Cold Region Hazards and Risks

Cold Region Hazards and Risks

Colin A. Whiteman

School of Environment and Technology, The University of Brighton, Brighton, England, United Kingdom

A John Wiley & Sons, Ltd., Publication

This edition first published 2011, © 2011 by John Wiley & Sons, Ltd

Wiley-Blackwell is an imprint of John Wiley & Sons, formed by the merger of Wiley's global Scientific, Technical and Medical business with Blackwell Publishing.

Registered office: John Wiley & Sons Ltd, The Atrium, Southern Gate, Chichester, West Sussex, PO19 8SQ, UK

Other Editorial Offices:
9600 Garsington Road, Oxford, OX4 2DQ, UK

111 River Street, Hoboken, NJ 07030-5774, USA

The Atrium, Southern Gate, Chichester, West Sussex, PO19 8SQ, UK

For details of our global editorial offices, for customer services and for information about how to apply for permission to reuse the copyright material in this book please see our website at www.wiley.com/wiley-blackwell

The right of the author to be identified as the author of this work has been asserted in accordance with the UK Copyright, Designs and Patents Act 1988.

All rights reserved. No part of this publication may be reproduced, stored in a retrieval system, or transmitted, in any form or by any means, electronic, mechanical, photocopying, recording or otherwise, except as permitted by the UK Copyright, Designs and Patents Act 1988, without the prior permission of the publisher.

Designations used by companies to distinguish their products are often claimed as trademarks. All brand names and product names used in this book are trade names, service marks, trademarks or registered trademarks of their respective owners. The publisher is not associated with any product or vendor mentioned in this book. This publication is designed to provide accurate and authoritative information in regard to the subject matter covered. It is sold on the understanding that the publisher is not engaged in rendering professional services. If professional advice or other expert assistance is required, the services of a competent professional should be sought.

Library of Congress Cataloguing-in-Publication Data

Whiteman, C. A. (Colin A.)
  Cold region hazards and risks / Colin Whiteman.
      p. cm.
  Includes bibliographical references and index.
  ISBN 978-0-470-02927-5 (cloth) — ISBN 978-0-470-02928-2 (pbk.)
 1. Cold regions. 2. Environmental risk assessment–Cold regions. I. Title.
  GB642.W45 2010
  363.700911—dc22
                                                            2010028090

A catalogue record for this book is available from the British Library.

This book is published in the following electronic formats: ePDFs: 978-0-470-97318-9

Set in by 10.5/12.5pt, Times Ten Roman by Thomson Digital, Noida
Printed in Singapore by Ho Printing Singapore Pte Ltd

First Impression   2011

Dedication

To Mother and Father

Contents

Preface · xiii

1 Introduction · 1

1.1 Concept and rationale · 1
1.2 Scope and classification of the hazards · 2
1.3 Hazard awareness · 5
1.4 Physical properties of ice · 7
1.5 Hazard and risk · 8
1.6 Summary · 10

2 Arctic Sea Ice · 13

2.1 Introduction · 13
2.2 The Arctic Ocean · 15
2.3 Sea ice · 17
   2.3.1 Definition and formation · 17
   2.3.2 Thickness and age · 17
   2.3.3 Distribution and extent · 21
2.4 Impacts · 23
2.5 Direct impacts · 24
   2.5.1 Entrapment · 24
   2.5.2 Access to Arctic sea ice · 24
   2.5.3 Shore ice override (ivu) · 27
2.6 Indirect impacts · 29
   2.6.1 Coastal erosion · 29
   2.6.2 Access to the Arctic Ocean space following ice loss · 34
   2.6.3 Iceberg severity · 35
   2.6.4 Ice sheet stability · 35
   2.6.5 Ecosystems · 35
   2.6.6 Climate change and the MOC · 37
2.7 Mitigation · 38
   2.7.1 Introduction · 38
   2.7.2 Sea-ice presence · 39

		2.7.3 Sea-ice loss	41
2.8		Summary	42

3 Ice Sheets – Antarctica and Greenland   45

3.1	Introduction	45
3.2	Ice sheet systems	47
	3.2.1 Ice domes	48
	3.2.2 Ice streams and outlet glaciers	49
	3.2.3 Ice shelves	49
	3.2.4 Sea ice	51
3.3	Greenland	52
3.4	Antarctica	56
3.5	Impacts of ice sheet loss	62
	3.5.1 Sea-level rise	62
	3.5.2 Iceberg flux	65
	3.5.3 Ecosystem change	66
	3.5.4 Climate and ocean circulation change	67
3.6	Mitigation	68
3.7	Summary	69

4 Icebergs   71

4.1	Introduction	71
4.2	Iceberg characteristics	72
	4.2.1 Composition	72
	4.2.2 Size	73
	4.2.3 Shape	74
	4.2.4 Sources	75
	4.2.5 Distribution	76
	4.2.6 Life expectancy	79
4.3	Iceberg impact and risk	81
	4.3.1 Ship–iceberg collisions	82
	4.3.2 Iceberg–fixed installation collisions	86
	4.3.3 Sea bed scouring	88
	4.3.4 Impact on ecological systems	88
	4.3.5 Impacts on global climate	89
4.4	Iceberg mitigation	90
	4.4.1 Detection, monitoring, databases and research	91
	4.4.2 Threat evaluation and prediction	95
	4.4.3 Ice management	96
	4.4.4 Avoidance	97
4.5	Summary	98

5 Glaciers   101

5.1	Introduction	101
5.2	Inherent glacier hazards	102

	5.2.1	Crevasses	102
	5.2.2	Seracs	104
	5.2.3	Ice avalanches	104
	5.2.4	Complex avalanches	113
	5.2.5	Ponded lakes	116
5.3	Glacier mass balance changes		116
	5.3.1	Advancing glaciers	117
	5.3.2	Surging glaciers	118
	5.3.3	Retreating glaciers	121
5.4	Mitigation measures		129
5.5	Summary		130

6 Glacier Lake Outburst Floods (GLOFs) — 133

6.1	Introduction		133
6.2	The glacial meltwater system		134
6.3	GLOFs		134
6.4	Trigger mechanisms		136
6.5	Risk		142
6.6	Mitigation		150
	6.6.1	Hard measures (engineering)	150
	6.6.2	Soft measures	151
6.7	Summary		154

7 Permafrost — 157

7.1	Introduction		157
7.2	Permafrost distribution and characteristics		159
7.3	Permafrost hazardousness		163
7.4	Lowland permafrost hazards		167
	7.4.1	Mitigation measures	173
	7.4.2	Climate change	174
	7.4.3	Mapping	175
	7.4.4	Geotechnical engineering	178
7.5	Mountain permafrost hazards		189
	7.5.1	Mitigation strategies	195
7.6	Summary		201

8 Snow Avalanches — 203

8.1	Introduction		203
8.2	Definition, classification and motion		204
8.3	Factors promoting avalanches		207
	8.3.1	Meteorology	207
	8.3.2	Terrain	211
8.4	Impacts of avalanches		212
	8.4.1	Damage	214

		8.4.2 Fatalities	214
8.5		Mitigation methods	218
		8.5.1 Information	219
		8.5.2 Modification and control	225
		8.5.3 Avoidance	227
8.6		Summary	236

9 River Ice – Ice Jams and Ice Roads — 237

9.1	Introduction	237
9.2	Ice jams	238
	9.2.1 Introduction	238
	9.2.2 Ice formation and freeze-up	241
	9.2.3 Ice breakup	246
	9.2.4 Ice jam processes and sites	248
	9.2.5 Ice jam impacts	250
	9.2.6 Ice jam mitigation	253
9.3	Ice roads	259
	9.3.1 Introduction	259
	9.3.2 Ice-forming processes and ice types	260
	9.3.3 Construction	262
	9.3.4 Loading	263
	9.3.5 Road use	265
	9.3.6 Hazards	267
9.4	Summary	272

10 Winter Storms — Ice Storms and Blizzards — 273

10.1	Introduction	273
10.2	Definitions	274
10.3	Weather systems and processes	276
	10.3.1 Air pressure patterns	276
	10.3.2 Air temperature	280
10.4	Impacts	282
	10.4.1 Fatalities and injuries	285
	10.4.2 Physical damages	286
	10.4.3 Economic losses	286
	10.4.4 Transport and traffic disruption	287
	10.4.5 Power loss	289
10.5	Mitigation	289
	10.5.1 Forecasting	290
	10.5.2 Forward planning	291
	10.5.3 Effective procedures	293
	10.5.4 Public response	294
	10.5.5 North-east Snowfall Impacts Scale (NESIS)	295
	10.5.6 Impact of climate warming	296
10.6	Summary	298

11	Conclusions – The Future	299
	References	303
	Glossary	335
	Acronyms	341
	Index	345

Preface

Sometimes, in the Arctic, you just have to sit and wait. It is the nature of tundra travel that the weather plays a key role. If the fog is down the Cessna doesn't fly; but then there is time to sit and think, or catch up with reading, which is how I came to be in the library of the Aurora Research Centre, in Inuvik, NWT, Canada. A thick volume on Arctic contaminants (not actually included in this text) sparked a train of thought that led to cold hazards and risk, and the idea that this would make an attractive subject for undergraduate students. However, with the exception of avalanches, ice-related hazards very rarely feature in existing geohazards texts. There may be some geomorphological coverage of particular topics in specialist glacial and periglacial books but nothing that brings the full range of cryogenic hazards together within one cover. At the same time as this idea arose, climate change was receiving wide coverage in the media and the subject of environmental hazards was gaining popularity amongst students. Polar and alpine regions appeared to be most vulnerable to climatic warming and ice-related events were attracting more media coverage. Dramatic pictures of disintegrating Antarctic ice shelves were projected into our living rooms for the first time, an Alpine glacier received an insulating blanket to preserve it for skiing, and stirring tales of the 'Northwest Passage' were revived as Arctic sea ice retreated. With popular coverage of these issues growing, it seemed an appropriate moment to provide a broad physical and human geographical background to these and related cold region hazards.

This has not proved to be an easy task. At the same time as popular coverage was rising, scientists in all areas have been exceptionally busy seeking answers to these accelerating environmental changes in an effort to understand what is happening and provide some advice on ways to reduce the potential impacts. Consequently the amount of literature on most topics is daunting. While adopting a global outlook in terms of the scope of examples, it has been necessary to be selective and apologies are due to those who feel that their work has been overlooked or inadequately handled. Nevertheless, I hope that there is sufficient scope and detail in each area to provide an understanding of the nature of the hazard, its potential impacts and the different approaches that have been taken to mitigate the hazard and limit risk.

For many reasons I must acknowledge a number of individuals and groups. Jim Rose is responsible for my interest in cold region geomorphology and geology. Julian Murton invited me to spend research time in the Arctic and, through our Inuvialuit assistants, Fred Wolki, Enoch Pokiak and Raymond Cockney, enabled me to gain a deeper understanding of Arctic life and environments. My colleagues in the Geography Division of the School of Environment and Technology at the University of Brighton kindly supported the cold region hazards module proposal, and some have even raised my awareness of human issues behind the physical environment. Their friendly rivalry provided a stimulating working environment under Roger Smith's able leadership. I have also received willing support from technical staff in the School and research assistant, Margaret Allen, has spent many hours in the search for appropriate material, as well as arranging two successful popular conferences related to polar environments which have raised the profile of cold environments within the School. At Wiley, I am grateful to Rachel Ballard for inviting me to submit the initial proposal and to unnamed colleagues who either reviewed the proposal or, subsequently, parts of the completed text, though I am responsible for any remaining errors. Also at Wiley, I am indebted to Izzy Canning, Chelsea Cheh, Aparajita Srivastava and, especially, to my project editor, Fiona Woods, who has provided encouragement and the necessary drive to see this project to completion. Finally, I thank Jill for her patience and long-suffering support.

Colin A. Whiteman

1
Introduction

1.1 Concept and rationale

This book is about ice, and its associated hazards. Ice is implicated in the 1970 Huascarán disaster, which killed about 18 000 people in the Andes of Peru. In 1912 over 1500 people died when the *Titanic* struck an iceberg and sank during her maiden voyage across the Atlantic. Each year, about 150 people are killed by avalanches and in bad snow seasons, such as 1950–1951 in Europe, this number can nearly double. Occasionally, catastrophic avalanches, such as the one that struck Galtür, Austria, in 1999 killing 31 people (Keiler *et al.*, 2006), are widely reported, but most geohazard textbooks neglect geocryogenic hazards. Instead, attention is focused on tsunamis, earthquakes, volcanic eruptions, landslides and floods, hazards which, more frequently, cause larger numbers of fatalities.

A number of reasons can be suggested for this neglect of cold region hazards. One has already been mentioned: ice-related hazards, with the exception of some avalanches, are rarely dramatic. A second reason is that cold regions are inhospitable to many people and generally support only low densities of population. Consequently, large centres of dense population are rare, except in the European Alps, and few people are perceived to be at risk. Thirdly, the remoteness of many polar and alpine regions means that some hazards pass unrecorded or are at best underrecorded, until a more newsworthy event brings the hazard to a wider audience. For example, such an event occurred in the Himalaya in 1985 when a glacial lake burst beyond its confining moraine and destroyed a small hydro-power plant that was about to be commissioned (Ives, 1986). In the context of this less-developed region, the ice avalanche that initiated this flood dealt a devastating blow to the aspirations of a financially poor community. On the other hand, this event was clearly instrumental in raising both national and international awareness of a Himalayan problem that

Cold Region Hazards and Risks, First Edition. Colin A. Whiteman.
© 2011 John Wiley & Sons, Ltd. Published 2011 by John Wiley & Sons, Ltd.

has since received substantially more attention (see Chapter 6). Lack of awareness is a fourth reason geocryogenic hazards generally have such a low profile. Obviously there are exceptions, such as densely populated Switzerland which in many respects leads the world in its study of geocryogenic hazards (e.g. Haeberli *et al.*, 2009). Otherwise, even scientists (Solomon *et al.*, 2007) have been surprised at the rapid rate at which some geocryogenic systems are changing and becoming more hazardous. Until scientists realized what was happening, there was little chance that others – the media, the general public and politicians – would be aware of the increasing risks associated with geocryogenic hazards. If the number of fatalities is the only measure of hazard importance, then perhaps ice does not warrant any special attention. However, in the present context of rapid climate change, ice-related hazards assume a much greater significance. In this text a broader, longer-term approach to hazard and risk will be taken, involving the social, cultural and economic well-being of human societies, and the biodiversity of natural ecosystems.

1.2 Scope and classification of the hazards

Most people are familiar with *snow* avalanches (Chapter 8), even though they may never have seen one, or even travelled through alpine terrain where these phenomena are common, but how many people have heard of *ice* avalanches (Chapter 5)? Likewise, many individuals will know of icebergs (Chapter 4), if only through the medium of film – blockbusters such as *Titanic* have seen to that – but how many people are familiar with other types of floating ice such as sea ice (Chapter 2) and ice shelves (Chapter 3)? Some ice-related hazards are unobtrusive because the ice is mostly out of sight below the ground surface. Permafrost (Chapter 7) falls into this category. Those of you reading this in a temperate maritime environment, or in the tropics or subtropics, may not even have heard of river ice jams (Chapter 9) and ice storms (Chapter 10). However, these hazards are familiar to the inhabitants of continental interiors (the northern states of the USA and the southern provinces of Canada, for example) and more northern latitudes, where subzero winter temperatures are normal. The television programme, *Ice Road Truckers*, has brought another hazard of polar and northern regions, ice roads (Chapter 9), to the attention of a wider audience. Two other hazards will be considered. One is the ice sheets (Chapter 3) which have the potential between them to raise sea levels by a massive 70 m, though hopefully not all at once! The other is the blizzard (Chapter 10) which may constitute the most widespread of cold region hazards as it depends on the global atmosphere.

At this stage, it is worth attempting some degree of hazard classification (Table 1.1 and Figure 1.1), in order to make sense of their different characteristics and to clarify their interrelationships with each other and with the human context. Any system of classification is likely to be a simplification of

Table 1.1 Hazard characteristics

Hazard type	Hazard process	Impact Direct	Impact Indirect	Timing Sudden	Timing Gradual	Term Short	Term Medium	Term Long	Global context	Latitudinal context
Arctic sea ice	Impact, crushing, erosion, surface failure, albedo change	x	x	x	x	x	x	x	Hydrosphere	Polar
Ice sheets	Sea level rise, albedo change, climate change		x		x	x	x	x	Lithosphere, hydrosphere	Polar, sub-polar
Ice shelves	Iceberg impact, glacier acceleration	x	x	x	x	x	x	x	Hydrosphere	Polar, sub-polar
Icebergs	Impact	x		x		x			Hydrosphere	Polar, sub-polar
Glaciers	Impact (seracs), land burial, slope instability	x	x	x	x	x	x	x	Lithosphere, hydrosphere	Polar to tropical Alpine
GLOFs	Flood		x	x		x			Lithosphere, hydrosphere	Polar to tropical Alpine
Permafrost	Ground instability	x		x	x	x	x	x	Lithosphere	Polar, Alpine
Snow avalanches	Impact and burial	x		x		x			Lithosphere, (atmosphere)	Polar to tropical Alpine
Ice avalanches	Impact and burial	x		x		x			Lithosphere	Polar to tropical Alpine
Ice jams	Impact and flood	x	x	x		x			Hydrosphere	Polar to temperate
Ice roads	Surface failure	x		x		x			Hydrosphere	Polar to temperate
Blizzards	Excessive snow deposits, gale force winds	x		x	x	x			Atmosphere	Polar to tropical Alpine
Ice storms	Excessive ice deposits, structural collapse	x		x	x	x			Atmosphere	Polar to temperate

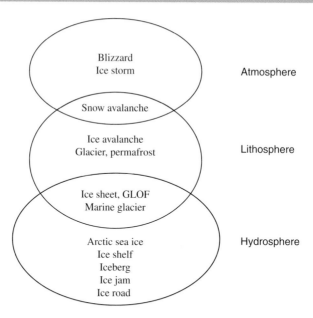

Figure 1.1 Global context of cold region hazards

reality: some hazards are extremely complex. For example, the large terrestrial ice sheets have fast-moving components (ice streams) as well as what might be described as normal, slow-moving areas. Additionally, it is now apparent that the behaviour of these ice sheets cannot be adequately modelled without also considering adjacent ice shelves and the surrounding, seasonal sea ice. Nevertheless, for the purposes of this exercise, ice sheets and glaciers will be considered generally as large, slow-moving, often land-based (terrestrial) masses of solid ice, possibly 3–4 km in thickness in the case of ice sheets, and with the potential to alter the global distribution of land and sea significantly. They can be contrasted with floating ice, such as sea ice and river ice which may average only 3–4 m in thickness and be seasonally very variable in extent. However, while the hazards of river ice are very localized and seasonal, the potential effects of sea-ice hazards may be felt much more extensively in both space and time. Ice shelves and icebergs are other forms of floating ice but in both cases these have a link to land-based ice. Ice shelves are largely an extension of land-based ice sheets, while icebergs are free floating, mobile ice bodies that have calved from land-based ice or ice shelves, and are potentially more difficult to deal with. In contrast to the hazards mentioned so far, which are terrestrial or water-based, ice storms and blizzards are essentially atmospheric in context and closely controlled by weather systems. Consequently, their impacts can be sudden and the distribution of their impacts in both space and time difficult to predict. Another sudden and often unpredictable hazard is the snow avalanche. Although it occurs on land, atmospheric conditions are the essential prerequisites of this hazard. Weather systems deliver the snow and may then metamorphose it into a susceptible snowpack. Two glacier-related hazards not mentioned specifically so far are ice avalanches and glacier lake outburst floods (GLOFs). Unlike snow avalanches, ice avalanches depend on the structure and position of the glacier from which they originate, although their impacts can achieve the same ends as snow avalanches – instantaneous death and destruction of property. GLOFs are not geocryogenic per se but they derive from glacial contexts and are often very difficult to predict. Their impacts are confined to glacial valleys but may be very variable in extent. The final hazard to be mentioned is permafrost, a unique hazard in the sense that it occurs *beneath* the surface, usually on land but also under some coastal seas. Depending on the content of ice in the permafrost, this hazard could directly impact up to 25% of the Earth's surface, perhaps more than any other cold region hazard, if the impacts of other hazards on the global atmosphere are discounted.

1.3 Hazard awareness

Probably nothing has done more to enhance our awareness of the hazards of cold regions than climate change, because by definition cold region hazards are climate and weather dependent, and ice depends on temperature for its behaviour and, indeed, for its very existence. The Earth's environment has always been in a state of flux, ever since it was a cloud of dust and gas 4.6 billion years ago. Some periods during this time have been relatively stable; others have experienced rapid change on a massive scale. The Earth system is often subdivided into the lithosphere (rock), hydrosphere (water), atmosphere (air), biosphere (fauna and flora) and *cryosphere*, the latter composed of ice, and the component least familiar to the majority of people. However, the cryosphere is arguably the single component of the global environment that has changed most noticeably. During the last few million years the Earth has fluctuated between glacial periods, when ice sheets and glaciers advanced across the globe, and interglacial periods, when ice has been forced to retreat. The last major ice retreat phase began about 15 000 years ago and for the last 11 500 years the Earth has experienced an interglacial phase. There have been minor fluctuations during this interglacial period, such as the Little Ice Age, when alpine glaciers presented a new hazard as they advanced down their valleys, but for over a century that trend has been reversed. The last 30 to 40 years have seen a period of rapidly rising temperature, especially in the Arctic, but also in alpine regions. Most glaciers are in retreat mode, sea ice on the Arctic Ocean has thinned and its area decreased, and permafrost has warmed. Even the great Greenland Ice Sheet has shown signs of increased melting.

Although climate change waxes and wanes as an issue in the media, it is now difficult to avoid, in one context or another. High profile reports of the Intergovernmental Panel on Climate Change (IPCC), the latest in 2007, and the Arctic Council's Arctic Climate Impact Assessment (ACIA, 2004) have achieved wide publicity. Consequently, the narrow, traditional view of geohazards, largely excluding the cryosphere, is slowly beginning to change. Television and newspaper correspondents are increasingly filing their reports from the decks of polar ice-breakers or the holds of helicopters as they seek to explain the decay of Arctic sea ice, or the disintegration of yet another Antarctic Peninsula ice shelf. Today, 8 April 2010, before I sat down to write this, BBC TV Breakfast News (BBC NEWS, 2010b) informed viewers that the replacement satellite (Cryosat–2) for the one that blew up during take-off in 2005, had been launched. The BBC statement was accompanied by a simple animated diagram illustrating how the satellite would record Arctic sea ice thickness, a crucial function in tracking changes to this important component of the cryosphere. If Arctic sea ice completely disappears, it must have an impact on climate because its current function as a reflector of solar radiation would be lost. Of more immediate concern is the effect it would have on Arctic inhabitants, especially the indigenous populations. Until recently Inuit traditional knowledge served a vital purpose in keeping hunters safe on the sea ice. Now, it is not uncommon to hear complaints that traditional knowledge is no longer 'fit for purpose' in some areas of Arctic sea ice (Ford, Smit and Wandel, 2006). Rapidly thinning and retreating ice is forcing a rethink, and the Inuit are facing steep 'learning curves' to keep up with changes in their once familiar territory. Winter hunting seasons are becoming shorter as access to the sea ice becomes less easy and more dangerous.

In August 2003, during the European Permafrost Conference in Zurich, the convener of the meeting, Professor Wilfried Haeberli, was called to the television studios to explain possible reasons for a massive rockfall from the slopes of the famous Matterhorn mountain in Switzerland (ETH Life, 2008), where several climbers were trapped until they could be airlifted to safety. About the same time, climbers following a route to Mont Blanc, in France, were less fortunate when they were hit and killed by another rockfall (Summitpost, 2003). It is likely that these headline-catching events were induced by the melting of mountain permafrost. An increasing number of landslides and rockfalls in the European Alps are close to the lower margin of mountain permafrost where the decay of this hazard is concentrated as the global temperature rises (ETH Life, 2008). For economic reasons, glaciers have been covered with insulating membranes to reduce summer melting and help preserve the valuable winter skiing industry (BBC NEWS, 2005).

Some ice-related hazards provide the media with spectacular footage for their prime-time news broadcasts, so we are now able to see massive tabular icebergs drifting away from their parent ice shelf in Antarctica (BBC NEWS, 2002). It is extremely unlikely that anyone will die as a *direct* consequence of the disintegration and melting of an ice shelf, but it has been shown that land-based ice streams within the West Antarctic ice sheet (WAIS) have accelerated and calved into the ocean once the protective ice shelves have gone. As these ice streams are land-based, their melting causes global sea level to rise. Although this sea-level rise will almost certainly be too slow to be *directly* life-threatening, it nevertheless has the potential to achieve this result *indirectly* through its impact on vulnerable economies such as Bangladesh.

These news stories, from which the general public derive so much of their information about scientific issues (Allan, 2002), are themselves derived largely from the press releases of research organizations, from individual scientists, or from scientific journals and other publications. In recent years a great deal of this new information on cryospheric subjects has come from the rapidly advancing, remote sensing capabilities of satellites, such as Landsat 1–5, ERS 1–2, RADARSAT 1–2, ICESat and the recently-launched Cryosat-2 (*The Satellite Encyclopaedia*, 2010). Regular monitoring of the cryospheric system, since the emergence of scientific satellites in the 1970s, has greatly facilitated the research effort, especially in remote and difficult terrain. However, this constant monitoring generates unprecedented quantities of data and this is increasingly being organized into accessible databases on a wide range of subjects such as glaciers, permafrost, ice jams, icebergs and GLOFs (glacial lake outburst floods). One important outcome of this constant effort is a new awareness of, and deep concern about, the *rapid* rates of change of some geocryogenic systems such as ice sheets, glaciers and Arctic sea ice. This has enabled the IPCC (Solomon *et al.*, 2007) to raise the probability of some of its predictions about the cryosphere from 'likely' to 'very likely' and has undoubtedly stimulated the interest of governments and grant-awarding bodies in new cryospheric research. In turn, this should enhance understanding of cold region hazards and the dangers associated with ice.

1.4 Physical properties of ice

Ice is a complex material. Under certain conditions it is a relatively hard, strong solid, capable of smashing buildings and sinking ships. However, under stress it can exhibit elastic, ductile and brittle behaviour (Knight, 1999). This means that glacier and sea ice can deform or crack under different conditions. Tensional stress, as glaciers accelerate on steep slopes, causes cracking and the formation of crevasses. These are a hazard for unwary mountaineers and may cause masses of ice to avalanche from the end of a glacier, or marine glaciers to calve an iceberg. In other circumstances, cold ice can form strong bonds with rock and maintain the stability of rock faces. If, however, the temperature rises *above* 0 °C, ice melts and takes on the properties of water, a relatively weak, mobile fluid. Melting ice-rich permafrost can cause slope instability resulting in rockfalls, landslides or debris flows. In the right circumstances, snow, composed of ice crystals, may be stable on a slope but if the conditions change the snowpack can become unstable and detach from the slope as an avalanche.

In addition to these changes in material strength, there is also a significant volume change across the 0 °C temperature threshold: ice occupies approximately 9% more space than the equivalent amount of water. Consequently ground can heave as its temperature falls past zero degrees and moisture freezes, and can subside as the temperature rises past the same threshold and ground-ice melts. This volumetric change is a major hazard of permafrost regions. Given that the *average* temperature of the Earth is around 13 °C, often only a small fluctuation in actual temperature is required to bring about some quite dramatic changes in the environment.

Under cold polar conditions the whole of the sea surface can freeze to form a layer of ice a few metres thick. This happens over most of the Arctic Ocean and around the continent of Antarctica. As will become clear in the following chapter, stable sea ice can be an asset to humanity, but as it moves under the influence of winds and sea currents, and as its extent increases and decreases seasonally, it can create hazardous conditions. Ice also forms on rivers where it can again be hazardous, especially if it breaks up and forms jams of ice blocks, or if it is too thin to support the movement of vehicles and people.

On land, similar cold conditions in both high-latitude and high-altitude regions can cause snow to accumulate which gradually transforms into glacier ice. Under the influence of gravity ice moves, either slowly by the deformation of individual crystals (creep), or more rapidly by basal shearing, basal sliding or with the aid of deforming subglacial sediments (Knight, 1999). The usual rate of flow of glaciers and ice sheets (exceptionally ~19 m per day (e.g. Jakobshavn Isbrae, Greenland) but generally much less) is not normally hazardous. When the input (snow accumulation) and output (ice ablation or melting) of a glacier are in balance the main dangers to be aware of are crevasses and falling seracs on the glacier surface. The bigger problems arise, either directly or indirectly, when glacier margins advance or retreat as mass balance increases or decreases. Providing glacier ice remains in an impermeable state with limited crevassing it is capable of damming the flow of water derived either from the melting of the glacier itself or from normal surface run-off. Lakes on and around glacier margins can lead to disastrous floods unless their behaviour is seasonally predictable. In coastal areas ice can be transferred from the land onto the sea, as either floating ice shelves or icebergs. Under normal conditions ice shelves remain attached to the land ice and present relatively little danger. They create problems when they disintegrate by forming massive tabular icebergs and by allowing the terrestrial ice to accelerate. Icebergs, on the other hand, float free and are a constant direct hazard to both shipping and fixed installations until they melt.

1.5 Hazard and risk

Without doubt ice in its various environmental manifestations can be a dangerous material and give rise to disastrous events. When these ice-related features have the potential to impact negatively on humans, their possessions and their general environment, they become *hazards*. Potential hazards are always present in the landscape but not necessarily a constant threat. The more that is known about how hazards function, the easier it is to predict hazardous events. As will become clear when considering individual hazards, monitoring, mapping and interpretation are key requirements for assessing the level of *risk* attached to a particular hazard. Monitoring may vary from the daily, practical measurement of snowpack on a hillside to remote sensing by satellite. Remote sensing, in particular, has enabled significant progress to be made in the fields of mapping and monitoring, and the results of this work suggest that the risk of some of the hazards considered here is increasing significantly.

Risk is a combination of the probability that an event will occur, and the likely cost of the event. Essentially, risk is the product of frequency and magnitude. It can be defined by a simple equation:

$$\text{Risk} = \frac{\text{Hazard probability} \times \text{Expected loss}}{\text{Loss mitigation}}$$

The probability of a hazardous event is usually determined from historical data. Longer time series enable more accurate predictions to be made. Risk managers often refer to a design event; that is an event of a particular magnitude and frequency around which they design their mitigation response. Normally, larger events recur less frequently than small events. Recurrence interval, probability and annual frequency of an event can be calculated as follows:

$$\text{Recurrence interval,} \quad T_r \text{ (years)} = (n+1)/m$$

where m = event ranking(largest to smallest) and n = number of events in the time period.

$$\text{Probability of event,} \quad P(\%) = 100/T_r.$$
$$\text{Annual frequency,} \quad \text{AF} = 1/T_r \text{ (years)}.$$

For example, in a list of 50 (n) events the largest event ($m=1$) will have a recurrence interval of 51/1 which is 51 years. The smallest (50th) event will have a recurrence interval of 51/50 which is 1.02 years, that is it occurs almost annually. The same largest and smallest events would have a probability of 1.96% and 98% and an annual frequency of 0.02 and 0.98 years respectively.

Expected losses from a hazardous event can be calculated from a knowledge of the number of people living within the expected range of the event and the value of their property. One factor that is contributing to risk in many areas is increasing population density coupled with the migration of residents and tourists into more remote and hazardous areas. Natural environmental costs may also be taken into account. The total value of loss can be reduced providing appropriate *mitigation* measures are put in place to reduce impact. Effective mitigation relies on a sound knowledge of hazard mechanisms. For example, in avalanche-prone areas, structures can be modified and buildings restricted by a system of land-use planning or zoning. The avalanche events themselves can sometimes be modified by inducing the avalanche before the snowpack becomes excessively deep and the potential volume of snow dangerously large. Mitigation of a hazard can also be achieved by forecasting, based on historical records, and effective warning procedures. Sometimes evacuation may be necessary. Mitigation can also be achieved through the education of both the community involved and responsible authorities at both national and international levels. Communities where mitigation is poorly developed, for reasons of inadequate wealth or ignorance, will obviously be more vulnerable to the impacts of hazards than those living in more favourable circumstances. Wealth and knowledge will also influence the adaptive capacity of communities, as will their resilience. At present, some of the hazards, such as ice-sheet melting, appear to be changing slowly relative to the capacity of communities and individuals to adapt. However, whether the long-term consequence of this impact – a rise in sea level – will always be equally manageable, remains to be seen. Arctic coastal communities, as noted earlier, are already being forced to respond to another massive hazard, the decline in the extent and thickness of sea ice. Many countries with cold regions are reviewing the level of risk presented by their geocryogenic hazards, redesigning their building and land-use regulations or making plans to resettle vulnerable communities.

1.6 Summary

Although technical specialists may have been aware of the hazards of particular cold environments, for a number of reasons details have rarely been included in geohazards texts, although maybe this is about to change. While the direct effects of avalanches and icebergs may be familiar, the impacts of many other hazards of cold regions are less direct and less well known. However, with the well-publicized possibility that global temperature could rise by as much as 3 $^\circ$C within the next few decades (Solomon *et al.*, 2007), more people are becoming aware of potential hazards, especially those associated with melting ice. As ice changes into water, the volume of material decreases, ground subsides, slope stability declines and lake levels rise (or lakes disappear altogether if the base of permafrost is penetrated). Also, coasts become increasingly exposed to storms as sea ice retreats, shrinking glaciers and rising snow lines reduce potential stores of water for irrigation and hydroelectric power, and the scope of traditional alpine tourism decreases. Many people, far beyond the limits of conventional polar and alpine 'cold regions', will feel the effects of these changes in the cryosphere as global sea levels rise. Given this rapidly changing global context it seems appropriate to bring geocryogenic hazards to a wider audience. With this in mind, each of the following chapters addresses the scientific background of a different set of geocryogenic hazards, assesses the risks faced by both humanity and natural ecosystems and considers appropriate methods for mitigating the impacts of each hazard.

2
Arctic Sea Ice

2.1 Introduction

Under long-term, north-polar-average conditions, Arctic Ocean water is cold enough to freeze. In winter the ocean is largely covered by at least 2–4 m of ice (Figure 2.1), except for the region north of Scandinavia which is warmed by the Gulf Stream (North Atlantic Drift) (Figure 2.2). The area covered by sea ice contracts during the summer to a minimum around mid-September, as the Arctic warms, and then expands again during the winter to a maximum in March. In the past, and still today, Arctic sea ice presents a hazard to shipping and other off-shore activities as its mass and movement is capable of crushing vessels. However, in 2007 Arctic sea ice made headlines in newspapers around the world for a different reason (BBC NEWS, 2007). A dramatic reduction in the area of ice on the Arctic Ocean, at its annual September minimum, produced speculation by Wieslaw Maslowski that an ice-less Arctic Ocean would occur as early as the summer of 2013. More realistic estimates by Peter Wadham ("earlier than 2040") and Mark Serreze ("2030") are not unreasonable following three more years when the summer ice minimum has failed to reach the 1979–2000 average (Figure 2.3). As will become clear, sea ice can be a hazard when it is present, and also when it is not present.

Many expeditions, some searching for the elusive 'Northwest Passage' as a short cut between Europe or north eastern North America and Asia, have become trapped and sometimes crushed by the build-up of sea ice in winter (Fleming, 1998). Indigenous communities, adapted to a marine-based subsistence culture, have lost property and occasionally there have been fatalities due to override of sea ice onto the coast (in Alaska a process referred to as 'ivu'). The surface relief of sea ice can hinder movement but generally, from the Inuit perspective, sea ice is seen as an asset for travel purposes.

Cold Region Hazards and Risks, First Edition.   Colin A. Whiteman.
© 2011 John Wiley & Sons, Ltd. Published 2011 by John Wiley & Sons, Ltd.

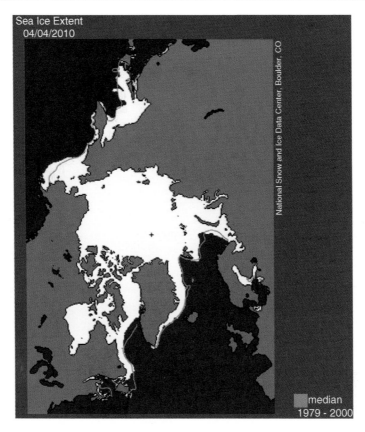

Figure 2.1 Sea-ice extent in the Arctic Ocean on 4 April 2010; the orange line shows the median extent for the period 1979–2000 (NSIDC, 2010)

However, in recent years the hazards traditionally associated with Arctic sea ice have begun to change. Instead of excess sea ice hazards are now more likely to arise from a paucity of ice, either its complete loss or a significant thinning. The ice that used to provide the Inuit and polar bears with a reliable travelling platform now sometimes fails to support skidoos or to accumulate sufficiently early to provide an adequate hunting season for both humans and polar bears. Thinning of the ice cover alters light transmission to the ocean beneath the ice and melting freshwater ice changes the salinity of marine ecological systems. Reduction of albedo, as the ocean surface transforms from light-coloured, reflective ice and snow to dark, solar radiation-absorbing ocean water, leads to a warmer ocean which enhances the rate of sea ice loss. The exposed ocean releases more heat to the atmosphere which in turn melts more ice in a positive feedback. It is too early to be certain exactly how the widespread loss of Arctic sea ice will impact on regional and

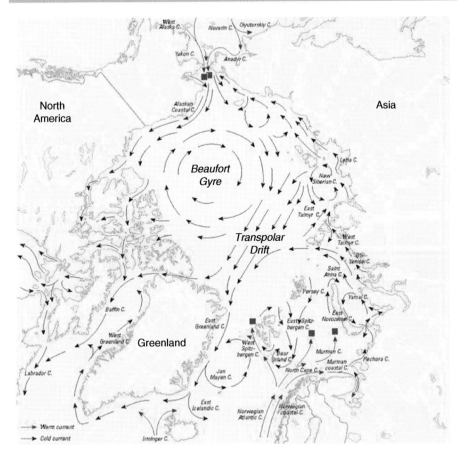

Figure 2.2 Arctic Ocean circulation (source: Arctic Monitoring and Assessment Programme (AMAP), Figure 3.29, AMAP (1998))

global climates, but what is certain is that climate and weather patterns will be altered under a regime of severely reduced or even absent Arctic sea ice. One can only speculate on the impacts of the increased access for shipping, mineral exploration and ecotourism, on an ice-free Arctic Ocean.

2.2 The Arctic Ocean

The Arctic Ocean (Figure 2.4) occupies two deep basins, separated by the Lomonosov Ridge, and extends across shallow shelves to the northern coasts of Eurasia (Scandinavia, Russia (including Siberia)), Alaska, Canada (Yukon, Northwest Territories, Nunavut and the islands of the Canadian Archipelago) and Greenland. Narrow straits (the Bering Strait between Alaska and Siberia

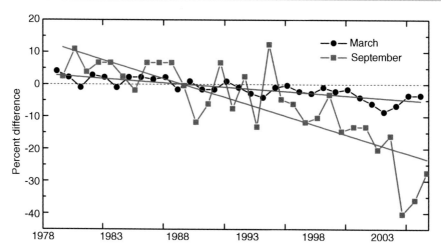

Figure 2.3 Percentage difference in ice extent in March (month of maximum ice extent) and September (month of minimum ice extent) relative to mean values for the period 1979–2000. Rate of decrease for the March and September ice extents is -2.5% and -8.9% per decade, respectively (Reproduced from Perovich et al., 2009, Fig. S2)

and the Fram Strait between Baffin Island and Greenland) and the North Atlantic (between Greenland and Svalbard and Svalbard and Scandinavia) provide access for the ingress of relatively warm water to the Arctic basin and egress of cold water out of the basin to the North Atlantic and Pacific Oceans. This exchange of water is important for the distribution of ice around the Arctic.

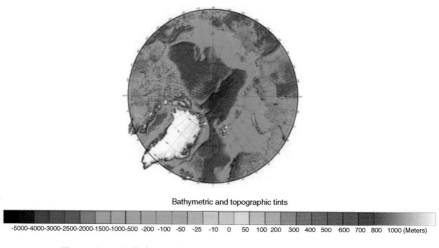

Figure 2.4 Relief map of Arctic Ocean basin (source: NGDC, 2008)

2.3 Sea ice
2.3.1 Definition and formation

Sea ice is ice that is formed at sea by the freezing of sea water. During formation, most brine is gradually expelled from the ice so that old sea ice is fresh and floats easily on the denser, saline ocean water. Formation of sea ice is a complex process with a number of different stages. It has been found that due to the different configurations of the Arctic Ocean (an ocean largely surrounded by land) and Antarctica (broadly, a continent surrounded by water) the processes and characteristics of sea ice in these opposing polar regions differ (Hansom and Gordon 1998; NSIDC, 2009). This chapter will concentrate on sea ice in the Arctic, where it has a strong relationship with human activity. Antarctic sea ice will be discussed in the next chapter, on ice sheets.

Cold air above the ocean reduces the sea temperature. As the sea temperature approaches freezing point its density increases and surface water sinks. Warmer water replaces it at the surface and is then cooled by the air above until a layer of water near the ocean surface becomes cold and the formation of ice crystals can begin. Dense, saline ocean water typically freezes at about $-1.8\,°C$. Initially, small needle-like ice crystals (frazils), usually 3–4 mm in diameter, form in the freezing ocean water before floating to the surface and bonding together (see Canadian Ice Service web site for details of different types of sea ice, http://www.ec.gc.ca/glaces-ice/default.asp?lang=En&n=D32C361E-1). Under calm conditions frazil crystals form a very thin smooth grey sheet, with the appearance of an oil slick, called grease ice. This may gradually thicken into, first, dark Nilas and then light Nilas. Further thickening forms congelation ice. Winds and current may cause overlap or rafting of these thin sheets. Eventually a more stable sheet of ice will form. If the sea surface is agitated by winds, the frazil crystals will be jostled together and accumulate into slushy circular discs known as pancakes. Pancake ice typically develops raised edges as individual pancakes impact each other. As with congelation ice, rafting may occur if sea surface motion is sufficiently strong. Eventually the pancakes freeze together into a more stable sheet. The result of these thermodynamic processes is a layer of ice averaging about 4 m in thickness.

Ice adjacent to the land which is fixed in place rather than moving is referred to as landfast ice (George *et al.*, 2004; Figure 2.5). There is usually a transition zone to floating pack ice which is on the move at about 20 km/month in winter and 80–100 km/month in the summer. The seaward limit of landfast ice is commonly indicated by grounded pressure ridges, known as stamukhi, formed in waters 15–40 m deep.

2.3.2 Thickness and age

The rate of thickening of an ice cover depends essentially on the surface temperature of the ice (or snow) and the thickness of the ice and snow (Maykut

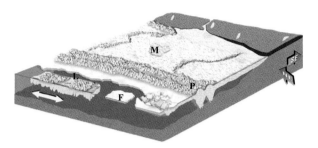

Figure 2.5 Landfast ice showing pressure ridges (P), multiyear ice (M), floe ice (F) and an open lead (L) (modified from George *et al.*, 2004)

in French and Slaymaker, 1993). Ice thickness is related to how cold the temperature is and for how long, commonly referred to as freezing degree days (FDD). For example, if the temperature remains at $-3.8\,°C$, that is $2\,°C$ below normal sea water freezing point of $-1.8\,°C$, for five days, this is equivalent to 10 (2×5) FDDs. A further series of six days at a temperature of $-6.3\,°C$ is equivalent to 27 FDDs, that is $4.5\,°C$ multiplied by 6. This gives a cumulative total of 37 FDDs. An empirical formula for ice thickness, incorporating the FDD concept, was devised by Lebedev in 1938:

$$\text{Ice thickness} = 1.33 \times \text{FDD}^{0.58}.$$

However, the increasing thickness of sea ice is not a limitless process. More ice forms at the base as heat is transferred away from the ocean through the ice to the cold air above. As ice thickens (and snow accumulates) it insulates the ocean from the atmosphere and the transfer of heat takes longer. Eventually the system reaches a thermodynamic equilibrium in which the ice is sufficiently thick to completely inhibit heat transfer to the atmosphere. In the Arctic, this equilibrium occurs when the ice reaches a thickness of about 3 m, a process that may take several seasons of freezing and melting. (In the Antarctic, by contrast, this thermodynamic equilibrium occurs when the ice is 1–2 m thick.)

Sea ice thickness is related to its stage of development. Currently forming *new* ice is defined as ice that is less than 10 cm thick. *Young* ice is 10–30 cm thick and may be described as grey (10–15 cm thick) or grey–white (15–30 cm thick) ice. *First-year* ice is generally between 30 and 200 cm thick but has not survived a summer melt season. Ice that does survive into the next freezing season is termed *multiyear* ice. This is thicker (typically 2–4 m) and generally more complex than younger ice, with ridges, individual ice floes, and linear areas of open water (leads) all forming a complex mosaic of different thickness and colour which is difficult to map (Thomas and Dieckmann, 2003; Figure 2.6). In relation to hazards, older ice tends to be thicker and more resilient than younger ice (Perovich *et al.*, 2009).

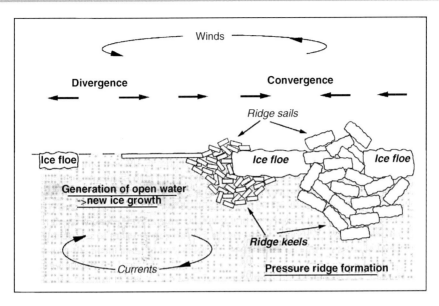

Figure 2.6 Dynamic modification by divergence and convergence of sea-ice thickness distribution (source: Thomas and Dieckmann, 2003, Fig. 3.1)

Although ice thickness and age vary across the Arctic from year to year as weather conditions fluctuate, some general patterns are discernable. For example, multiyear ice is common to the north of the Canadian Archipelago (Figure 2.7), where persistent anticyclonic conditions with low temperatures and temperature inversions in winter favour ice formation, and low cloud and fog in summer inhibit melting. In contrast, Baffin Bay and the Davis Strait, and the areas north of Siberia and around the Bering Strait annually lose most of the sea ice so that conditions are not suitable for the development of multiyear ice.

In the NOAA Arctic Report Card for 2009, Perovich *et al.* (2009) provide some interesting details about how Arctic sea ice has changed (Figure 2.8 (a) and (b)). Recently, the area of multi-year sea ice has reduced rapidly at a rate of $1.5 \times 10^6 \, \text{km}^2$ per decade, three times faster than during the three previous decades (1970–2000). Satellite altimetry measurements indicate substantial overall thinning of ∼0.6 m in sea ice thickness between 2004 and 2008. The volume of winter multiyear ice shows a net loss of >40% in the period 2005–2009. These decreases in volume and average thickness reflect thinning and loss of multiyear sea ice due to melting and ice export from the Arctic Ocean. Seasonal ice has become the dominant Arctic sea ice type in terms of both area and volume. These recent satellite estimates have been supplemented by longer records of declassified, sub-ice, sonar measurements from US Navy submarines which cover ∼38% of the Arctic Ocean (Figure 2.6). Within this area the overall mean winter thickness of 3.6 m measured by submarines in

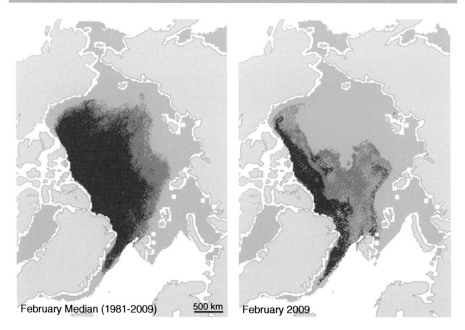

Figure 2.7 Median age of February sea ice from 1981–2009 (left) and February 2009 (right). Dark colours show older, thicker ice. As of February 2009, ice older than two years accounted for less than 10% of the ice cover. Data provided by J. Maslanik and C. Fowler, University of Colorado, Colorado Center for Astrodynamics Research. (Reproduced from NSIDC, 2010e)

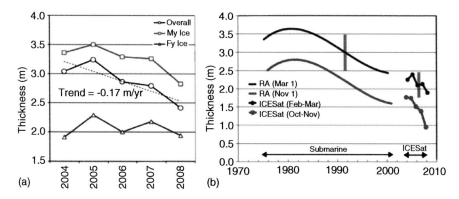

Figure 2.8 (a) Winter Arctic Ocean sea-ice thickness from ICESat (2004–2008). The black line shows the average thickness of the ice cover while the red and blue lines show the average thickness in regions with predominantly multiyear and first-year ice, respectively. (b) Interannual changes in winter and summer ice thickness from the submarine and ICESat campaigns within the data release area spanning a period of more than 30 years. The data release area covers approximately 38% of the Arctic Ocean. Blue error bars show the uncertainties in the submarine and ICESat data sets (figures and caption fom Perovich et al., 2009, Fig. S4, after Kwok et al., 2009 and Kwok and Rothrock, 2009)

1980 is down to a mean of 1.9 m in 2008 according to the ICESat record, a reduction of over 47%.

Under dynamic sea and weather conditions, sea ice moves in response to currents, tides and wind and tensions and pressures may develop within the ice. Sheets of ice may buckle and fracture, and where two ice floes converge and impact, pressure ridges of broken ice may form (Hansom and Gordon, 1998; Figure 2.6). Thermal expansion of newer sea ice is also capable of creating pressure ridges (Environment Canada, 2003a). Under stormy conditions, pressure ridges can grow to 20 m or more in height with equivalent keels extending down below the water surface. Near the coast these keels may extend to the sea bed and help to stabilize coastal ice and facilitate access onto the ice from the shore (Figure 2.5), but ice ridges at the surface can be a hindrance to travel, as many North Pole explorers have found. Conversely when sea ice is under tension it splits and open water, referred to as a lead, appears. Other recurrent areas of 'open' water (that is open water *plus* new ice) are called polynyas. These are common in the Canadian Archipelago. They may be due to (a) mobile ice caused by turbulent motion in narrow channels or persistent offshore winds in the lee of islands, or (b) convective overturning which brings warmer water to the surface. Polynyas can be of great biological importance, providing open water, and hence places for marine mammals, such as Narwhals, to breathe during the winter.

2.3.3 Distribution and extent

Under favourable winter conditions the whole of the Arctic Ocean is covered with ice, except for parts of the Barents Sea and the Greenland Sea, between Norway, the island of Björnöya and Svalbard, where the relatively warm North Atlantic Drift (Gulf Stream) current penetrates into these Arctic waters (Figure 2.2) and maintains ice-free conditions throughout the year. On the western side of the North Atlantic Ocean ice forms along the eastern seaboard of Greenland, and extends through the Fram Channel into Baffin Bay and the Davis Strait on the western side of Greenland, facilitated by cold ocean currents. Elsewhere ice spills out into the Northern Pacific Ocean through the Bering Strait.

The distribution and extent of sea ice at any particular time is a response to several different variables including temperature, wind, ocean currents, tides, the Coriolis Force, sea surface tilt and the inherent stress properties of the ice. However, annual climatic cycles (NSIDC, no date, a) are responsible for the prominent seasonal pattern of Arctic sea ice change. The maximum winter extent of Arctic sea ice occurs in March and averages around 15.8 million km^2. During the summer, melt conditions reduce the area of sea ice to a minimum of about 6.7 million km^2 by mid-September, although, in recent years, this minimum has been achieved several days earlier than the average for the twentieth century.

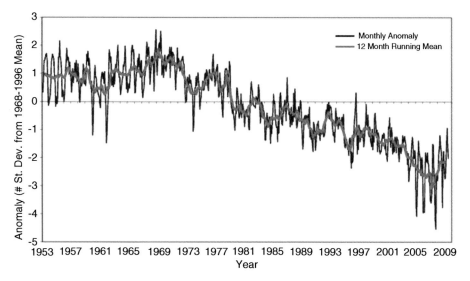

Figure 2.9 Arctic sea-ice extent standardized anomalies, January 1953 – July 2009 (source: http://nsidc.org/sotc/sea_ice.html accessed 06 04 10)

The pattern of change of Arctic sea ice has become clearer since 1979, when satellite monitoring of sea ice, using microwave radiometer measurements, was introduced. Although there is considerable annual variation a downward trend in the March maximum and the September minimum of sea ice extent during the last 30 years is clear (Figure 2.9). During the 1990s, September minima set four records for the smallest aerial extent of Arctic sea ice ever observed. Since 2000 further records have been set: 2007 was exceptional and the famous 'Northwest Passage' between the Atlantic and Pacific Oceans became navigable throughout its length. Substantial areas of multiyear ice were lost. Stroeve *et al.* (2007) suggested that change is occurring faster than the IPCC 2007 (Solomon, *et al.*, 2007) models predict because the models do not adequately incorporate atmospheric variations. Correlation of Arctic ice with general circulation models (GCMs) was also inhibited because only ice *extent* could be measured effectively whereas GCMs are more accurately related to ice *volume*, that is the product of both ice extent *and* ice thickness. Only recently have remote sensing techniques been developed which are able to provide an acceptable degree of precision in the determination of ice thickness and therefore ice volume. This should give more confidence in explaining the reduction of sea ice and predicting the scale of likely hazards related to Arctic sea ice changes.

Lenton *et al.* (2008) suggest that a 'tipping point' for *complete* loss of summer Arctic sea ice may be reached within the next 10 years. This does not mean that the Arctic will become ice-free by 2018 but that by 2018, feedbacks in the environmental system will make it *inevitable* that the Arctic will experience an ice-free summer. Wieslaw Maslowski's strongly contested, reported suggestion (BBC NEWS, 2007) was that this could occur as soon as 2013. Mark Serreze's more conservative view (BBC NEWS, 2007), just after the spectacular minimum of 2007, was 2030. The basis for this prediction is that there is more thin first-year ice vulnerable to summer melting, a longer melt season enhanced by the albedo feedback and more warmth in all seasons making recovery during a series of cold years less likely. It remains to be seen whether the downward September minimum trend continues and what the scale of change turns out to be.

Whatever the long-term future of Arctic sea ice turns out to be, the Arctic is a hazardous place to live, work and travel and has been since it was first occupied by humans. Extreme temperatures combined with cyclical and fluctuating ice conditions have ensured that life around sea ice is never risk-free. Although a great deal of traditional knowledge and experience concerning the nature and behaviour of Arctic sea ice has been acquired over the centuries, recent changes to the system are unprecedented in recorded history. Rapid climate change is becoming increasingly obvious and scientists are forced to respond equally quickly in order to keep track of these changing circumstances.

Whatever the eventual outcomes, the effects of these changes are already creating new hazards for Arctic communities (ACIA, 2004, 2005). Their traditional knowledge no longer seems to be as reliable as it once was because ice arrives later, is thinner and melts or retreats earlier and ecological systems no longer respond in familiar ways. Not all hazards are related to present changes in the natural system. Today, tourist ships have largely replaced exploration vessels looking for the 'Northwest Passage', but their personnel must be as aware of traditional hazards, such as ice entrapment or collisions with thick multiyear ice and ice ridges, as were their eighteenth and nineteenth century counterparts. The next section will consider the impacts of Arctic sea ice hazards in more detail.

2.4 Impacts

Hazards associated with sea ice are due not only to its presence (*ivu*, entrapment, obstruction) but, perhaps more significantly these days, to its reduced extent and thickness (access, ice sheet stability, coastal erosion). If, in the foreseeable future, it is partially or completely absent for at least part of the year, regional and global climate is likely to be affected, although considerable uncertainty surrounds any predictions in the absence of perfect global climatic models. Some impacts of sea ice are felt directly, for instance, override of ice onto the shore (*ivu*), or failure of thin ice to support a load. Other impacts are less direct. For example, when ice fails to form or withdraws further from the shore this extended wave fetch may expose coasts to increased erosion under stormy conditions. Some impacts are short-term and are already being felt, such as a shortened hunting season. In contrast, climate change is a longer-term, larger-scale issue and less predictable. The sea ice hazards, summarized below, will now be considered in more detail.

Direct impacts:

- entrapment,
- reduced access,
- *ivu.*

Indirect impacts:

- coastal erosion,
- increased tourist shipping, exploration and exploitation,
- ecology including UV,
- iceberg severity,
- ice sheet stability,
- climate change and the MOC.

2.5 Direct impacts

2.5.1 Entrapment

From a historical, European perspective sea ice gained a reputation as a destroyer of ships and an obstruction to exploration and commerce. Besides the 'Northwest Passage', the location of the north magnetic pole was a key objective of exploration, and whaling was an important economic activity during the eighteenth and nineteenth centuries. There are numerous accounts (see for example Fleming, 1998) of ships having to overwinter, trapped in sea ice, sometimes by design but also due to the rapid early onset of winter. Some of these ships survived until the spring melt season released them, but others were crushed by stress caused by the mass and movement of the ice. Eventually, heavily reinforced ships actively sought entrapment for scientific purposes. For example, the Norwegian vessel, *Fram* spent three years trapped in Arctic sea ice in order to determine the pattern of drift. Today this activity still continues, though often with the aid of remotely sensed buoys rather than ships. A recent example of sea ice entrapment occurred in the Antarctic when the tourist vessel Kapitan Khlebnikov was trapped for several days by sea ice in the Weddell Sea (The Guardian, 2009). As the next section will show sea ice can move around very rapidly and unexpectedly, and will remain a hazard to shipping even if, as predicted, the Arctic Ocean does become ice-free for a time during the summer.

2.5.2 Access to Arctic sea ice

Although, for some, sea ice only presents problems, for many indigenous communities of the Arctic it is an integral part of their life; it provides transport connections, and a platform for harvesting culturally and economically important resources (Ford, 2009). Ideally, the ice needs to be smooth, continuous and thick, but these properties appear to be less common and this is creating problems for Arctic people, as their own testimony shows (see Box 2.1 for details). Broken ice with leads requires both land-based and water-borne transport. Broken ice caused by ice-floe impacts often generates rougher surfaces with pressure ridges. These can help to maintain ice stability but hinder progress. Sea ice is perfectly capable of supporting considerable weight, as the presence of large aircraft at the North Pole (when it is not ice-free!) testifies. However, there are limits to its strength (Table 2.1), and the reduction in its thickness, has already forced Arctic travellers to become more wary of the risk of movement across sea ice. One man at rest requires 13 cm of sea ice. 18 cm is needed by a 0.4 ton load moving slowly. A fully-laden skidoo and trailer probably needs twice this thickness at least. Snowmobile travel carries greater risk (Ford, Smit and Wandel, 2006). Indigenous communities are facing significant impacts on their traditional economies and cultures due to these problems of access.

As mentioned earlier, shorefast ice is stabilized by different means of grounding (George *et al.*, 2004). If these effects do not occur, shorefast ice is vulnerable to winter and early spring break-off ('calving') events. For example, in 1997, 12 whaling camps broke away requiring 142 people to be rescued by helicopter. Furthermore, break-offs tend to leave shorefast ice thinner and weaker. The increased frequency of these 'calving' events reflect shorter sea-ice seasons, which provides less time for thick ice to form and multiyear floes to freeze in and less time for stabilizing grounded pressure ridges to form. These changing conditions are obviously having an effect on existing sea-ice knowledge and require a new information base to be developed from new experiences, although the rate of change is likely to

---

Box 2.1 Quotations and comments from Inuit elders regarding sea ice in the Arctic

The Arctic indigenous population maintains a strong connection with the marine environment through hunting and fishing. Access to food supplies is often related to travel access and safety. However, in many parts of the Arctic, the increasing unpredictability of sea ice as a travel platform, and its more broken nature, makes travel dangerous. Less thick multiyear ice means spending more time on first-year ice which is less safe, not only for humans but for important mammal species including the ringed seal and polar bear. Reduced ice cover in summer enhances the effect of storms and increases the dangers of boat travel.

The following quotations, concerning the state of sea ice, are from ACIA (2004, p. 97).

'Long ago, there was always ice all summer. You would see the [multiyear ice] all summer. Ice was moving back and forth this time of year. Now, no ice. Should be [multiyear]. You used to see that old ice coming from the west side of Sachs. No more. Now between Victoria Island and Banks Island, there is open water. Shouldn't be that way.'
Frank Kudlak, Sachs Harbour, Canada, 1999

'I know that today that seals, it might be because of early spring break-up or that they are out on the ice floes, that the seals are nowhere.'
Man aged 62, Kuujjuaq, Canada

'When there is lots of ice, you don't worry too much about storms. You get out there and travel in between the ice [floes]. But last few years there has been no ice. So if it storms, you can't get out...'
Andy Carpenter, Sachs Harbour, Canada, 1999

The following quotations, concerning the approach to hunting by young Inuit, are from Ford, Smit and Wandel (2006):

'[The younger generations] are not out there hunting.'
Tommy Tatatuopik

'It is more dangerous for [the younger generation] because they don't know the conditions, what to avoid.'
Kautak Joseph

'We don't share as much as before.'
David Kalluk

The following are comments from the 'Inuit Observations of Climate Change' project, a joint exercise by the Inuvialuit of Sachs Harbour (Ikaahuk), Banks Island, Canada and the International Institute for Sustainable Development (ACIA, 2004; IISD, 2010).

- Multiyear ice no longer comes close to Sachs Harbour in summer.
- All summer no ice in water.
- Less sea ice in summer means that water is rougher.
- Less ice cover in summer means rougher, more dangerous storms at sea.
- Open water is now closer to land in winter.
- Too much broken ice in winter makes travel dangerous.
- Unpredictable sea ice conditions make travel dangerous.
- Less multiyear ice means travelling on first-year ice all winter; this is less safe.
- It is more difficult to hunt seals because of a lack of multiyear ice.
- Hunter cannot go out as far in winter because of a lack of firm ice cover.
- Earlier break-up. Freeze-up later. Last year freeze-up was 3–4 weeks later due to warm water.
- Fewer polar bears are seen in the autumn because of lack of ice.
- Last year, early July, skinny young seals. Ice goes too quick and mother seals go.

Table 2.1 Ice thickness versus ice strength –fresh ice includes lake and river ice (source: http://ice-glaces.ec.gc.ca/App/WsvPageDsp.cfm?Lang=eng&lnid=10&ScndLvl=no&ID=10167– accessed 08 03 10)

Load	Operation	Fresh ice	Sea ice
One person	At rest	8 cm	13 cm
0.4 ton	Moving slowly	10 cm	18 cm
2-ton vehicle	Moving slowly	25 cm*	40 cm*
10-ton tracked vehicle	Moving slowly	43 cm	66 cm
13-ton aircraft	Parked	61 cm	102 cm

* Estimated numbers, not provided in original table.

make adjustment difficult. Hunters often have other full- or part-time jobs and are less likely not to hunt after time has been set aside even if the conditions are not ideal. An added factor is the introduction of new technology, including remote-sensing information, regional weather forecasting, faster snowmobiles, GPS (global positioning systems) two-way radios and the availability of rescue services (helicopters and boats). Changing technology is modifying risk perception and related behaviour, not always to the benefit of individuals, who may be encouraged to become overconfident and take more risk than is justified by increasing hazards (Ford, Smit and Wandel, 2006).

2.5.3 Shore ice override (ivu)

Although the perils of sea ice were, first, widely publicized in urban Europe and America, its impact had been felt, especially by coastal indigenous populations throughout the Arctic for millennia. Shore ice override (called *ivu* by the Inupiaq of Alaska) occurs wherever sea ice is driven onshore during storms and high wind events. Kovacs and Sodhi (1980) provide an extensive historical review of this phenomenon and discuss its technical aspects. The incentive for their report was concern for the safety of artificial islands and other structures built to support exploratory drilling activities in the Beaufort Sea hydrocarbon field as examples had been reported of ice shearing the top off barrier islands (Shapiro, Bates and Harrison, 1978 quoted in Kovacs and Sodhi, 1980). *Ivu* tend to occur in autumn and early winter before the shorefast ice is too strong and when storms are most frequent, but they can occur at any time of the year and are not necessarily associated with strong winds. The scale and character of an *ivu* depends on the magnitude of the driving forces (ice sheet size, wind, currents, ice thickness, ice strength) and the resistance of the beach (friction between ice and beach, beach slope and height).

In most cases *ivu* affect gently sloping beaches and usually penetrate only 10–20 m inland, but sea ice has overridden the steep, 9 m high bluff at Barrow,

Alaska and destroyed structures and taken lives (Kovacs and Sodhi, 1980). Exceptionally, ice override has been recorded up to 0.8 km inland (Ross, 1835 quoted in Kovacs and Sodhi, 1980).

Ice impacts shores and beaches in different ways. During ice *shove*, the beach is ploughed up to form a ridge of sediment which may be 5 m or more in height. When ice *pile-up* occurs, ice is stacked in layers or jumbled masses on the beach as the sea ice breaks and shears over the ice in front. Ice pile-up can reach 20 m but rarely extends more than 10 m inland. Ice *ride-up* refers to the situation when the shore ice moves as a single layer, frequently more than 50 m inland of the beach. It has been estimated that the distributed force required during ice-piling or ride-up varies from 10 to 350 kPa. Field observations revealed that shore ice pile-up or ride-up appears to occur within a period of less than 30 min (Kovacs and Sodhi, 1980).

Ivu regularly affect the north coast of Alaska where there are several settlements and the Prudhoe oil field. In the 1880s a bluff-top ice override trapped and killed a man in a subterranean entryway to a traditional-style sod house at Utkiave (Barrow) (Spencer, 1968). Brower (1960, quoted in Kovacs and Sodhi, 1980) described many boats being crushed and buried at Barrow in 1937. The same author also claimed that on one occasion several Barrow houses were crushed with everyone inside but Kovacs and Sodhi, (1980) believe that this may refer to an abandoned village west of Barrow. The village of Nuwuk on the spit of Point Barrow, further east, may also have been abandoned because of ice override (Hume and Schalk, 1964). Power poles were knocked down along a 700 m stretch of beach near Barrow on 2 January 1978 by a 1.3 m thick ice override (personal communication from L. H. Shapiro (1978) to Kovacs and Sodhi, 1980). In another incident at Kotzebue, Alaska, a summer cabin was pushed off its foundations and smashed, and a 25–30 cm diameter wooden pile supporting the cache platform (for keeping food away from bears) was sheared off (personal communication from A. G. Francis (1979) to Kovacs and Sodhi, 1980). A notable event occurred at Point Barrow, Alaska in 2006 when a 6–12 m high ridge of ice piled up along the shore (Figure 2.10(a)). The State of Alaska Division of Homeland Security and Emergency Management monitored the event as the North Slope Borough authorities were concerned that city pipelines, utilities and roads along the shoreline may be threatened (DHS&EM, 2006). In the event the vulnerable infrastructure was not damaged but one lane of the coastal Stevenson Road was blocked and ice had to be moved by a bulldozer (Associated Press, 2006; Figure 2.10(b)) to clear the road and allow access for hunters and others to and from the shorefast ice along the coast. An *ivu* may have been the cause of the prehistoric collapse of a house at Utqiagvik, Barrow, Alaska, that resulted in the preservation of the 'Frozen Family' (five females) probably some time between 1500 and 1800 (Lobdell and Dekin, 1984). However, it is usually damage rather than death that results from an *ivu* event.

Figure 2.10 Ice pile-up at Barrow, Alaska, January 2006: (a) source: IARC (2006), photo: Alice Brower; (b) source: http://www.worldproutassembly.org/archives/2006/01/arctic_ocean_ic.html - accessed 27 03 10 The Associated Press, Friday, January 27, 2006 10:23 PM EST

2.6 Indirect impacts
2.6.1 Coastal erosion

While some erosion is achieved by the advance of sea ice *onto* the coast, more erosion of the coast is possible when the extent of sea ice declines. Withdrawal from the coastal zone exposes a longer sea fetch (the distance across which waves are generated by wind) which can be whipped into larger waves by storm-force winds. The Beaufort Sea coastlands in northwest Canada are especially vulnerable because of the high content of ground ice which melts on exposure to warming Arctic Ocean water and the problem is exacerbated by sea level rise. The presence of significant settlements, such as Tuktoyaktuk (NWT, Canada) and Barrow (Alaska, USA), and existing (Prudhoe Bay) and potential hydrocarbon exploration and exploitation has provided the incentive for several investigations of coastal erosion.

Using 1955 USGS topographic maps and 1985 and 2005 Landsat 5 Thematic Mapper data, Mars and Houseknecht (2007) recorded a doubling of coastal erosion along the National Petroleum Reserve, from $0.48\,km^2\,year^{-1}$ during the period 1955–1985 to $1.08\,km^2\,year^{-1}$ during the period 1985–2005. Jones *et al.* (2009; Figure 2.11 (a) and (b)) found retreat along 60 km of the Alaskan Beaufort Sea coast had increased from $6.8\,m\,year^{-1}$ between 1955 and 1979 to $8.7\,m\,year^{-1}$ between 1979 and 2002 and to $13.6\,m\,year^{-1}$ between 2002 and 2007. During the record Arctic sea ice loss year of 2007 up to 25 m of erosion occurred in some parts of this coast even without a significant westerly storm event.

(a)

(b)

Figure 2.11 Erosion along the Arctic coast of Alaska, USA – photos: (a) Christopher Arp, (b) Benjamin Jones, (source: Jones *et al.*, 2008 http://geology.com/usgs/alaska-coastal-erosion/ accessed 29 03 10)

One concern is the possible erosion of toxic drilling wastes from old exploratory wells in the National Petroleum Reserve (NFIC, 2010). Another cause for concern is the loss of Inuit cultural and historical sites (Jones *et al.* (2008). Esook, an early trading post has already been lost to erosion and Qalluvik, an abandoned Inupiaq village site is vulnerable to coastal retreat. In 2008, archaeologists were working at Point Barrow to record an ancient Inupiaq cemetery, with grave goods including a harpoon point dated to between 820 and 1020, and relocate it inland.

Further west in Alaska, the settlement of Shishmaref (Figure 2.12 (a) and (b)) is located in a very vulnerable position on a gravel bar. During October

Figure 2.12 (a) Shore protection; (b) erosion at Shishmaref, Alaska, USA (source: Kawerak, no date)

Figure 2.13 View of Tuktoyaktuk, 26 June 2001. Note sea ice in harbour. A is the area of protective concrete slabs and B is the zone of erosion where protection has been ineffective. This is the area where buildings are most vulnerable to erosion (source: http://www.assembly.gov.nt.ca/_live/pages/wpPages/maptuktoyaktuk.aspx accessed 07 04 10)

1997, a severe storm eroded over 9 m of the north shore, requiring 14 homes and the National Guard Armoury to be relocated (Kawerak, no date). Five more homes were relocated in 2002. Other storms have continued to erode the north shore an average of 1 to 1.5 m year^{-1}. In July 2002, residents voted to relocate the community although this has yet to happen. At Unalakleet, the 2009–2013 Development Plan, including proposals for a revetment to save coast and water supply lines, has been submitted to the Community of Unalakleet and the Bering Strait Development Council (Kawerak, 2009).

In Canada, the largest (956 at the 2007 census; LANT, no date) Arctic settlement with erosion problems is Tuktoyaktuk, built on a spit of land extending into the Arctic Ocean near the mouth of the Mackenzie River (Figure 2.13). It is adjacent to the best natural harbour in the western Arctic, an essential base for Beaufort Sea hydrocarbon exploration. The vulnerability of Tuktoyaktuk stems from a complex of factors: (a) a maximum cliff height of only 5 m; (b) an ice-rich erodible permafrost substrate; (c) a long-term rate of sea-level rise between 1 and 4 mm year^{-1}; (d) a variable wave fetch depending on the distance of the summer pack-ice from the shore; (e) onshore west to north-west autumn storms creating storm surge tides up to 2.5 m. Average coastal retreat rates have been about 1 m year^{-1} during the second half of the twentieth century (Harper, 1990), but much of the erosion occurs periodically during large events. (see Box 2.2 for further details concerning erosion at Tuktoyaktuk, and the measures that have been taken to mitigate the problem).

Box 2.2 Coastal erosion at Tuktoyaktuk, NWT, Canada

Tuktoyaktuk occupies a prime position beside an excellent harbour on the coast of the Beaufort Sea near the mouth of the Mackenzie River (Figure 2.13). However, the low-lying, ice-rich, permafrost topography on which the settlement is located is very vulnerable to erosion. Significant erosion events have often occurred when strong storms have coincided with periods of exceptional fetch (Figure 2.14). One such event occurred between 28 and 31 August 1987, when pack ice in the Beaufort Sea retreated creating an ice-free fetch of almost 500 km. Prolonged northwesterly winds of over $50\,km\,h^{-1}$ at Tuktoyaktuk generated a wave height of 4 m at an oil well drilling site 45 km offshore in 32 m of water, and a storm surge of 1.4 m at Tuktoyaktuk on both days, compared to the normal wave amplitude of a few tens of centimetres. Similar conditions applied between 21 and 23 September 1993. This time an ice-free fetch of over 300 km and gusts of $80\,km\,h^{-1}$ and sustained winds of $50\,km\,h^{-1}$ for 30 h, generated a near-record positive storm surge of 2.17 m above chart datum. The storm overtopped protective sandbagging and eroded up to 8 m. This was serious but not as bad as a storm in 1970 that caused 13.7 m of cliff-top erosion in one day. An oil exploration-related analysis (McGillivray et al., 1993) predicted an increase in the length of open water seasons in the Beaufort Sea and the increased likelihood of severe storms coinciding with sufficient fetch to generate large waves.

Johnson et al. (2003) provide a useful review of the problem and suggest an adaptation strategy. There have been several studies on erosion at Tuktoyaktuk (e.g. Government of Canada, 1976; Aveco, 1986; UMA, 1994) to determine the best method of mitigation. These included beach nourishment, Longard tubes, groins, longshore protection and an offshore breakwater. A combination of Longard tubes (geotextiles) and groins built in 1976 lasted only 5 years before it was destroyed, partly by vandalism. Beach nourishment and sandbagging lasted little better and 50% of the area was destroyed in the 1993 storm. In 1998, 40 monolithic slabs were installed (Trillium, 1997) over geotextile and a gravel pad, but the budget only

Figure 2.14 Sketch diagram of coastal erosion at Tuktoyaktuk (yellow circle) showing retreat of sea ice and increase in fetch which enhances wave height and power

extended to 100 m of coastline near the end of the spit. A prediction of future shoreline positions for the next 10 and 25 years (Solomon, 2002) and a risks assessment (EBA, 2002) indicated that 10 buildings would be impacted after 10 years and 15 after 25 years. Given the difficult physical conditions of the site, mentioned elsewhere, coupled with likely longer open water seasons, accelerated sea-level rise and more rapid permafrost degradation the probability of long-term protection for Tuktoyaktuk is not high. The most appropriate options, according to the UMA (1994) study are:

- annual replenishment of the bank with sand and gravel – capital cost Can$600,000 and a life cycle cost of Can$2.8 million;
- stacked overlapping gravel bags costing Can$5.5 million and a life cycle cost of Can$9.6 million;
- concrete mats joined by chains with a Can$8.1 million capital cost and a life cycle cost of Can$9.1 million.

The last of these options is considered to be the best, but trials have not been fully monitored. Interestingly, Shaw *et al.* (1998) judged the Beaufort Sea coast generally to be one of the three Canadian coastal areas most sensitive to sea-level rise and recommended that societal response to changes in sea level should favour retreat and accommodation strategies. Apart from some redevelopment within the exiting community but away from the vulnerable coastline, complete resettlement is another adaptation strategy that will have to be assessed.

2.6.2 Access to the Arctic Ocean space following ice loss

It is perhaps too soon to judge the precise effects of an increase in shipping in Arctic waters but already, in anticipation of a substantial decrease in Arctic sea ice, countries are attempting to establish sovereignty over portions of the Arctic Ocean area for mineral rights. The USGS has estimated that there are considerable hydrocarbon resources (25% of known reserves?) in the Arctic basin. Russia recently deposited its flag on the sea bed at the North Pole! Canada and the USA are in dispute over access to waters around the Canadian Archipelago. The EU has published guidance on likely problems associated with the new circumstances in the Arctic region (EU, 2008; The Guardian, 2008). North-west and north-east shipping routes will be established between the Atlantic and Pacific Oceans. Tourism is increasing as sea ice extent diminishes. Tourism will be encouraged by programmes such as Billy Connolly's 'Journey to the Edge of the World', which saw him in the Canadian Arctic. All these developments are likely to impact the Arctic environment and its inhabitants. In *After the Ice* Alun Anderson (2009) provides an interesting commentary on several issues surrounding sea ice depletion in the Arctic, including potential effects of greater access by commercial and tourist interests. If access improves, *The Scramble for the Arctic* (Sale and Potapov, 2010) may become a reality and is unlikely to be achieved without some adverse impacts on the environment.

2.6.3 Iceberg severity

The relationship between sea ice and icebergs is discussed in more detail in Chapter 4. Essentially, sea ice helps to protect icebergs from the erosive effects of waves. Marko *et al.* (1994) showed that the severity (survival and maximum extent) of an iceberg season around the Grand Banks off the coast of eastern Canada is related to sea ice density. There is a connection between annual iceberg severities and the preceding January sea ice extent in the Davis Strait between Baffin Island, Canada and Greenland. Consequently the risk of iceberg impact on shipping and fixed installations increases when icebergs are given the protection of more extensive sea ice.

2.6.4 Ice sheet stability

This is another issue that is discussed elsewhere in relation to ice sheets (Chapter 3). Harwood, Winter and Srivastav (1994) concluded that a coherent girdle of sea ice surrounding the Antarctic continent is critical in helping to maintain the stability of both the West and East Antarctic Ice Sheets by influencing air and water temperatures. Sea ice may also play a protective role in shielding ice shelves and marine glaciers from wave attack, as it does in the case of icebergs. Rennermalm *et al.* (2009) found a covariability between patterns of sea ice and ice sheet melt, suggesting that less sea ice may enhance melting over the Greenland Ice Sheet.

2.6.5 Ecosystems

Two creatures above all others share iconic status in polar regions: the penguin in Antarctica and the polar bear in the Arctic. They share a reliance on sea ice as a base for at least some of their life-cycle activities. Penguins are considered in relation to ice sheets. Here, more attention will be paid to polar bears and other members of Arctic ecosystems. Simpkins (2009) provides a useful review of the situation with regard to marine mammals. These animals are considered to be in a difficult situation because they are adapted to life where seas are at least seasonally ice-covered and the general extent of Arctic sea ice is progressively diminishing (NSIDC, 2010d; Moore and Huntington, 2008). Animals such as polar bears live long and reproduce slowly and, although they have survived previous glacial and interglacial changes, the key issue seems to be whether they can adapt sufficiently quickly to adjust to *present* rapid rates of change in their habitat. Impacts of sea ice loss on different animals have recently been reviewed (Richter-Menge and Overland, 2009). Impacts vary depending on specific ecological requirements and the strength of the link between the animal and sea ice (Moore and Huntington, 2008; Table 2.2).

For example, polar bears and ringed seals may, in the future, survive longer in refugia in the Canadian Arctic Archipelago where sea ice is thicker and more persistent. Other, sub-Arctic, migratory species may be able to access areas from which sea ice had previously excluded them. However, if this requires travelling further and expending more energy as the summer sea ice edge retreats it may turn out to be a negative impact (Freitas *et al.*, 2008; Durner *et al.*, 2009). While some polar bear groups are currently stable or even increasing others, such as those in western Hudson Bay, show lower body condition and reduced survival rates as reduced sea ice forces bears to fast on shore longer (Stirling, Lunn and Iacozza, 1999; Regehr *et al.*, 2007). Jay and Fischbach (2008) reported that Pacific walruses hauled out on the shores of Alaska and Russia in unusually large numbers and in new locations during the record low summer sea ice extent of 2007. This redistribution resulted in an increase in trampling deaths as walruses on shore stampeded in response to terrestrial disturbances. There is concern that they could deplete nearshore

Table 2.2 Relationship of Arctic marine mammals to sea ice (source: Simpkins, 2009)

Species	Primary diet	Relationship with sea-ice habitat
Bowhead whale	Zooplankton (filter feeder)	Forage in productive marginal ice zone
Beluga whale	Diverse fishes and invertebrates	Refuge from predation? Access ice-associated prey
Narwhal	Ice-associated and benthic fishes (deep diver)	Forage in areas of very dense ice
Ringed seal	Diverse fishes and invertebrates	Resting and nursing platform Access ice-associated prey
Bearded seal	Benthic invertebrates	Resting and nursing platform Access to benthic foraging grounds
Walrus	Benthic invertebrates	Resting and nursing platform Access to benthic foraging grounds
Polar bear	Seals (primarily ringed) and other marine mammals	Hunting platform

benthic resources if they are forced to maintain this pattern of behaviour. A study of Narwhals (Laidre and Heide-Jørgensen, 2005) highlights the importance of regional variations in sea-ice extent. These authors detected a reverse of the normal decreasing sea-ice trend in parts of Baffin Bay. Here, an increase of winter sea ice during the last 50 years meant a decrease in the amount of winter open water, and an increase in the risk of entrapment of these sea mammals. On land, observations of warmer temperatures and increased green vegetation in the summers with lower sea-ice concentration, confirm models that predict this correlation (Richter-Menge and Overland, 2009). In the atmosphere, ultraviolet (UV) radiation in Antarctica has risen as a result of the 'ozone hole' above that continent, and there is also potential for future ozone depletion over the Arctic. However, while ozone depletion will probably remain the major influence on UV radiation over Antarctica, in the Arctic, loss of sea ice is likely to be the main cause of an increase in the impact of underwater UV (Vincent and Belzile, 2003).

Besides these natural reactions to sea-ice retreat, there is the broader issue of human access and impact on the Arctic Ocean as it becomes more accessible. If current trends continue, not only will the iconic Northwest Passage become ice-free on a regular basis during the summer, but it becomes possible to envisage a much more extensive ice-free area of the Arctic Ocean, perhaps within a couple of centuries (Lenton et al., 2008). For some (e.g. the USA and Russia (former Soviet Union)), the Arctic has been a focus of interest for many years from the military perspective. However, there are other reasons (fishing, shipping, hydrocarbon and mineral exploitation, in particular) why the Arctic is of wider international interest and these were highlighted in a recent report by the WWF (Huebert and Yeager, 2008). One concern is that more extensive fishing activities will lead to overfishing, inappropriate practices such as bottom trawling and a loss of biodiversity. Increased shipping, in the form of international bulk carriers, northern resource development support vessels, fishing fleets, tourist cruise ships and surface naval vessels (submarines already use the Arctic Ocean), introduces the threats of oil spillage, introduction of alien species and general environmental disturbance such as the disruption of fish foraging areas or migration routes. Hydrocarbon exploitation brings with it a familiar list of potential impacts such as oil spillage, air pollution and environmental disturbance. Apparently, new ice-capable vessels are under construction as shipping companies anticipated the changes in sea-ice extent.

2.6.6 Climate change and the MOC

Modification of existing climate regimes is probably the least predictable impact of decreasing Arctic sea ice, due to the scale and complexity of the climate system. Nevertheless, some changes to atmospheric and associated oceanic components have already been measured, although models are not yet

Table 2.3 Albedo of some surfaces associated with Arctic sea ice

Surface	Albedo
Snow on ice	0.90
Ice	0.50–0.70
Shallow melt ponds on ice	0.20–0.40
Deep melt ponds on ice	0.15
Ocean	0.06

capable of predicting what might be the long-term outcome. One critical positive feedback loop exists between snow, ice and ocean with respect to albedo or surface reflectivity (Table 2.3).

White snow reflects most radiation, but ice is also a good reflector. When the ice goes, there is also nowhere for the snow to rest and the open ocean will reflect much less radiation and consequently absorb much more heat. In turn, this will cause more melting, making it difficult for ice to reform in the winter. This means that more warmth will escape to the atmosphere before melt-back. Surface air temperature will rise and more heat will be available to slow down freezing onto the base of the sea ice. The ice will thin, melt more easily and expose more ocean to solar radiation heating and so on in a large-scale, positive ice–albedo feedback, in which the initial temperature change is amplified, an effect referred to as Arctic amplification (Holland and Bitz, 2003; Serreze and Francis, 2006). At present sea-ice cover is sufficient to insulate the ocean from the atmosphere effectively, but as more ice is removed from the system this will no longer apply and the Arctic will move into a new state. Although there is still uncertainty about when this is likely to be achieved (e.g. Winton, 2006), as climate models continue to show considerable inter-model scatter (Holland., Serreze and Stroeve, 2010), there is some evidence that summer sea-ice loss is affecting general atmospheric circulation in the following winter season (Overland, Wang and Walsh, 2009). By raising upper atmosphere air temperatures, the normal north-to-south pressure gradient that drives west-to-east airflow is weakened, thereby contributing to changes in atmospheric circulation.

Changes to the MOC are discussed more fully in relation to ice sheets. The export of fresh sea ice from the Arctic is likely to add to any tendency for the MOC to change but this, like climate change, is very uncertain and not a short-term hazard.

2.7 Mitigation

2.7.1 Introduction

During the last few years Arctic sea ice has received probably more media attention than any other component of the cryosphere, with front page headlines in *The Independent* (16 December, 2008) declaring 'Arctic at tipping point', supported by two pages of colour photographs and diagrams illustrating the explanatory text. This followed the suggestion of Lenton *et al.* (2008) that Arctic sea ice may be close to a threshold beyond which it moves rapidly to a mode with complete absence of sea ice during the summer. Ironically, such a tipping point is the aspect of Arctic change which is least amenable to any form of direct mitigation, as current projections for the significant reduction of carbon dioxide in the atmosphere appear to extend well beyond the date proposed for complete loss of summer sea ice in the Arctic. With the Arctic expected to receive more additional warmth than any other area of the planet (Solomon *et al.*, 2007), substantial changes seem inevitable, and a range of mitigation strategies will be required to cope with the impacts of a decreasing Arctic sea-ice cover (ACIA, 2004, 2005). Strategies available for the mitigation of the Arctic sea-ice hazard, related to both the presence and absence of sea ice, will now be considered in more detail.

2.7.2 Sea-ice presence

Several methods are available to mariners, and others who travel in the Arctic, for addressing problems that arise when the ocean is covered with ice. These include: (a) applying knowledge of where the ice is located, and its characteristics; (b) removing ice from the line of travel; (c) reinforcing vessels and 'going with the flow/floe'. The Norwegian *Fram*, amongst other vessels, successfully achieved option (c) and acquired good scientific information about the circulation of ice in part of the Arctic during the course of its three year ice-bound drift between 1893 and 1896 (Nansen, 1897). For many years other specially constructed vessels, known as ice-breakers, have been employed to keep open routes through sea ice for other vessels to follow (option (b)), or have been used, in their own right, to service ice-bound exploration sites, for example. However, the work of these ice-breakers, the safe passage of other vessels and the movement of travellers over the ice are all facilitated initially by option (a) – application of knowledge of the ice.

Most Arctic countries have an organization responsible for providing information about sea ice. One such is the Canadian Ice Service (CIS), initiated as a response to the iceberg hazard. The CIS produces maps of sea ice, colour-coded for age and type, which are supported by a further coding system, known as the 'Egg Code' (Figure 2.15(a)), so-called because of the shape of the line enclosing the data for each site. The data refer to important ice characteristics such as thickness, ice concentration and floe size. The information on sea ice is contained in a manual accessible on the CIS web site (MANICE, 2005). Providing the map user understands the codes, it is easy to find out where the thickest, most concentrated ice is located or where the ice is thinnest and least concentrated. The first area with the thick, concentrated ice can then be

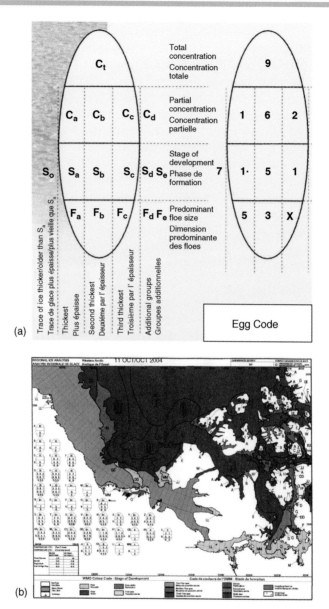

Figure 2.15 (a) Egg code; (b) egg code map of ice conditions in the Canadian Western Arctic Region for 11 October 2004 (source: Canadian Ice Service MANICE, 2005)

avoided if you are in a vessel, or used to your advantage if you are travelling on the ice, while those crossing the ice will want to avoid the second type of area, which may be made even more inaccessible by the movement of vessels seeking an easier passage through less concentrated ice. A short time spent studying Figure 2.15(b) and the CIS web site (CIS, 2003a) should give the reader a preliminary understanding of the benefits to be gained from the 'Egg Code'. NSIDC (2006) and the Arctic and Antarctic Research Institute (AARI) provide similar services for the USA and Russia respectively. At one particular location, Barrow, Alaska, a sea-ice webcam, mounted on a four-storey building in the city, has been set up to illustrate and record the changes to the sea ice that take place in the near-shore area (Barrow webcam, 2010). This provides the local community with some warning of ice movements that may affect the coastline or any people who are active on the ice.

2.7.3 Sea-ice loss

The mitigation methods discussed above are employed against the presence of ice, but increasingly Arctic sea-ice problems are related not to its presence but to its absence, or at least to its thinning or retreat. The eventual scale of sea-ice loss will influence the type of mitigation measures that are taken to reduce impacts. A number of mitigation methods have been applied, depending on the nature and scale of the problem.

Travel on the ice, especially for hunting, is an essential aspect of traditional Inuit culture (ACIA, 2004, 2005). For many Inuit, hunting provides essential foods and other useful products such as clothing. Without access to sea ice, Inuit culture, as currently practised, will be forced to change significantly. This, of course, it has done many times in the past but what may be different this time is the rapidity of the rate of change. Consequently, reducing or reversing the increase in GHGs would seem to be an obvious mitigation strategy to employ on behalf of indigenous populations wherever they occur in the Arctic. However, alternatives to traditional hunting are already in place in less remote settlements, in the form of 'western' stores, for example, and these are patronized especially by younger Inuit (Fred Wolkie, personal communication, 1999; see also Box 2). It is conceivable that the younger generation of Inuit, influenced by a wage-based economy, easier travel and different expectations, coupled with the external imposition of harvesting quotas (Ford, Smit and Wandel, 2006), will adopt a lifestyle that is no longer significantly dependent on access to sea ice. However, for others in the community, the capacity to adapt to the current rate of change may be limited. For these people and indeed for all Inuit 'the Inuit Circumpolar Conference ... framed the issue of climate change in a submission to the United States Senate as an infringement on human rights because it[s effects] restrict access to basic human needs as seen by the Inuit and will lead to the loss of culture and identity (Watt-Cloutier, 2004 quoted in Parry *et al.*, 2007). Only time will tell whether such mitigation strategies have been successful.

Another key impact related to sea-ice loss is coastal erosion. This is not a simple issue as sea-ice retreat is only one of a number of factors, including ground-ice content, sea level and weather conditions, which contribute to the problem. Reducing climate warming would help to solve the problem of sea-ice loss but judging by past attempts to agree methods, this is a long-term aspiration. Meanwhile, erosion continues. Shorter-term solutions are coastal protection (Box 2.2) and population relocation. As demonstrated at Tuktoyaktuk, the protection methods are not equally successful. The most effective methods are generally most expensive, and this can be a serious difficulty for small communities. Both Tuktoyaktuk in Canada and Shishmaref in Alaska, USA, have contemplated moving to alternative, less vulnerable sites as a mitigation strategy against coastal erosion. Shishmaref residents actually voted for this option in 2002 (Kawerak, no date), but those in Tuktoyaktuk seem keen to protect their present position, although this will almost certainly depend on external funding.

Finally, there is the question of how to deal with direct human impacts on the Arctic Ocean, once it is opened up more extensively to different types of shipping and the use of fixed installations. A number of national and international organizations have already been established with the interests of the Arctic in mind. One of these is the Conservation of Arctic Flora and Fauna Committee (CAFF) which came into being in 1991 and evolved into a wider organization, the Arctic Council in 1996 (Huebert and Yeager, 2008). The Arctic Council commissioned CAFF and the Arctic Monitoring and Assessment Programme (AMAP) together with the International Arctic Science Committee (IASC) to undertake a study of the impact of climate change on the Arctic, and these bodies subsequently produced the Arctic Climate Impact Assessment Scientific Report (ACIA, 2004), and a synthesis volume, *Impacts of a Warming Arctic* (ACIA, 2005), designed for more general consumption. This was a valuable initiative in bringing the problems of the Arctic, including sea-ice loss, to the attention of a wider scientific and general public community. However, the Arctic Council is focused on environmental issues and is not financially independent. In the opinion of WWF (2008) this organization and others charged with monitoring ocean conditions and regulating its use will require additional powers in order to see that potential impacts of the changing conditions in the Arctic Ocean are mitigated effectively.

2.8 Summary

Unlike avalanches, the direct impacts of Arctic sea ice are rarely quantifiable. Occasionally fatalities can be directly attributed to sea ice, perhaps when there has been a misjudgement of the precise conditions of the sea ice. Ivu events have caused fatalities, but in practice only rarely. Most sea-ice impacts are less direct in terms of their effects on humanity and natural systems, although not necessarily less significant in the longer term. Coastal erosion is a complex problem. In some places like Tuktoyaktuk and the north coast of Alaska it is severe and progressive and may require abandonment of some settlements.

Even more significant, because a whole culture depends on it, is decreasing access to the ice by the Inuit, and the fauna on which they have depended for so long. As sea-ice extent and thickness declines, Arctic inhabitants become less confident in their environment. The climate is an even larger and less predictable issue. Climate is already warming more quickly in the Arctic than elsewhere and contributing to ice loss. Through the positive albedo feedback, climate itself is likely to be changed, though exactly how remains to be seen.

3
Ice Sheets – Antarctica and Greenland

3.1 Introduction

'Losing Greenland: Is the Arctic's biggest ice sheet in irreversible meltdown? And would we know it if it were?' (Witze, 2008, p. 798). This provocative headline appeared in the journal, *Nature*, in April 2008 in response to the prediction of a tipping point around the year 2300 for substantial, irreversible loss of the Greenland and West Antarctic ice sheets (Lenton *et al.*, 2008). The accompanying editorial argued that 'the entire polar community needs to come together to monitor Greenland's meltdown on a comprehensive scale' (Nature, 2008a, p. 781). However, ice sheet loss is not a new issue, as another *Nature* paper, entitled 'West Antarctic Ice Sheet and CO_2 greenhouse effect: a threat of disaster' (Mercer, 1978) makes clear but the response to Mercer's early warning was more muted. In 1978 global warming had hardly emerged as a problem; now it is high on most environmental agendas as evidence for climate change and rising polar temperature mounts (Solomon *et al.*, 2007). The concern implicit in the titles of the articles mentioned above is therefore well-founded. If the Greenland and Antarctic ice sheets both melt completely, the surface of the global ocean will rise by at least 64 m (IPCC, 2001). Even allowing for the apparent stability of East Antarctica, by far the largest ice sheet component, global sea level is still expected to rise by 12 m, if only the Greenland Ice Sheet and the West Antarctic Ice Sheet melt. It is not surprising then that scientific expertise is currently being directed towards a full understanding of these two great cryogenic systems. Recent innovations in remote sensing have substantially increased our ability to assess the whole of these

---

Cold Region Hazards and Risks, First Edition.   Colin A. Whiteman.
© 2011 John Wiley & Sons, Ltd. Published 2011 by John Wiley & Sons, Ltd.

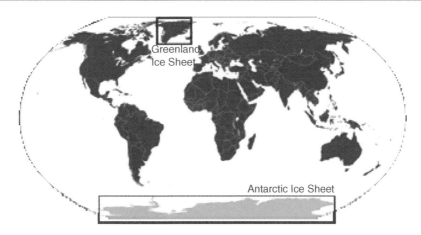

Figure 3.1 Location of Greenland and Antarctic ice sheets (Reproduced courtesy of The National Snow and Ice Data Center, University of Colorado, Boulder.)

massive ice sheets regularly, rather than occasionally at only a few isolated sites, as was the case before the 1970s. What has emerged from the remotely sensed data is that these complex multicomponent systems, are not as stable and slow to change as they appeared but alter their behaviour surprisingly quickly.

The Earth currently hosts two large ice sheets: Greenland (GIS) in the northern hemisphere between latitudes 60°N and 83°N and Antarctica, in the southern hemisphere between 64°S and 90°S (the South Pole) (Figure 3.1). Their latitudinal spread and position in relation to other landmasses and oceans means that, although they possess broadly similar components – ice domes, ice streams and valley glaciers and shelf ice – these two great ice sheets, are different glaciological systems and will be treated separately. Antarctica is to a large extent isolated from the rest of the world's oceanic and atmospheric systems by a strong westerly circumpolar circulation that restricts heat transfer by ocean and atmosphere into the Antarctic region. In contrast, Greenland is influenced by oceanic and atmospheric circulations derived from more temperate regions.

Greenland is broadly a unified system peaking at 3340 m at its summit, whereas Antarctica is usually divided into two distinct subunits, the West Antarctic Ice Sheet (WAIS) including the Antarctic Peninsula, and the more massive East Antarctica Ice Sheet (EAIS) that peaks at 4200 m. East Antarctica is located on land mostly above sea level and is generally considered to be stable over the long term. In contrast, the base of West Antarctica rests extensively on land that is below sea level. As the WAIS thins it will be able to float free of the land which will adjust isostatically more slowly than the ice. This is likely to cause the ice of the WAIS to melt rapidly as relatively warm ocean water gains access to its base. This vulnerability to rapid change explains why the WAIS is currently receiving a great deal of research attention. The same applies to Greenland, where recent investigations have revealed significant increases in near-coastal melting and in ice flow velocity leading to a net loss of ice mass. The timescale for these changes now seems to be measured in hundreds rather than thousands of years. Economic systems will need to adjust to this massive physical change. Hard choices between maintaining the coastal status quo, managing a dignified retreat and total abandonment of coastal defences will be required. Migration may be part of the solution.

Although some parts of the mass of these ice sheets may be moving very rapidly by glacial standards and some outlet glaciers and ice streams do surge periodically, in the current global climate regime neither GIS nor WAIS appear to present a significant hazard due directly to ice *advance*. At present, these ice sheets are much more likely to create problems for humanity due to ice *retreat*, the most obvious hazard being sea level rise, though this is not the only concern. Other indirect impacts related to retreat include increased iceberg flux, reduced ocean salinity, and changes to ecological systems, the Meridional Overturning Circulation (MOC) and global climate itself. The probability of some of these impacts, sea level for example, is high and well understood. Other potential hazards, such as changes to the MOC and global climate, are much less predictable, given our current understanding of these systems. They are, nevertheless, issues for serious consideration now, if the precautionary principle is applied within the context of a longer time perspective. It is noteworthy that the IPCC, in 2007, expressed '*very* high confidence' in some of its statements about likely impacts of changes in Antarctica and Greenland instead of its previous 'high confidence' rating. In IPCC terms this is a significant escalation of concern for the future of these ice-sheet systems and their associated hazards.

3.2 Ice sheet systems

Over the last 30 to 40 years, since the advent of satellite remote sensing and global positioning systems, the complexity of the Earth's remaining ice sheets has become increasingly apparent. It is no longer valid to treat these huge masses of ice as uniform entities: that has become clear with the recognition of fast flowing 'ice streams' for example. The ice associated with ice sheets can here be conveniently divided into four broad categories, central ice domes (divide), ice streams and valley glaciers, ice shelves and sea ice, depending on its position within the overall system (Figure 3.2). Although much sea ice is seasonal rather than permanent, and is not strictly part of the ice sheet in the same way as the other components, it nevertheless contributes significantly to ice sheet dynamics and so will be discussed here as part of the overall cryogenic system.

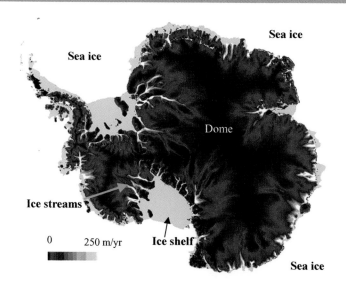

Figure 3.2 Ice sheet structure and flow pattern exemplified by Antarctica (Reproduced with permission from Bamber J.L., Vaughan D.G., Joughin I. (2000). "Widespread complex flow in the interior of the Antarctic Ice Sheet". Science 287: 1248–1250. Fig. 2a doi:10.1126/science.287.5456.1248. © The American Association for the Advancement of Science)

3.2.1 Ice domes

At the core of the ice sheet system is a large dome of ice, typically above 2000 m in altitude, which forms a reservoir of thick, cold, slow-moving, mostly internally deforming ice, located within the area of net ice accumulation. On this central dome, ice accumulates relatively slowly due to the paucity of precipitation in these relatively arid regions (annual precipitation on the polar plateau of Antarctica is less than 100 mm and in central Greenland probably less than 200 mm, which means that both of these dome areas are technically polar deserts (Briggs and Smithson, 1992). As these central domes are high and cold (mean annual temperature on the polar plateau of Antarctica is −50 °C (Hansom and Gordon, 1998) and at the summit of the Greenland plateau, −30 °C (Box, 2002)), they develop little surface melt water. Geothermal heat from below the ice sheet can produce some basal melting to facilitate movement, but ice flow rates are generally low and ice moves only slowly, largely by internal creep, away from the ice divide towards the ice sheet margin. Ice diverges from domes and intervening ice divides and converges toward ice streams nearer the margin. In contrast, marginal areas are lower and warmer, but there is a contrast between GIS and Antarctica (Hansom and Gordon, 1998). In Greenland marginal zones are mainly land based and temperatures are sufficient to cause melting, loss of volume and therefore a decrease in general velocity. In Antarctica the margin is still within the net accumulation zone above the equilibrium line so there is little melting, and both accumulation and flow rates increase with ablation mainly by calving into the sea. However, the pattern is complex. The simple concentric flow pattern is obscured by a complex subglacial topography (Hughes. Denton and Fastook, 1985, Fig 2.2), especially in the WAIS, and in some areas the ice is almost stagnant where it is obstructed by mountains. In these marginal zones ice is frequently evacuated from the ice sheets via rapidly moving ice streams or outlet glaciers.

3.2.2 Ice streams and outlet glaciers

Ice streams are distinctive linear features within the marginal zones of ice sheets and can be clearly distinguished in remotely sensed images or aerial photographs by their obvious flow structure. They begin in strongly convergent areas of the ice sheet especially where ice is concentrated in subglacial troughs. They may be up to 50 km wide, 2000 m thick, hundreds of kilometres long and flow at speeds of over 1000 m year^{-1} (NERC-BAS, 2007). Their rapid movement contrasts with the surrounding ice and at the edges of ice streams deformation causes ice to recrystallize making it softer and concentrating the deformation into narrow bands or shear margins. Flow velocity is dependent on basal sliding and relies heavily on basal melt water or deformable wet sediment slurries for lubrication. Although ice streams account for only 10% of the volume of the Antarctic ice sheet, they may account for 90% of the discharge of this ice sheet through a network of tributaries that penetrate up to 1000 km from their grounding line to the relatively stable and inactive areas of the dome (Bamber, Vaughan and Joughin, 2000; Figure 3.2). The ice streams will therefore be responsible for most of Antarctica's contribution to sea level rise. In Greenland, most ice streams flow into narrower ice tongues or calve directly into the ocean forming icebergs (see Chapter 4). In Antarctica, by contrast, most ice streams and outlet glaciers flow into ice shelves, the third major component of the ice sheet system.

3.2.3 Ice shelves

Ice shelves, as their name implies, are tabular masses of ice that float on the ocean but are attached to and closely integrated with the rest of the ice sheet especially through its ice stream component. Characteristically, ice shelves can be distinguished by their relatively level upper and lower surfaces and their cliff-like outer margins (Figure 3.3) where they calve. Their average thickness is more than 400 m and they calve where the ice typically thins to less than 250 m in thickness (ACECRC, 2009). Antarctic ice shelves, cover an area of about 1.5 million square kilometres in total, and form no less than 11% of the entire ice

Figure 3.3 Larsen Ice shelf showing a lead (open water) where the ice shelf is splitting apart (Reproduced courtesy of The National Snow and Ice Data Center, University of Colorado, Boulder, photo: Ted Scambo, NSIDC.)

sheet (Solomon et al., 2007).The world's largest ice shelves, the Ross and Filchner–Ronne ice shelves, occupy massive embayments between East and West Antarctica. Other large southern hemisphere ice shelves occur around the Antarctic Peninsula, though several, such as the Wordie and Larsen A and Larsen B shelves have catastrophically disintegrated during the last two decades. The Wilkie Ice Shelf was also on the point of disintegration in April 2009. Collapse of these ice shelves is a complex phenomenon (Cook and Vaughan, 2010) probably related to thinning by basal melting, to fragmentation due to surface melting which induces ponding and incision of the shelf by relatively warm melt water and to instability induced by loss of contact with grounding points. Ice shelves around the GIS are much smaller than those in Antarctica, occupying only a few thousand square kilometres and often constituting little more than floating glacier tongues.

As ice shelves float the classical 'Archimedes Principle' implies that they already displace a volume of water equivalent to their own weight and so do not *directly* contribute to sea level rise when they melt. However, ice shelves do have a significant *indirect* effect on sea level rise, in that they obstruct and retard the flow of land-based ice and therefore slow its contribution to global sea level rise. Proof of this comes from the acceleration of glaciers and ice streams formerly inland of ice shelves that have disintegrated. Morris and Vaughan (2003) have suggest that there is a thermal limit to ice shelf life and Cook et al. (2005) have placed this between -5 and $-9\,°C$, as no ice shelves survive in a warmer environment than $-5\,°C$ and none have disintegrated in a colder environment than $-9\,°C$.

3.2.4 Sea ice

The fourth component of the ice sheet system, as defined here, is floating sea ice (see Chapter 2). There are significant differences between the sea ice of Antarctica and that of the Arctic (NSIDC, 2009). In the Arctic sea ice varies from 15×10^6 km^2 in March to 7×10^6 km^2 in September. In Antarctica most sea ice is lost annually and the extent of thicker and deformed multiyear ice is much more restricted than in the Arctic (Table 3.1). The SCAR Report (Turner et al., 2009) predicts that Antarctic sea ice could reduce by a third within the next century.

As already noted, sea ice is not part of the ice sheet, as it forms and melts seasonally, independent of the ice sheet. However, this does not mean that it makes no contribution to the existence of ice sheets and their potential sea level impact. Harwood, Winter and Srivastav (1994) suggested that the girdle of sea ice surrounding Antarctica is a critical component of the ice sheet's stability. Sea ice may assist in buffering the interior ice sheet from the heat of the ocean. Sea ice has a higher albedo than the open ocean and therefore reduces air temperature. The presence of sea ice means that wave activity at the sea surface is reduced which in turn reduces erosion of ice shelves and, in turn, tidewater glacier margins. This has been demonstrated in the Davis Strait and Baffin Bay, between Greenland and Baffin Island, where sea ice has been shown to prolong the survival of icebergs (see Chapter 4). Sea ice is also an important component of polar ecological systems.

This broad outline of the structural components of ice sheets does not differentiate between the GIS and Antarctica. However, in the next two sections these ice sheets will be discussed separately because their very different locations mean that their relationships to other global environmental systems – atmosphere, oceans, continents – are significantly different and, as mentioned above, this imposes different patterns of behaviour on the two ice masses.

Table 3.1 Differences between Arctic and Antarctic sea ice characteristics (Reproduced courtesy of The National Snow and Ice Data Center, University of Colorado, Boulder.)

	Arctic	Antarctic
Average maximum areal extent	15 000 000 km^2	18 000 000 km^2
Average minimum areal extent	7 000 000 km^2	3 000 000 km^2
Typical thickness	~2 m	~1 m
Geographical distribution	Asymmetrical	Symmetrical
Snow thickness	Thinner	Thicker
Trend 1979–2008	Significant decrease of 4.1% per decade	Small increase of 0.9% per decade

3.3 Greenland

The Greenland Ice Sheet is 2.9 million km^3 of ice which, in the words of Alexander Witze (2008, p. 798), 'brings with it an inherent sense of stability'. The traditional view of ice sheet stability is one in which heat and melt water penetrate only slowly through dense ice to lower layers where most of the ice deformation, basal sliding and deformation of basal till take place. Similarly, surface mass balance only slowly alters ice thickness and, in turn, the driving stresses in these deforming layers. This means that the impact of a warming atmosphere takes a long time to change the behaviour of the ice sheet significantly. Much of the ice sheet would be frozen to its bed, subglacial melt water would be negligible, and the whole mass would move slowly, dominantly by internal ice crystal creep driven by gravity. Under such conditions, ice loss would be slow and ice sheet decay could be viewed as a long-term (thousands of years) hazard, implying a low level of risk. In effect, the ice sheet could be considered stable and not worthy of serious consideration as a hazard at the present time, except for its regular supply of icebergs (see Chapter 4) and a very small gradual contribution to sea level rise amounting to only fractions of a millimetre over decades or centuries. Consequently, the 2001 IPCC Report laid little stress on its prediction that Greenland's contribution to sea level rise up to 2100 would be insignificant. However, Greenland *could* raise global sea level by as much as 7 m, if it melted completely. Studies have shown (e.g. Cuffey and Marshall, 2000; Willerslev *et al.*, 2007) that large areas, currently beneath the GIS, have been vegetated in the past and the result of this exposure must have been a substantial rise in sea level. In fact, Greenland *has* raised global sea level by between 2 and 5 m in the past, during the Eemian (Ipswichian) Interglacial some 125 000 years ago, when global temperatures were at least 1 °C higher than they are today (Tarasov and Peltier, 2003). This means that the potential for a substantial impact on global coastal communities clearly exists. The problem is whether the GIS poses a threat that requires immediate attention or whether investigation of this system can be addressed at a later date with no risk of short- or even medium-term impact.

The latter, more relaxed, scenario would probably be appropriate if increasingly accurate remote sensing technology, supplying regular detailed data on a large scale, had not become available from the late 1970s onwards. Since that time the use of data derived from passive microwave satellite platforms has revolutionized our understanding of ice sheet dynamics. Data from 1979 to 1999 (Abdalati and Steffen, 2001) revealed an increasing melt trend for the GIS, interrupted only by the eruption of Mount Pinatubo in the Philippines that reduced global temperatures by ejecting dust and ash into the upper atmosphere. Steffen *et al.* (2004) reported that the year 2002 exceeded melt extent records for the previous 23 years, with a maximum melt area of 685 000 km^2 and an average increase in melt area across the GIS of 16%

between 1979 and 2002. Of particular note was the melting in the north and northeast (Figure 3.4) which experienced extreme melting up to a height of 2000 m. The satellite data was confirmed by automatic weather stations on the ice sheet. A later study (Tedesco, 2007) supported the rising melt trend by showing that the 'melting index' (area × time or km^2 × days), an indication of where melting is occurring and for how long, increased significantly in 2007 for high altitude areas over 1930 m. The melt season in these areas lasted 25–30 days longer than the 19 year average. The study also indicated that the 2007 melt index for lower parts of the GIS was 30% higher than average. Even in the

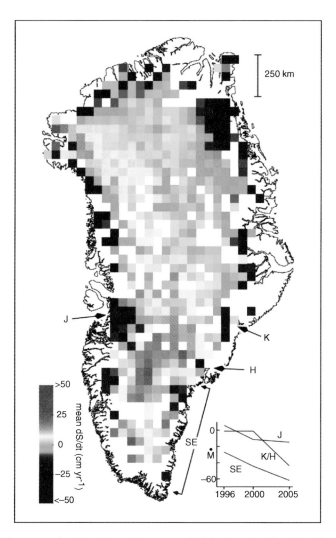

Figure 3.4 Mean rate of surface change (cm per year) of the Greenland Ice Sheet (source: Climate Change 2007: The Physical Science Basis. Working Group I Contribution to the Fourth Assessment Report of the Intergovernmental Panel on Climate Change, Figure 4.17. Cambridge University Press.)

northern parts of Greenland record snowmelt was reported (Tedesco, 2007). A long-term study of the GIS mass balance (Rignot *et al.* (2008) shows an accelerating loss of ice mass, from around 30 gigatons during the 1970s and 1980s to 97 gigatons in 1996 and up to 305 gigatons in 2007, the latter figure approaching the equivalent of 1 mm of global sea level rise. A key point from these melt studies, especially those at a higher altitude, is that refrozen snow can absorb up to four times more energy than fresh, unthawed snow. Wet snow absorbs more solar radiation than dry snow. Melting provides water vapour which enhances cloud formation which then traps more outgoing long-wave radiation.

The knock-on effect of this increased melting is an increase in the volume of melt water on the surface of the GIS. These streams and lakes are relatively common in marginal areas of the ice sheet (Echelmeyer, Clarke and Harrison, 1991) and Zwally *et al.* (2002) suggested that this surface melt water plays an important part in the dynamics of the ice sheet. Once formed lakes can be enhanced by a positive albedo feedback, the lower albedo of the lake water enhancing radiative melting (Lüthje *et al.* 2006). Although lake area provides some indication of the scale of the lakes, their glaciological significance is better appreciated if water volume can be determined. In 2007, Sneed and Hamilton reported a method, using multispectral ASTER satellite imagery, for estimating lake depth by assuming the albedo of the bottom surface of the lake and the optical attenuation of its water. From area and depth, lake volume can be calculated.

The possible significance of this lacustrine surface melt water was highlighted by Zwally *et al.* (2002), who proposed a link with the acceleration of marginal Greenland glaciers. GPS readings at the Swiss Camp (69.6°N, 49.3°W), near the equilibrium line in the southwest sector of the GIS, showed changes in ice velocity between winter and summer. From a constant winter base velocity of 31.33 cm/day velocities increased during the four summers of 1996–1999 to 32.8, 35.1, 40.1 and 38.6 cm/day, respectively, and these increases correlated with the intensity and timing of summer melting. This seasonal pattern was attributed to the rapid migration of surface melt water through *moulins* (vertical or subvertical drainage channels through ice) to the ice–bedrock interface where it enhanced glacial sliding and the deformation of wet deformable sediment (Figure 3.5). Although lakes usually drain during the summer, their drainage tends to be episodic, whereas surface streams are considered to provide widespread and continual drainage at this time. Therefore, according to this hypothesis, the probable cause of summer acceleration of the ice sheet is an increase in water pressure at the base of the ice sheet due to the through-glacier transfer of surface melt water. However, this hypothesis may not be the complete explanation, as Joughin *et al.* (2008) reported a glacier flow response to the drainage of a massive lake down a *moulin*, of only 1 m. Catania, Neumann and Price (2008) reported that *moulins* are concentrated in the *marginal* areas of the ice sheet and are far less

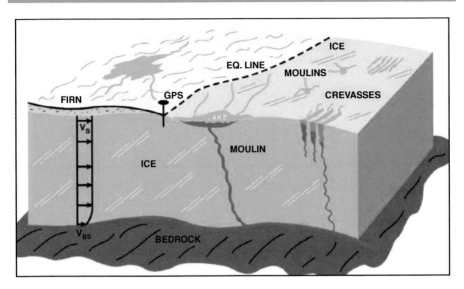

Figure 3.5 Diagram of proposed link between surface meltwater system and the glacier bed (source: Zwally *et al.*, 2002)

common inland near the equilibrium line where the data of Zwally *et al.* (2002) came from. However, Price *et al.* (2008) demonstrated that the seasonal acceleration observed by Zwally *et al.* (2002) could have been initiated closer to the margin. Here, the ice is generally thinner and more crevassed, ablation, the melt-water flux and ice temperature are higher, and the seasonal velocity doubles. According to their flow model, flow in the marginal zone is coupled to flow further inland and can therefore explain what Zwally *et al.* (2002) found without the need for widespread melt-water supply via *moulins*.

However, a further complication is that the velocity of some of Greenland's fastest moving glaciers (e.g. Jacobshavn Isbrae) fluctuates and is not a simple increase reflecting the reported melt increases (Rignot *et al.*, 2008). Holland *et al.* (2008) related the acceleration of Jacobshavn Isbrae to warm subsurface ocean water. This seems reasonable as the floating tongue of Jacobshavn Isbrae survived 1950s air temperatures, equivalent to those associated with its recent thinning and collapse. Retreat of the grounding lines of tidewater glaciers and acceleration of glacier calving facilitates faster upstream flow and ice evacuation. The fact is that existing ice flow models are not yet capable of accurately representing the type of changes represented by the behaviour of Jacobshavn Isbrae. It is likely that such changes reflect atmospheric temperatures, oceanic temperatures and circulation and the dynamics of the ice sheet itself (Solomon *et al.*, 2007, Chapter 4.6).

3.4 Antarctica

With 30 million km^3 of ice, Antarctica is 10 times larger than Greenland, and can be divided into two distinct parts, separated by the Transantarctic Mountains (Hansom and Gordon, 1998). The East Antarctic Ice Sheet (EAIS) is approximately circular, rises to 4200 m above sea level and appears to be very stable. The interior of the EAIS is too cold to be significantly affected by the currently projected amount of atmospheric warming. Apart from a few small peripheral areas where satellite altimetry measurements show that the altitude of the surface is decreasing (Davis *et al.*, 2005), the EAIS has a positive mass balance due to enhanced precipitation (snow) from more humid air induced by the higher temperatures. This suggests that at the present time there is no reason to see the EAIS as a cause for concern in terms of its contribution to sea level rise and other potential hazards such as iceberg flux. On the contrary, Davis *et al.* (2005) suggested that the positive mass balance of the EAIS actually reduces sea level rise.

The same cannot be said for the WAIS and the Antarctic Peninsula (AP). Here, there have been significant changes involving the loss of substantial quantities of ice, and even speculation about the 'collapse' of the whole ice sheet. As mentioned earlier, Lenton *et al.* (2008) have suggested that a tipping point for the WAIS may occur in around 300 years time. This does not mean that all the ice would disappear then, but that an irreversible pattern of change would have been established, which would lead to significant, if not total, loss of ice on these parts of Antarctica. Some changes have already been sufficiently spectacular, to warrant prime time news headlines in the broadcast media (e.g. BBC, 2010a).

One such change over the last 50 years has been the almost total loss, or significant retreat, of seven out of the 12 ice shelves around the AP (Table 3.2), often involving the production of massive tabular icebergs. Over 30 years ago, Mercer (1978, p. 325) concluded that 'one of the warning signs that a dangerous warming trend is underway in Antarctica will be the breakup of ice shelves on both coasts of the Antarctic Peninsula...[and that they] should be regularly monitored by LANDSAT imagery'. They have been; and Cook and Vaughan (2010) have extensively reviewed the spectacular changes to the AP ice shelves. The _9 °C isotherm more or less separates surviving AP ice shelves from those that have been lost, and represents the thermal limit of viability for ice shelves on the AP (Vaughan and Doake, 1996; Morris and Vaughan, 2003). Over 28 000 km^2 of ice shelf have been lost during the last half century, in some cases by progressive retreat (e.g. Müller) but sometimes by abrupt collapse (e.g. Larsen A and B). During the last 50 years the AP has warmed at a rate of 3.7±1.6 °C per century (Vaughan *et al.*, 2003). The loss of the Larsen A ice shelf (Skvarca *et al.*, 1998; Table 3.2) coincides with this warming on the east coast of the peninsula. However, although ice shelf loss has often been attributed primarily to climate warming (e.g. Doake and Vaughan, 1991; Mercer, 1978),

3.4 ANTARCTICA

Table 3.2 Loss of ice from Antarctic Peninsula ice shelves (Reproduced courtesy of The National Snow and Ice Data Center, University of Colorado, Boulder.)

Ice shelf	1950s	1960s	1970s	1980s	1990s	2000-2	2008/9	Total change (km^2)	% remaining
Müller	78	69	60	64	45	44	40	−38	51
Jones	29	31	36	26	21	10	0	−29	0
Wordie	1420	1917	1538	827	906	312	139	−1281	10
Wilkins	16 577			15 986	14 694	13 663	11 144	−5434	67
George VI	25 984	25 806	25 249	24 707	24 260		24 045	−1939	93
Bach	4798	4721	4825	4685	4582	4562	4487	−311	94
Stange			8286	8148	8030	7949	8022	−264	97
Prince Gustav	1632	1299	1328	1019	665	276	11	−1621	1
Larsen A	4021	3736	3873	3394	926	638	397	−3624	10
Larsen B		11 573	11 958	12 190	8299	4429	2407	−9166	21
Larsen C		56 131	58 036	50 241	51 246	51 593	50 837	−5295	81
Larsen D		21 716		22 372	22 345	21 851	22 602	884	104
Total area	152 246	151 862	153 483	143 661	136 020	129 589	124 128	−28 117	82
Change between periods		−384	1621	−9823	−7640	−6432	−5461		

the analysis of Cook and Vaughan (2010) shows that the distribution and timing of the AP ice shelf loss is not a simple matter.

Alongside increases in summer melting, large ponds formed on the ice shelf and these drained into and wedged open surface crevasses that extended down to basal crevasses filled with sea water (Scambos *et al.*, 2000). Prior to the collapse of the Larsen B Ice Shelf, several ice shelves along the Antarctic Peninsula had been thinning by 1–5 m per year between 1992 and 2001 (Zwally *et al.*, 2002), probably, in part, due to basal ice melting, induced by warming sea temperatures (Shepherd *et al.*, 2003; Shepherd, Wingham and Rignot, 2004). Increases in sea temperatures due to climate change are likely to be enhanced by the reduction of surface albedo as light-coloured ice converts rapidly to dark ocean. In addition, Turner *et al.* (2009) reported that waters of the Antarctic Circumpolar Current (ACC) have warmed more rapidly than the global ocean as a whole at depths between 300 and 1000 m. Temperature increases averaged 0.06 °C per decade from the 1960s to the 2000s and by 0.09 °C per decade since the 1980s. These changes are considered to be consistent with enhanced greenhouse gas effects forcing a southward shift of westerly winds and, in turn, a southward shift of the ACC. The stronger winds are associated with the more positive Southern Hemisphere Annular Mode (SAM) which increased westerly winds over the Southern Ocean by 15–20% and drove warm Circumpolar Deep Water up against the western AP coast and the Amundsen Sea coast where the most significant changes are taking place.

Ice shelf dynamics are clearly very complex, and changes are likely influenced by a range of internal and external factors. Nevertheless, although more detailed modelling is required to enable accurate predictions of future ice shelf loss to be made, there can be no doubt about the relevance of ice shelf loss to other components of the Antarctic cryosphere. Several analyses (e.g. Rott *et al.*, 2002; Scambos *et al.*, 2003; Pritchard and Vaughan, 2007; Hulbe *et al.*, 2008; Pritchard *et al.*, 2009) have shown that ice shelf loss leads to the acceleration and thinning of marine glaciers flowing from the Antarctic Peninsula plateau, suggesting that ice shelves act as a buttress against glacier flow. This vital function was clearly demonstrated by the spectacular fragmentation of the Larsen B Ice Shelf on the eastern side of the Antarctic Peninsula. During five weeks early in 2002 (Scambos *et al.*, 2003) most of the Larsen B ice shelf disintegrated, exposing many of its tributary glaciers and allowing them to accelerate. Behind the surviving fragments of the ice shelf little change in velocity of the glaciers was observed but where protection of the shelf had been removed glacier speeds increased by up to eight times (Rignot *et al.*, 2004; Scambos *et al.*, 2004). Although this acceleration may not persist indefinitely, it does imply a contribution to sea-level rise. Cook *et al.* (2005) summarized changes in the position of 244 AP glacier margins since the late 1940s. Overall, 87% of the glaciers show a net retreat, while 13% show a generally small advance during this time. Detailed analysis indicates a

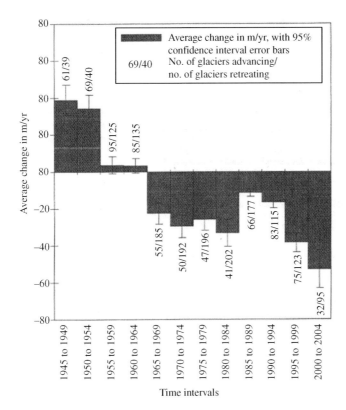

Figure 3.6 Average change (advance or retreat) in m per year with 95% error bars, of Antarctic Peninsula marine glaciers (source: Cook et al., 2005, Science, 308, 541–544.)

progressive, though fluctuating change over time (Figure 3.6). From 1945 to 1954, 62% of glaciers were advancing but by 2000–2004 this figure had fallen to 25% and most glaciers were retreating. There had been a gradual change in the timing of the transition between advance and retreat. At first, in the late 1940s, only glaciers north of 64°S were retreating. Subsequently, glaciers further south began to retreat, those between 68°S and 70°S not until 1965. The very large ice shelves, the Ross and Filchner–Ronne, associated with the WAIS, require sea temperatures to increase by 10 °C, rather than the 4 °C recently measured in that area. Consequently the largest marine glaciers that are located behind these big ice sheets are relatively safe for the foreseeable future, but the same cannot be said of those on the AP.

Like the ice shelves, marine glacier retreat broadly reflects the atmospheric warming across the peninsula, but, again, the relationship is complex and involves other factors. The 25% of the glaciers that were advancing in 2004 are scattered along the length of the peninsula. This means that increasing atmospheric temperature cannot be the complete cause of the pattern of retreat. Nor does sea-ice concentration match the retreat pattern (Cook et al., 2005).

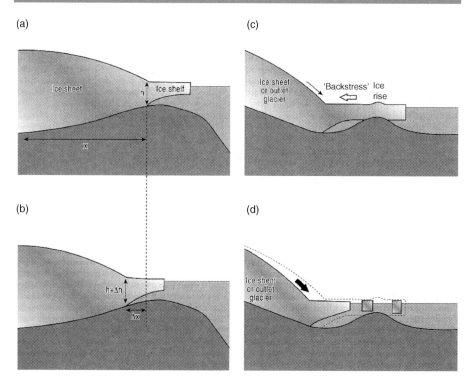

Figure 3.7 Schematic diagram of marine glacier instability in relation to grounding line change – see text for explanation (Reproduced with permission from Bentley, 2010, Journal of Quaternary Science, 25, 5–18. Fig. 2 Wiley-Blackwell.)

Grounded, tidewater glaciers are known to respond in a complex manner to changes in mass balance (e.g. van der Veen, 2002). Normal climatic forcing is supplemented by subglacial topography and oceanic effects, that is, by the conditions at its base and forward margin. Heat from global warming is stored at the approximate depth in the ocean of the grounding line of terrestrial ice in Antarctica. Where an ice sheet is grounded below sea level and on ground that slopes back towards the glacier (a reverse slope) it is inherently unstable (Schoof, 2007). Bentley (2010; Figure 3.7 (a)–(d)) has illustrated the process. Retreat (a) to (b) brings the grounding line (where the glacier changes from rock-based to floating) into deeper water. The glacier at the grounding line is therefore thicker, which increases the flow rate and thins the glacier. This dynamic thinning causes floatation and further retreat until a positive slope is reached. In the case of the glacier buttressed by an ice shelf (c), there is a backstress protecting the glacier. If the ice shelf collapses (d), this backstress is removed and the glacier will accelerate, thin and retreat slightly upslope. If, however, this happens on a reverse slope the retreat will be unstable.

Rignot et al. (2008) used satellite interferometric synthetic-aperture radar observations to demonstrate that ice sheet loss from WAIS along the

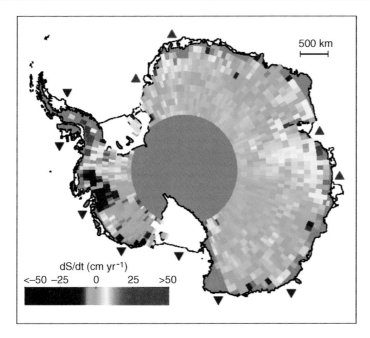

Figure 3.8 Mean rate of surface change on the Antarctic Ice Sheet (source: Climate Change 2007: The Physical Science Basis. Working Group I Contribution to the Fourth Assessment Report of the Intergovernmental Panel on Climate Change, Figure 4.19. Cambridge University Press.)

Bellingshausen and Amundsen sea coast increased by 59% in 10 years to reach 132 ± 60 Gt per year in 2006. (Interestingly, the Bellingshausen Sea is the one area of Antarctica where sea ice extent has been significantly reduced.) In the AP, ice losses increased by 140% from 1992 to 2006 to reach 60 ± 46 Gt per year. In another study (Pritchard *et al.*, 2009), high-resolution ICESat (Ice, Cloud and land Elevation Satellite) laser altimetry showed that some glaciers are thinning at more than 9.0 m per year in the Amundsen Sea embayment of the WAIS (Figure 3.8). According to Turner *et al.* (2009) Pine Island Glacier is moving 60% faster than in the 1970s and Pine Island and adjacent glacier systems are now more than 40% out of balance: 280 ± 9 Gt of ice per year are discharging, but only 177 ± 25 Gt per year of new snow is falling. However, when Payne *et al.* (2004) modelled perturbations at the grounding line they concluded that the Pine Island Glacier would reach a new equilibrium after ~150 years.

The losses detailed above are located in areas where Joughin *et al.* (2009) used models constrained by remotely sensed data to infer the basal properties of the Pine Island and Thwaites Glaciers. Strong basal melting in areas upstream of the grounding lines of both glaciers was indicated, where the ice flow is fast and the basal shear stress is large. Weak sub-glacial areas along much of Pine Island Glacier's main valley could destabilize the glacier if it retreats past the band of strong bed just above its current grounding line.

These discussions of the recent behaviour of the Greenland and West Antarctic ice sheets show that considerable uncertainty still surrounds the reasons for the observed changes, and this limits confidence in the prediction of future developments (Parry et al., 2007, AR4, WGII 15.3.4). Even so, these same authors accept that at least the Antarctic Peninsula is showing a clear response to contemporary climate change, and that other parts of Antarctica *may* also be showing climate change-related changes. Interestingly, Turner *et al.* (2009) highlight the fact that the ozone hole over Antarctica has been shielding the continent from much of the effect of global warming for 30 years, but ozone concentrations are expected to recover over the next century and enhance the effect of greenhouse gases. The possibility that change could take place on century to millennial timescales (Vaughan, 2008), rather than within a longer time frame of several millennia, has certainly stimulated research and debate involving important national organisations such as BAS (British Antarctic Survey) and NOAA (National Oceanographic and Atmospheric Administration, USA). Some impacts of the changes have already been reported, others have been predicted and both will now be discussed.

3.5 Impacts of ice sheet loss

It should now be clear that the world's two remaining ice sheets are massive and complex environmental systems, with the capacity to create significant problems for human societies and natural ecological systems, if not immediately, then in the foreseeable future. The main impacts can be summarized as follows:

- sea level rise,
- iceberg flux,
- ecosystem change,
- climate and ocean circulation change.

Some of these impacts, such as those relating to the ecosystems of Antarctica, are already being reported. Other impacts are probably happening, although almost imperceptibly at the present time. For example, sea level rise due to ice sheet loss is currently very limited, but its longer-term effects could be substantial and global in extent. Iceberg flux is also likely to vary over time and will probably be different in the two hemispheres. Less predictable at this stage are large-scale changes to climate and ocean circulation, which have happened in the past in natural circumstances, but not as a result of significant human-driven forcing.

3.5.1 Sea-level rise

Sea-level rise is a global problem that a few years ago was perceived as a relatively long-term problem of only modest proportions. With rare exceptions (e.g. Mercer, 1978) both Antarctica and Greenland were seen as relatively stable, with the net positive mass balance of Antarctica actually reducing the scale of sea level rise (Solomon *et al.*, 2007). Now there is a greater sense of urgency as even Antarctica is contributing to global ocean volume.

The potential impact of ice sheet loss, on human societies in particular, was clearly recognized by the IPCC (Solomon *et al.*, 2007) in their last assessment of key vulnerabilities and the risk from climate change. The Fourth Assessment Report (FAR) expressed the view that since the Third Assessment Report (TAR; IPCC, 2001)

> 'the literature offers more specific guidance on possible thresholds for partial or near-complete deglaciation for the [GIS] and [WAIS]. There is *medium confidence* (author's italics) that at least partial deglaciation of the [GIS] and possibly the [WAIS] would occur over a period of time ranging from centuries to millennia for a global average temperature increase of 1–4°C (relative to 1990–2000), causing a contribution to sea level rise of 4–6 m or more'(Solomon *et al.*, 2007, Ch 9, p. 797). This is 'a key impact that creates a key vulnerability due to its magnitude and irreversibility, in combination with limited adaptive capacity and, if substantial deglaciation occurred, *high levels of confidence* (author's italics) in associated impacts' (Solomon *et al.*, 2007, Ch 9, p. 793).

Only possibly, in the distant future, might widespread deglaciation be reversible. Before then, near-total deglaciation of GIS and WAIS would have raised sea level by up to 12 m and substantially reconfigured global coastlines. The Amundsen Sea sector of the WAIS alone would contribute 1.5 m of sea level rise.

The scale of impact of this change is related to the ability to adapt, which in turn depends on the rate of change and the wealth of the community. Estimates of rates vary from 1 m to only a few centimetres per century, depending on the model used for the calculation. The 2007 IPCC Report (Solomon *et al.*, 2007) estimated conservatively that sea level would rise by 0.59 m by the end of the twentyfirst century. However, this has since been challenged in a number of journal articles (e.g. Rahmstorf, 2007; Pfeffer *et al.*, 2008; Grinsted, Moore and Jevrejeva, 2010) and reports, largely because remote sensing technology has indicated an accelerating rate of change of these huge cryogenic systems. The Dutch Delta Commission, for example, expects sea level to rise 0.65 to 1.3 m by 2100 and 2 to 4 m along the Dutch coast by 2200, in comparison with only 0.2 m during the twentieth century. The SCAR Report (Turner *et al.*, 2009) predicts an increase of sea level from melting ice sheets of 1.4 m by 2100, enough to displace 17 000 000 people in Bangladesh. A 30% probability that ice loss from the WAIS could raise sea level at 2 mm per year has been suggested. Cazenave and Llovel (2010) show that accelerating ice sheet loss has contributed towards raising land ice melt contribution to sea level from 55% to 80%. However, both the latest IPCC Report (Solomon *et al.*, 2007) and the SCAR Report (Turner *et al.*, 2009) acknowledge that existing ice sheet models are not capable of reproducing the observed behaviour of ice. For example, ice sheet degradation processes, basal lubrication, grounding line changes, the influence of ice shelf loss on the flow of marine glaciers and ice streams and dynamic thinning (Pritchard *et al.*, 2009) due to accelerated flow require further analysis before their effects can be confidently included in models.

These knowledge gaps are of concern. Accurate sea-level-rise forecasts are essential in order to avoid inadequate preparation due to underestimation, or unnecessary waste due to overestimation (Pfeffer *et al.*, 2008). The expanding output of research papers on these two ice sheets clearly demonstrates growing scientific and political concern about their potential impacts, but until the models are refined, the precise impacts due to ice sheets will remain uncertain although it is likely to be sufficient to elicit a response from at least 10% of the global population who inhabit coastal regions. The fact that sea level rise is not spatially uniform only exacerbates the problem. Minimum sea-level rise is expected in the Southern Ocean while the Arctic Ocean is likely to experience maximum sea level rise.

Lenton *et al.* (2008) concluded that these ice sheet systems would reach a tipping point for deglaciation in only 300 years. Current models (Solomon *et al.*, 2007) suggest a threshold of 3.2–6.2 °C of local warming, equivalent to a global warming of 1.9–4.6 °C, relative to pre-industrial temperatures, for near total deglaciation of Greenland. According to these models Antarctica should gain mass through increased snow accumulation but not all studies confirm a significant continent-wide trend in accumulation and it is often assumed that such an increase would be outweighed by greater ice discharge. If the increase in global temperature exceeds 5 °C, mean summer temperatures over the big ice shelves of the WAIS have an even chance of exceeding melting point, but some studies, according to the IPCC, consider that they may disintegrate at lower temperatures aided by basal or episodic surface melting. Recent unpredicted behaviour of the GIS and WAIS in terms of local acceleration and mass loss highlights the inadequacy of existing ice-sheet models for projecting future change. The relative importance of parameters such as mass balance, surface melting, meltwater flow, ocean temperatures, grounding line retreat, back stresses and ice calving rate in ice sheet systems is still not fully understood. Available time series for these parameters are often short and fail to capture inherent system variability, giving limited scope for forward projection and prediction. Nevertheless, palaeoclimatic evidence from the last interglacial suggests a global mean temperature only slightly warmer than today, polar temperatures 3–5 °C warmer and a sea-level rise of 4–6 m contributed by Greenland and possibly the WAIS. On the basis of this evidence and current understanding of ice sheet behaviour, the IPCC (Solomon *et al.*, 2007, p. 794), as noted above, has expressed *medium confidence* that partial deglaciation of the Greenland ice sheet, and possibly the WAIS, would occur within centuries to millennia given a global average temperature increase of 1–4 °C (relative to 1990–2000), which would cause sea-level to rise 4–6 m or more'. Consequently the sooner ice sheet models and predictions are improved, the higher confidence levels will rise and the better coastal communities will be able to plan for, and respond to, predicted changes in sea levels.

3.5.2 Iceberg flux

The flow of icebergs into polar seas, if not completely predictable, is probably the easiest of the recognized impacts to address, as monitoring is well established in the most hazardous area (Greenland) and has recently been introduced for very large icebergs in the Antarctic (NOAA, 2009). Nevertheless, the expected increase in iceberg flux, due to accelerating ice sheet margins, is likely to increase the pressure on monitoring, and risk assessment systems such as those discussed in detail in Chapter 4. This is especially so in the northern hemisphere where shipping is concentrated, but will no doubt apply in the southern hemisphere given the likely increase in tourist vessel movements. Already there has been a high-profile collision between a tourist ship and an iceberg (see Chapter 4), which led to the loss of the vessel, although not of lives. Over the short term the iceberg hazard will increase if present ice sheet dynamics persist or accelerate. In the long-term, if ice sheet loss continues it is conceivable that the iceberg hazard could disappear completely once all marine glaciers have retreated to a land-based position. Before that arises, the scale of the iceberg problem in relation to humans will depend largely on the effectiveness of monitoring and the skill of ships' captains and crews.

Unfortunately, these facilities and expertise are not available to penguins and there are reports (e.g. Kooyman *et al.*, 2007) of these birds having been severely impacted by icebergs. For example, in 2001 giant iceberg B15A collided with the Ross Ice Shelf at Cape Crozier, Ross Island and destroyed the Emperor penguins' nesting habitat with the result that the colony totally failed in 2001. Further collisions at Cape Crozier crushed incubating penguins or trapped them in ravines with the result that the colony was abandoned and a poorer habitat occupied. At Beaufort Island, 70 km NW of Cape Crozier, chick production declined to 6% of the 2000 count by 2004. B15A and a second iceberg, C16, obstructed the direct access of the Beaufort Island colony to its feeding and wintering areas via the Ross Sea Polynya. Due to the iconic status of penguins for the public, this was a high-profile event, but other less visible aspects of polar ecosystems are also vulnerable to the changes that are currently affecting polar ice sheet systems.

3.5.3 Ecosystem change

The extreme conditions of polar environments inhibit easy access to their ecosystems, and as these ecosystems are subject to multiple stressors, direct impacts of climate change are difficult to predict (Parry *et al.*, 2007, IPCC, 2007, AR4, WGII, 15.6.3). However, the growing awareness of climate change and recognition of the presence of ozone (O_3) holes (which allow increased ultraviolet (UV) light incidence) have stimulated considerable research in these areas. Parry *et al.* (2007; IPCC, 2007, AR4, WGII, 15.2.2) reported substantial evidence of change in warming Antarctic terrestrial and marine ecosystems. For example, Ainley *et al.* (2005) recorded declining abundances of krill, Adelie and Emperor penguins and Weddell seals, but increasing abundance of shallow-water sponges and their predators. A decrease in the extent of sea ice reduces the breeding platform area for seals and penguins, although more open leads and polynyas tends to favour seals requiring access to air.

The extent and timing of sea ice is obviously a key variable. A decrease in winter sea ice may inhibit the availability of krill and this will probably have knock-on effects on higher predators, such as albatrosses, seals, whales and penguins (Smith *et al.*, 2003). Autumn and winter sea ice development around Antarctica is directly associated with the production of cold, dense Antarctic bottom water that delivers nutrients to the ecosystems of the Southern Ocean. Later and thinner sea ice may create problems for phytoplanctonic blooms and the productivity of the seasonal ice zone because sea ice provides the stable conditions in which phytoplanktonic blooms can become established (Priddle, Smetacek and Bathmann, 1992). Consequently, a decline in primary production of the Southern Ocean has been predicted (Priddle, Smetacek and Bathmann, 1992), but thinner ice means more light and therefore perhaps the enhancement of phytoplankton growth (Marchant, 1992). A further complication, however, is that less ice exposes organisms to increased UV irradiance. This would benefit those with photorepair or photoprotective mechanisms. Clearly, determination of the actual impact of change on susceptible ecosystems is a complex problem and Vincent and Belzile (2003) advocate the development of improved, predictive models of biological UV exposure in the polar marine environment.

Another potential impact that has attracted considerable interest (e.g. Chwedorzewska, 2009) is the invasion of alien species into areas that have warmed or have been exposed due to ice retreat. Although some polar species may benefit initially from reduced environmental stresses, eventually colonization by lower latitude species with greater competitive ability may lead to substantial changes and possibly the loss of endemic species and unique landscape. The detrimental consequences of invasions for native species has already been recoded on sub-Antarctic islands (Frenot *et al.*, 2005, quoted in Parry *et al.*, 2007) and slow reproduction rates during rapid climate change may limit relocation of native species (Parry *et al.*, 2007). Hughes and Convey (2010, abstract) are concerned that

> 'the Protocol on Environmental Protection to the Antarctic Treaty, the primary instrument through which environmental management is addressed within the Antarctic Treaty System, says little about unintentional introduction of non-indigenous species to Antarctica', and argue that 'the measures described in most Antarctic Specially Protected Area (ASPA) and Antarctic Specially Managed Area (ASMA) Management Plans, by themselves, may not be sufficient to (1) minimize the possibility of introduction of plants, animals and microbes not native to the protected area or (2) adequately protect the many unusual assemblages of species, type localities or only known habitats of certain species found in Antarctica'.

3.5.4 Climate and ocean circulation change

The impacts discussed so far, are, to a large extent, confined to areas within or immediately surrounding the polar ice sheet systems. However, it is possible that the outcome of some changes may extend far beyond these areas and involve large-scale climatic systems and the Meridional Overturning Circulation (MOC) (often referred to as the Thermohaline Circulation (THC)), part of a global system of ocean currents responsible for global heat flux (Broecker, 1994). Projected rapid melting of the edges of the GIS, coupled with the retreat of Arctic glaciers and observed increases in river run-off (Peterson *et al.*, 2002, quoted in Parry *et al.*, 2007), will freshen the ocean surface in northern high latitudes. Together with rising temperatures, the influx of large quantities of lower-density fresh water into the North Atlantic from the GIS could interfere with the present ocean current circulation (Lemke *et al.*, 2007, quoted in Parry *et al.*, 2007, AR4, WGII 15.3.3) in which cold dense saline water sinks before returning southwards as ocean bottom water. There has been sufficient concern to justify a large joint venture of the Natural Environment Research Council, UK, the Universities of Southampton (UK) and Miami (USA), the Atlantic Oceanographic and Meteorological Laboratory, the Spanish Institute of Oceanography and the Max Plank Institute of Meteorology (Rapidmoc, 2010) to measure the MOC in the North Atlantic Ocean. A potential impact of a circulation change could be the reduction of heat transfer to the North Atlantic resulting in colder conditions in north west Europe and other climatic impacts elsewhere. However, the IPCC (Solomon *et al.*, 2007, p. 794) has expressed '*low confidence* in the scale of climate change that would cause an abrupt transition (of the MOC) or the associated impacts'. In their view, 'evidence from components of the MOC as well as uncertainties in the observational records, [during the span of] the modern instrumental record, [suggests that] no coherent evidence for a trend in the mean strength of the MOC has been found' (Solomon *et al.*, 2007, Box 5.1).

Potentially, changes in Antarctica could similarly affect the MOC, but the evidence is no more convincing than in the northern hemisphere. However, climate models do suggest changes in the stratification of the Southern Ocean which could alter the community structure of primary producers, the rates of draw-down of atmospheric CO_2 and its transport to the deep ocean (Parry *et al.*, 2007, AR4, WGII 15.3.1).

The caution expressed by IPCC in 2007 is appropriate, given the magnitude of the systems involved and the associated uncertainties. Nevertheless the IPCC is convinced (Parry *et al.*, 2007) that the effects of increased greenhouse gases, as well as decreases in stratospheric ozone which will exacerbate the problem, are already evident. If the current rate of change continues it will be remarkable for its speed. Broad estimates of how temperature, precipitation and sea ice extent might change, and their impact on marine and terrestrial biota, are possible. There is less confidence in the future of the Antarctic ice sheets and how they will respond, but the recent rapid changes do give cause for concern – especially for the stability of the WAIS.

3.6 Mitigation

The massive scale and complexity of ice sheet systems inhibits easy mitigation of related impacts. Although an increasing amount of research effort is being expended in seeking to understand how they function in relation to climate change, many details still require investigation before this relationship is fully understood. Consequently, apart from reducing the current rate of temperature increase to reduce melting, understanding processes in order to improve the accuracy of prediction is likely to be the best that can currently be achieved in terms of mitigating most ice sheet impacts.

Probably the easiest problem to resolve is iceberg flux. Chapter 4 details the measures that are taken in the most important iceberg area, around Greenland. Here both the USA, with its International Ice Patrol, and Canada, which runs the Canadian Ice Service, remotely sense icebergs and collate the information for the benefit of shipping and fixed installations. The rarity of collisions reflects the success of these North American mitigation systems. For the seas around Antarctica, the United States Geological Survey has established a scheme to monitor, as far north as 60°S, the very large icebergs that derive mainly from the breakup of ice shelves. However, a general scheme, similar to that around Greenland, does not operate in Antarctic seas as the density of shipping is too low to justify the expense. It therefore rests with the operators of vessels to be vigilant when traversing iceberg infested waters around Antarctica.

The problem that has attracted most attention, especially since the rate of ice sheet loss appeared to be accelerating, is rising sea levels. How much sea level change will occur, as different quantities of the ice sheets melt, is not too difficult to calculate, given an accurate estimate of ice volume and density. The uncertainty attached to sea level change is the rate at which change will occur. If the rate is known, impacts can be anticipated and plans established for their mitigation. The key unknown at present is *exactly* how the climate will change in the future, given that both human-induced and natural forcing is likely to influence the actual amount of change. The difficult component in the equation is natural forcing because that can only really be anticipated when the climate system as a whole is fully understood, which is probably a long way off. Unless something totally unforeseen happens, sea level is unlikely to increase very abruptly, giving humanity inadequate time to respond effectively. On the other hand, rising sea levels have already impacted so severely on some marginal systems in the Pacific and Indian Oceans that migration to safer ground has been contemplated (The Guardian, 2003).

With regard to the other impacts that have been discussed, ecological systems, global climate and the MOC, probably the best solution is to control the extent of climate change and understand what might happen in the long-term. For now, in an effort to restrict the unintentional introduction of alien species to Antarctica, following the rapid growth of tourism (27 000 visitors in the 2005/06 summer; IAATO, 2006, quoted in Parry *et al.*, 2007), stringent clothing decontamination guidelines have been implemented (IAATO, 2007) for tourist landings on the Antarctic Peninsula (Parry *et al.*, 2007, AR4 WGII, 15.7.2). However, in the long run, not much can be done to maintain existing ecological systems, even if this were considered desirable. The best that might be achieved is to record current species and understand their ecology. These are big issues with huge potential consequences. There are large uncertainties with limited possibilities for immediate mitigation. The IPCC recognized this in their latest report where they summarized their key uncertainties and set out related scientific recommendations and approaches (Parry *et al.*, 2007, AR4, WGII, 15.8; Table 3.3). Perhaps their most significant uncertainty is 'The adaptive capacity of natural and human systems to cope with critical rates of change and thresholds/tipping points.' Their recommended response to this is the '"integration of existing human and biological climate-impact studies to identify and model biological adaptive capacities and formulate human adaptation strategies'. Not much to do then!

3.7 Summary

Ice sheets represent one of the largest single hazards in terms of the number of people vulnerable to their potential impact, a sea level rise of 12 m. In seeking to mitigate this impact, one obvious difficulty is the enormous scale and complexity of ice sheets, and the fact that the GIS and the WAIS are different. Inevitably this has led to differences of interpretation of the evidence and consequently to uncertainty, not least regarding the direction and rate of change. Nevertheless, much information has been obtained, especially since the advent of satellites and remote sensing, not just about

Table 3.3 Key uncertainties and related scientific recommendations/approaches concerning Arctic and Antarctic landscapes from Parry et al., 2007, AR4, WGII, 15.8, Table 15.1

Uncertainty	Recommendation and approach
Detection and projection of changes in terrestrial, freshwater and marine Arctic and Antarctic biodiversity and implications for resource use and climatic feedbacks	Further development of integrated monitoring networks and manipulation experiments; improved collation of long-term data sets; increased use of traditional knowledge and development of appropriate models
Current and future regional carbon balances over Arctic landscapes and polar oceans, and their potential to drive global climate change	Expansion of observational and monitoring networks and modelling strategies
Impacts of multiple drivers (e.g., increasing human activities and ocean acidity) to modify or even magnify the effects of climate change at both poles	Development of integrated bio-geophysical and socio-economic studies
Fine-scaled spatial and temporal variability of climate change and its impacts in regions of the Arctic and Antarctic	Improved downscaling of climate predictions, and increased effort to identify and focus on impact 'hotspots'
The combined role of Arctic freshwater discharge, formation/melt of sea ice and melt of glaciers/ice sheets in the Arctic and Antarctic on global marine processes including the thermohaline circulation	Integration of hydrologic and cryospheric monitoring and research activities focusing on freshwater production and responses of marine systems
The consequences of diversity and complexity in Arctic human health, socio-economic, cultural and political conditions; interactions between scales in these systems and the implications for adaptive capacity	Development of standardised baseline human system data for circumpolar regions; integrated multidisciplinary studies; conduct of sector-specific, regionally specific human vulnerability studies
Model projections of Antarctic and Arctic systems that include thresholds, extreme events, step-changes and non-linear interactions, particularly those associated with phase-changes produced by shrinking cryospheric components and those associated with disturbance to ecosystems	Appropriate interrogation of existing long-term data sets to focus on non-linearities; development of models that span scientific disciplines and reliably predict non-linearities and feedback processes
The adaptive capacity of natural and human systems to cope with critical rates of change and thresholds/tipping points	Integration of existing human and biological climate-impact studies to identify and model biological adaptive capacities and formulate human adaptation strategies

the ice sheets themselves but also the surrounding sea ice and oceans. At least now the IPCC has recognized where major uncertainties exist and has been able to recommend appropriate ways in which research on ice sheets can be progressed.

4
Icebergs

4.1 Introduction

On the night of 14 April 1912 the 'unsinkable' RMS *Titanic*, travelling between Southampton, UK, and New York, USA, on its maiden voyage, struck an iceberg in the North Atlantic Ocean and sank within three hours with the loss of 1517 lives. Since then no iceberg impact events have come near the Titanic disaster in severity (Hill, 2004). This absence of major ship–iceberg accidents since 1912 is not due to chance. Without doubt, this iconic marine disaster focussed minds and raised awareness of the iceberg hazard amongst ship owners and captains as well as authorities with responsibility for marine safety; it provided the stimulus for better iceberg monitoring. Both the USA and Canada set up monitoring systems in response to that tragic event: the International Ice Patrol (IIP) in the USA (USCG-IIP, no date, Figure 4.1), and the Canadian Ice Service (CIS) in Canada (CIS, 2009a). Today, these organizations provide a service to shipping and coastal communities, that has largely succeeded in keeping the iceberg hazard off the front pages of newspapers, in spite of the fact that up to 40 000 icebergs are produced annually, mostly from the west coast of Greenland, and hundreds of them cross busy shipping lanes off the east coast of North America. However, icebergs have not been completely removed from the list of cold region hazards, as the sinking of the tourist ship, M/S *Explorer,* off the coast of Antarctica in November 2007 indicates, fortunately on this occasion without fatalities (The Guardian, 2007).

Although sophisticated monitoring systems and on-board technology are in place, icebergs remain an obstacle to shipping and other marine activities and systems. Vessels do plan routes and change course to avoid impacts and this increases journey distances and therefore costs to businesses. Some ship–iceberg collisions still occur. Fixed objects, like oil drilling platforms and

---
Cold Region Hazards and Risks, First Edition. Colin A. Whiteman.
© 2011 John Wiley & Sons, Ltd. Published 2011 by John Wiley & Sons, Ltd.

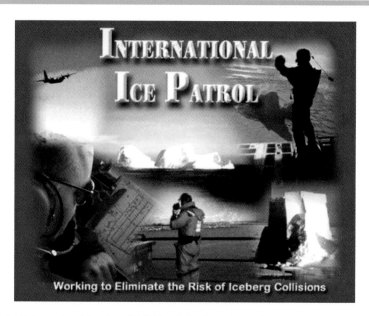

Figure 4.1   International Ice Patrol (USA) website logo (Reproduced from http://www.uscg-iip .org/cms/. Courtesy of the US Coast Guard. [accessed 1 March 2010].)

coastal installations are vulnerable to iceberg impact, and many attempts are made every year to tow icebergs away from valuable installations. Submarine pipelines and cables are vulnerable when icebergs run aground and scour troughs in the seabed. Icebergs introduce relatively cold, fresh water into warmer, more saline parts of the ocean and this *can* alter ecological systems and ocean current circulation. Very large tabular bergs in Antarctica have been blamed for disrupting penguin migration routes (NSF, 2002)! On a completely different space and time scale, it has been suggested that armadas of icebergs, from the Hudson Bay area of Canada, altered climate 20 000 or 30000 years ago during the so-called Heinrich events. Currently, there are no suggestions that a similar type of event could result from the very rapid decay of the Greenland Ice Sheet.

4.2   Iceberg characteristics

4.2.1   Composition

Icebergs are composed of fresh water because they originate from land-based tidewater glaciers and ice sheets, or from floating ice shelves, the off-shore component of ice sheets (see Chapter 3). Icebergs float because the density of ice (about $900 \, kg/m^3$) is lower than that of sea water (about $1025 \, kg/m^3$). The ratio of these densities means that 7/8 of an iceberg's mass must be below the water surface. Usually icebergs are 20 to 30% longer beneath the water surface than above, and generally not quite as deep as they are long at the waterline. Although ice forms the bulk of most icebergs, some bergs contain the eroded debris, often arranged in layers, from their existence as a terrestrial glacier. As they melt this debris falls to the ocean bed. Other icebergs may be unusually coloured due to the presence of algae.

4.2.2 Size

The CIS (Table 4.1) and IIP (Table 4.2) have both produced classifications of iceberg sizes to facilitate data collection, communication and impact risk assessment. Most icebergs are small to medium in size in the Labrador Sea and Grand Banks areas where they have been studied in most detail (Venkatesh and El-Tahan, 1988). The product of mass and velocity determines the momentum with which icebergs impact obstacles to their movement. In Canadian waters the largest icebergs are up to 8.5 million tonnes near their source, but 1 million tonnes is exceptional near major shipping lanes around the 48th Parallel (line of latitude), and most have reduced to less than 500 000 tonnes at this location. Robe, Maier and Russell (1980) have estimated that over 30 000 medium icebergs (5 million tonnes) are produced annually from West Greenland, mainly from nine major glacier sources.

Table 4.1 Iceberg size classification according to the CIS (source: CIS 2003a)

Type	Height above sea level (m)	Length (m)	Weight (megatons)
Growler	<1	<5	0.001
Bergy bit	1–<5	5–<15	0.01
Small iceberg	5–15	15–60	0.1
Medium iceberg	16–45	61–120	2
Large iceberg	46–75	121–200	10
Very large iceberg	>75	>200	>10.0

Table 4.2 IIP iceberg size classification, average values. Note: Type 1 = tabular; type 2 = non-tabular. (Reproduced with permission from Venkatesh, S. and El-Tahan, M, (1988) Iceberg Life Expectancies in the Grand Banks and Labrador Sea. Cold Regions Science and Technology, 15, 1–11, Table 1. © Elsevier.)

Code	Type	Size	Mass (tonnes)
1	2	Growler	450
2	2	Small	75 000
3	2	Medium	900 000
4	2	Large	5 500 000
5	1	Small	250 000
6	1	Medium	2 170 000
7	1	Large	8 230 000

Table 4.3 Shape classification according to IIP, ratios are derived from actual icebergs recorded by IIP south of Newfoundland (Reprinted from http://www.uscg-iip.org/pdf/AOS_2009.pdf. Courtesy of US Coast Guard.)

Type	Height:draft ratio	Description
Blocky	1:05	Flat-topped, steep vertical sides
Tabular	1:05	Flat-topped, horizontal banding
Wedged	1:05	Flat-topped, one steep side sloping to lesser sides
Domed	1:04	Smooth, rounded top
Pinnacled	1:02	Central spire with one or more surrounding spires
Dry-docked	1:01	Eroded with large U-shaped slot

4.2.3 Shape

Shape (USCG-IIP, 2009a; Table 4.3; CIS, 2003b; Figure 4.2) is also an important consideration as it may influence the stability of an iceberg, its movement in response to wind and water currents and the ease with which it

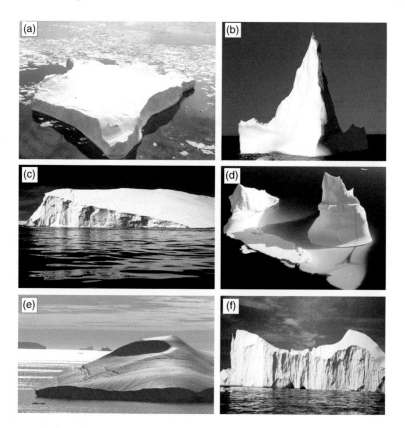

Figure 4.2 Iceberg shapes: (a) Tabular; (b) Pinnacle; (c) Wedge; (d) Drydock; (e) Domed; (f) Blocky (Source: Environment Canada, Canadian Ice service http://ice-glaces.ec.gc.ca/App/WsvPageDsp.cfm?Lang=eng&lnid=27&ScndLvl=yes&ID=239 (accessed 07 03 10))

can be towed. Most icebergs derived from ice shelves are tabular in shape and relatively stable. Icebergs calved from tidewater glaciers tend to have more irregular, less stable shapes. As melting progresses icebergs are undercut at water level. Subsequent tilting or even a capsize is revealed by the undercut at an angle to the present water surface.

Generally, water currents are the most important factor controlling iceberg movement because of the bergs' deep draft (most of the volume of an iceberg is below the water surface). Wind direction and velocity become significant only when an iceberg has a high 'sail' (height) to draft ratio, and as melting proceeds during the life of the berg.

4.2.4 Sources

On a global scale icebergs are limited in their distribution to relatively few polar, sub-polar and, rarely, temperate source areas (Figure 4.3). The best known of these is Greenland, especially its west coast (Figure 4.4), where up to 40 000 (average 30 000) icebergs per year are created from nine major glaciers, 25 000 from the coast between Smith Sound and Disko Bay and 10 000 from Melville Bay, between Cape York and Upernavik. Jacobshavn Isbrae, one of the world's fastest flowing glaciers, is a major single source. These icebergs transfer from Baffin Bay through the Davis Strait into the North Atlantic around the shallow Grand Banks off Newfoundland. Others enter the North Atlantic Ocean from the east coast of Greenland, from glaciers such as Daugaard-Jensen Gletscher and Vestfjord Glescher. The Arctic Ocean receives icebergs from small ice shelves around northern Ellesmere Island in the

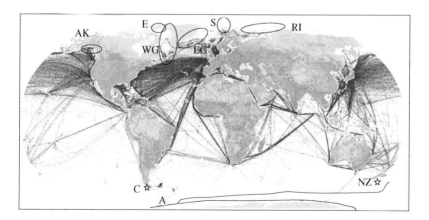

Figure 4.3 Main iceberg source areas superimposed on a map of shipping routes: AK = Alaska; E = Ellesmere Is.; WG = West Greenland; EG = East Greenland; S = Svalbard; RI = Russian Islands; C = Chile; NZ = New Zealand; A = Antarctica. Blue line shows possible drift of Antarctic icebergs to the vicinity of New Zealand (source of base map: http://www.sciencemag.org/content/vol316/issue5833/images/large/316_1866_F2.jpeg)

Figure 4.4 Icebergs calving from glaciers at Cape York, Greenland, September 2005. (source: http://en.wikipedia.org/wiki/File:Icebergs_cape_york_4.jpg [accessed 23 09 2010]) Photo: Mila Zinkova

Canadian archipelago, from Svalbard and from the Russian Arctic islands of Franz Josef Land, Novaya Zemlya and Severnaya Zemlya.

Elsewhere, tidewater glaciers, such as Columbia Glacier on the southern coast of Alaska, supply icebergs to the north Pacific Ocean, and in the southern hemisphere, similar glaciers contribute icebergs to the southeastern Pacific Ocean from the southern coast of Chile,and to the Indian Ocean from a few New Zealand glaciers. However, the extent of iceberg movements from these areas is relatively limited.

The last but not least source of icebergs is Antarctica. Here, most icebergs derive from ice shelves which, until recently, have buffered the terrestrial glaciers from direct contact with the Southern Ocean. During the last two or three decades, ice shelves on both sides of the Antarctic Peninsula, including the Wilkie, Larsen A and Larsen B shelves, have collapsed dramatically. Some of the fragments released by these disintegrating shelves form icebergs up to 200 km in length. These massive icebergs around Antarctica are named and tracked by the US National Ice Centre (NIC) until they pass north of latitude 60°S (NIC, 2009a).

4.2.5 Distribution

Iceberg distribution is concentrated in the seas around Greenland, Baffin Island, Devon Island, Bylot Island and Ellesmere Island in the Canadian Archipelago, southern Alaska, Nordaustlandet, Kuitaya, Edgeøya and West Spitzbergen in Svalbard, Novaya Zemlya, Savernaya Zemlya and Franz

Josef Land in the Barents and Kara Seas, southern Chile and Antarctica. Icebergs from Antarctica, have been sighted 400 km from the south coast of New Zealand (50°S) in 2006 and again near Macquarie Island (54°S) in 2009 (NIWA, 2009; Figure 4.3).

The general distribution, maximum extent and frequency of occurrence of icebergs depends on a variety of factors, including (a) calving (supply) rate of source glaciers and shelves, (b) trajectory (route taken), (c) flux or frequency and (d) life expectancy/deterioration. Initial rates of iceberg supply from their source depend on glacier flow velocity and the strength of the floating ice. Warm ocean temperatures will help to thin the ice and weaken it through basal melting. Stormy conditions and strong wave activity will hasten breakup of the floating ice apron. Ice shelf calving appears to be enhanced by the accumulation of surface water as this facilitates melting and thinning of the shelf in areas where the surface water occurs.

Once in the ocean, the trajectory of the icebergs is controlled dominantly by sea currents and wind but also, to a lesser extent, by the Coriolis force, by waves and by the surface slope of the ocean, although the last three of these are minor contributors. Given the current importance of the northwest Atlantic Ocean region in terms of iceberg density and the frequency of shipping movements, the following discussion will concentrate largely on iceberg flux in the area of the Grand Banks, the Labrador Sea, the Davis Strait and Baffin Bay. However, the potential increase in access to Arctic waters, due to a decrease in Arctic sea ice extent, may require a change in this emphasis in the near future. Governments, consultancies and the industrial community on both sides of the Atlantic and elsewhere are already studying this possibility with great interest (ANP, 2008).

Initially, most West Greenland icebergs move north-north-westerly in response to the West Greenland and Baffin Bay ocean currents (Figure 4.5). At the northern end of Baffin Bay they turn and flow south-south-eastwards along the west side of the Bay, eventually being picked up by the Labrador Current and carried across the Grand Banks of Newfoundland. Here they drift either eastwards north of the Flemish Cap or southwards between the Flemish Cap and the Grand Banks, an area often referred to as 'Iceberg Alley'. The southern limit of drift is generally defined by the northern edge of the warm North Atlantic Current (Gulf Stream). It is possible for icebergs to be transported across the warm current in cold water eddies (CIS, 2009b). Maximum rates may be 9.3 nautical miles per day. Total travel distance from West Greenland to the Grand Banks, Newfoundland can be anything from 2700 to 3700 km and take 2–3 years, though possibly less than 12 months if periods of grounding are avoided and the icebergs are not diverted into bays. Average drift rates have been measured at 1.7 nautical miles per day north of latitude 67°N where the sea ice cover persists for longer. Further south, from latitudes 60°N to 52°N, where there is less sea ice, drift rates may increase to an average of 7.6 nautical miles per day. Satellite tracking in

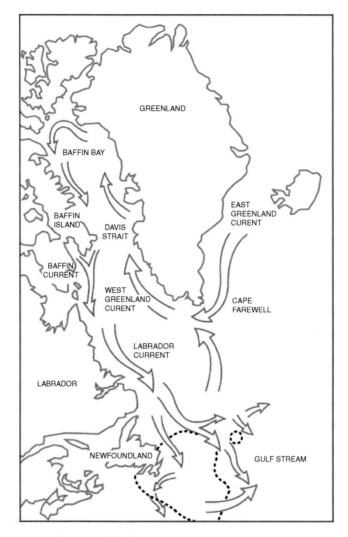

Figure 4.5 Mean trajectories of icebergs in between Greenland and Newfoundland, Canada: dashed lines outline the Grand Banks (large area) and Flemish Cap (small area) and 'Iceberg Alley' is the channel between these two areas (Description and format by kind permission Environment Canada © Environment Canada 2010.)

Baffin Bay shows that most icebergs move in a well-defined zone over the continental slope and follow the Baffin Current more or less along the 500 m isobath (submarine contour)(Marko, Birch and Wilson, 1982), but some follow the local bathymetry into bays and may become trapped in land-fast ice for 9–10 months.

4.2.6 Life expectancy

A key parameter in assessing iceberg risk is their life expectancy, which is dependent on their size and rate of deterioration. Factors determining survival to the Grand Banks include (a) intensity or volume transport rate of the Labrador Current; (b) direction, strength and duration of prevailing winds; (c) the extent of sea ice available to protect icebergs from wave erosion and (d) environmental conditions, especially air temperature, sea water temperature and wave action (amplitude and frequency). Venkatesh, Murphy and Wright (1994) recognized five different processes responsible for the deterioration and decay of icebergs:

- surface melting due to insolation,
- melting due to bouyant vertical convection,
- melting due to forced (air and water) convection,
- wave erosion,
- repeated calving of overhanging ice slabs (caused by wave erosion).

Of these processes, wave erosion (Table 4.4) and associated calving were considered to be by far, the most important, accounting for over 80% of total decay. Sea ice damps wave generation and propagation (Squire and Moore, 1980) and effectively shields the iceberg from wave action. Wave erosion for computational purposes is assumed to be effective when the ice concentration decreases to 2/10. Venkatesh and El-Tahan (1988) derived a life expectancy (L) equation for icebergs of different sizes in the Grand Banks and Labrador Sea area based on IIP data, such that $L = a\,m^b$ where $m =$ mass (tonnes), a and b are empirical constants for each month (Table 4.5). Table 4.6 gives some indication of the rate of iceberg melting.

Under present day climatic and oceanic conditions, it is rare for Greenland icebergs in the northwest Atlantic to travel beyond 44°N. Some exceptional movements have been recorded in the past but they are very rare. Only four are known to have travelled south of 40°N (USCG-IIP, 2009b):

- In 1883, a growler was located about 200 nm (nautical miles) south of the Azores.

Table 4.4 Rates of iceberg deterioration attributable to different melt processes (Reprinted from http://www.uscg.mil/lantarea/iip/home.html. Courtesy of US Coast Guard.)

Cause of melting	Deterioration	Percentage of total
Insolation	0.02 m/day	0.3
Buoyant convection	0.12 m/day	1.6
Wind-forced Convection	0.93 m/day	14.2
Wave induced	6.55 m/day	84.0

Table 4.5 Empirical constants for iceberg life-expectancy equation (Reproduced with permission from Venkatesh, S. and El-Tahan, M, (1988) Iceberg Life Expectancies in the Grand Banks and Labrador Sea. Cold Regions Science and Technology, 15, 1–11, Table 4. © Elsevier.)

Month	Grand Banks		Labrador Sea	
	a	b	a	b
January	0.159950	0.67	-	-
February	0.387270	0.67	-	-
March	0.415910	0.67	-	-
April	0.274150	0.67	-	-
May	0.159220	0.67	-	-
June	0.089125	0.67	0.832	0.6
July	0.053330	0.67	0.282	0.6
August	0.033884	0.67	0.159	0.6
September	0.019588	0.67	0.183	0.6
October	0.026546	0.67	0.349	0.6
November	0.036224	0.67	0.681	0.6
December	0.061944	0.67	-	0.6

- In 1907, an iceberg was sighted about 100 nm southwest of Fastnet, Ireland.
- In 1912, a growler was seen about 75 nm east of Chesapeake Bay, USA.
- In 1926, the southernmost known iceberg (a growler) reached 30° 20′ N, 62° 32′ W (about 150 nm from Bermuda).

Using ships' archives Newell (1993) found a downward trend in the size of the largest icebergs sighted. Large bergs were more common in the 1880s, reflecting colder conditions and more extensive protective sea ice during the Little Ice Age. This is a trend supported by McClintock, McKenna and Woodworth-Lynas (2007). Greenland is producing *more* icebergs as glaciers

Table 4.6 Number of days required to melt a 100 m iceberg at a given sea surface temperature, assuming: wave height = 1.8 m; wave period = 10 s; relative velocity = 0.25 m/s (Reprinted from http://www.uscg.mil/lantarea/iip/home.html. Courtesy of US Coast Guard.)

Sea surface temperature (°C)	Number of days
−1	179
3	20.5
6	12
10	8
15	5

accelerate under warming conditions (see Chapter 3), but iceberg *size* and iceberg flux are decreasing due to changing glacier dynamics and ocean water temperatures. Glacier acceleration results in dynamic thinning and warmer intermediate water is moving into fjords and decreasing grounding line depth, with the result that icebergs begin their existence with shallower drafts. Warmer ocean surface temperatures are increasing iceberg melt rate and decreasing sea ice concentration in Baffin Bay and the Labrador Sea is reducing protection from wave erosion.

However, while the overall size and flux trends may be downwards, there is still considerable variability in the number (severity) of icebergs reaching the Grand Banks area each year. This may be due, in part, to fluctuations between positive and negative phases of the North Atlantic Oscillation (NAO), the dominant pattern of winter atmospheric variability in the North Atlantic (Hurrell and Deser, 2009; USCG-IIP, 2009c). During a positive phase strong and persistent northwest winds blow along the Labrador coast bringing colder air temperatures and more sea ice to the Labrador and Newfoundland coasts which not only protects the icebergs but promotes their southerly movement. In contrast, the negative phase of the NAO tends to reduce iceberg survival and flux.

4.3 Iceberg impact and risk

The size and weight of icebergs, even growlers, is such that they pose a significant threat to vessels and other objects with which they collide. Even a glancing blow to a relatively thin ship's hull is capable of creating sufficient damage to sink an ordinary vessel, as M/S *Explorer* demonstrated in the Antarctic (The Guardian, 2007). Direct risks associated with icebergs are the product of iceberg numbers, their different characteristics and the vulnerability of the human systems with which they interact. High densities of icebergs in remote parts of the global ocean present little risk to humans directly, but may impose risk indirectly through changing ecosystems and even climate. Different aspects of iceberg risk, operating at different spatial and temporal scales, are summarized below:

- direct ship–iceberg collisions;
- direct impact of icebergs on fixed or moored installations
  - at sea (e.g. drilling platforms)
  - on land (e.g. harbours, piers);
- destruction of submarine installations (e.g. communication cables, pipelines) by sea bed scour;
- modification or disruption of ecological systems;
- disruption of global atmospheric and oceanic systems (e.g. Heinrich Events and climate change).

4.3.1 Ship–iceberg collisions

The sinking of RMS *Titanic*, with its record-breaking loss of lives, remains *the* iconic maritime disaster. The event implies that if a risk assessment was carried out for the journey, it was inadequate. At that time it would have been conducted without the benefit of three new databases collating iceberg sightings, iceberg management events and ship–iceberg collisions, which contribute to the mitigation of iceberg impacts at the present time (see below). A database of ship–iceberg collisions, extending back to the seventeenth century and covering the Northwest Atlantic, the Canadian Arctic and the coastal waters of southern Alaska, has been compiled at the Institute for Ocean Technology in the National Research Council, Canada (Hill, 2004). It illustrates the history of ship–iceberg collisions and their associated fatalities as far as they are currently known. The database shows that RMS *Titanic* is only one of at least 58 vessels since 1619 that apparently struck icebergs *and* sustained at least one fatality as a result. In total, the database records 3539 fatalities since 1619 (Table 4.7), although there may well have been additional losses that were not published or have yet to be researched for the database. The attribution of all the fatality events to iceberg impacts cannot be demonstrated beyond doubt (Hill, 2004), but the context in which most of these 58 ships were lost or damaged provides good circumstantial evidence for this conclusion to be valid.

In the six decades before *Titanic*, ship–iceberg collisions had become relatively commonplace, with 40 incidents and at least 1777 fatalities (some are recorded as 'unknown'). In contrast, for the period of almost a century since the *Titanic* foundered, the database contains only 11 events with 169 fatalities, of which just two events accounted for 82% of the fatalities. A number of reasons can be suggested for this significant decrease in ship–iceberg collisions and fatalities: improved reconnaissance, better on-board radar technology and fewer passenger vessel movements, for example (see next section). Nevertheless, although air transport may have relieved one type of trans-Atlantic travel anxiety, there are still good reasons why icebergs remain high on the list of risk assessment priorities of many organizations.

The US IIP decided at the outset that the 48°N line of latitude is the key boundary for the iceberg hazard related to shipping lanes. North of 48°N there may be more icebergs but the density of large ships substantially decreases. South of the line large-scale shipping increases but the density of icebergs decreases as the warmer waters of the Gulf Stream are approached. Consequently, the IIP has recorded the number of icebergs that pass south of 48°N (Figure 4.6), every year with the exception of the 1917–18 season during the First World War and the 1942–45 seasons during the Second World War. Annual figures are related to the standard ice-observing year, which begins in October and ends in September when the ice (sea ice and

4.3 ICEBERG IMPACT AND RISK

Table 4.7 Summary of database of ship–iceberg collisions since 1619 which resulted in fatalities (source: Hill, 2006)

Period	Events with number of fatalities																				Total fatalities	Number of fatality events
Post-2000																					0	0
1976–2000																					0	0
1951–1975	95	1	6																		102	3
1926–1950	2	1	43																		46	3
1901–1925	42	**1503**	1	5	2	11	2														1566	7
1876–1900	25	1	1	**14**	5	8	11	1	1	1											420	22
1851–1875	480	1	186	**135**	**35**	177	1	**1**	47	4	**43**	74	62	**12**	**1**	5	25	32	78	15	1063	9
1826–1850	1	30	1	1	**120**	**49**	1	**20**													252	8
Pre-1825	3	66	**11**	1	1	8															90	6
																			Totals:		3539	58

Note: figures in bold carry high confidence

icebergs) is generally at its minimum. Comparison between years must be made with caution in view of the considerable advances in reconnaissance techniques and the fact that some pre-IIP data have been obtained from a variety of other sources.

It is therefore difficult to decide whether the apparent increase in iceberg numbers south of 48°N since the early 1970s, especially during the last two

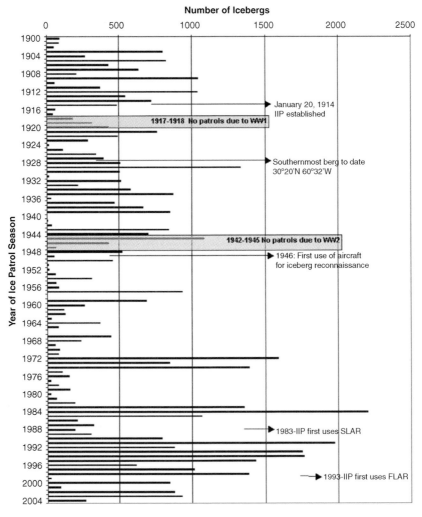

Figure 4.6 Number of icebergs recorded by IIP crossing latitude 48°N on the Grand Banks (source: USCG-IIP, 2009b)

Table 4.8 Ranking of the most severe ice seasons during the past 25 years (1983–2008) based on cumulative numbers of icebergs south of 48°N (Reprinted from http://www.uscg-iip.org/pdf/Annual_Report_2008.pdf. Courtesy of US Coast Guard.)

Rank	Year	Bergs south of 48°N
1	1984	2022
2	1991	1976
3	1994	1765
4	1993	1753
5	1995	1432
6	1998	1380
7	1983	1352
8	1985	1063
9	1997	1011
10	2008	976

decades of the twentieth century (Table 4.8), is attributable to climate change (global warming), reconnaissance developments (SAR, SLAR, FLAR and RADARSAT), natural variation or a combination of all these factors. Nevertheless, the frequency of icebergs crossing 48°N into the main hazard zone of the Grand Banks does vary significantly (Figure 4.6), with numbers fluctuating between 2202 in the very severe season of 1983/84 and zero in the 1965/66 season. Besides these annual differences, iceberg movements in the hazard zone south of 48°N are not evenly distributed across the year (Figure 4.7). The

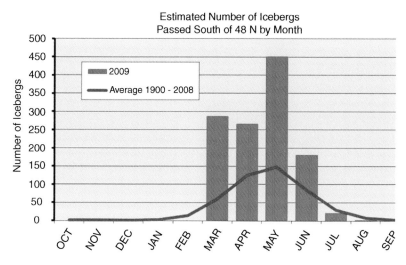

Figure 4.7 Estimated number of icebergs that passed south of 48°N each month during 2009 ice year (Reprinted from http://www.uscg-iip.org/pdf/Annual_Report_2009.pdf. Courtesy of US Coast Guard.)

main iceberg season south of Newfoundland is normally from March to July but a few icebergs occasionally appear in any month except October to December. About half of the annual total of icebergs is accounted for by the numbers in April and May (Marko *et al.*, 1994, Fig. 2b) when they are released from the melted sea ice. These annual and seasonal variations in iceberg numbers reflect the complexity of causal factors, and inter-annual variability remains a key component of risk assessment and ice management, especially in these North Atlantic waters (see below).

However, other oceans also experience problems with ship–iceberg collisions. Since about 1980 the Columbia Glacier in southern Alaska has been retreating very rapidly and the volume of icebergs calved from its terminus in Prince William Sound has increased by 500% (Tangborn and Post, 1998). This has significantly increased the risk to shipping along the southern Alaskan coastline, especially tourist vessels, as well as oil tankers linking with the trans-Alaska oil pipeline. This new situation has been monitored since July 1996 after a serious ship–iceberg collision in 1994 when the oil tanker, *Overseas Ohio*, struck a 4000 ton iceberg in Alaskan waters and badly dented the ship's bow (Figure 4.8). A mass balance model of Columbia Glacier has been developed to relate ice discharge to iceberg calving rate, and studies of fjord bathymetry and local tides and weather have been made in order to develop an iceberg prediction model for this region. In Antarctic waters, as mentioned earlier, the M/S *Explorer* sank following a collision with a small iceberg.

4.3.2 Iceberg–fixed installation collisions

Today, with so much off-shore hydrocarbon exploration and recovery taking place or planned in polar and sub-polar regions, iceberg management systems (see section on mitigation) have assumed a high profile in risk assessments associated with the hydrocarbon industry (Eik, 2008). In the area of the Grand Banks there are currently three operational oil fields, Hibernia, Terra Nova and White Rose, with drilling platforms of various types in place, all of which are potentially vulnerable to iceberg impact. The Hibernia field, discovered in 1979, began oil production in late 1997 and has an expected field life of 20 years, with estimated recoverable reserves of between 106 and140 million m^3. Production facilities include a fixed production platform (gravity base structure) incorporating a 15 m thick concrete wall designed to withstand the impact of a 6 million tonne iceberg with an estimated return period of 10 000 years. This may sound an unlikely occurrence given the 20 year lifespan of the oil field, but a 2.8 km long iceberg was sighted south of 50°N in July 1991 (Newell, 1993) and the recent disintegration of several north Ellesmere Island ice shelves (Copland and Mueller, 2009) provides a new source of big icebergs with the potential to

Figure 4.8 Iceberg damage to bow of the oil tanker Overseas Ohio, April, 2003 (source: DEC Alaska, 2003, Department of Conservation, Division of Spill Prevention and Response Photo Gallery)

enter the Grand Banks area (or the Beaufort Sea hydrocarbon exploration area in the Arctic Ocean – the recent breakup of the Ayles Ice Shelf in northern Ellesmere Island did send an ice island westwards towards the Beaufort Sea initially). Ice islands (exceptionally large icebergs) crossed the Grand Banks in the 1920s (Hill, 2008).

The iceberg hazard and drilling platform design are also issues for Russian cryoscientists and engineers. Icebergs can be expected in all deep-sea areas of the Russian Arctic with the exception of shallower coastal waters. The greatest concentration of icebergs occurs near Franz

Josef Land and Severnaya Zemlya in September, the season of maximum propagation from the source areas. Selection of ice-resistant drilling platforms for middle sea depth (200–400 m) has been considered by Chernetsov, Malyutin and Karlinsky (2008), who see this as a vital problem for the development of the Russian shelf oil reserves. Of particular concern is the Shtokman Gas-Condensate field in the northeastern part of the Barents Sea, north of Europe.

At the eastern end of Russia, in the Chukchi Sea, and in the Beaufort Sea, north of Alaska, a different ice hazard has recently emerged. This is the rapid disintegration of Arctic Ice Shelves, such as the Ayles Ice Shelf (Environment Canada, 2010a), along the north coast of Ellesmere Island where 23% of the area was lost in the single year of 2008. Although there has been no sign of regrowth and with the current warming of the Arctic this situation is not likely to change, ice shelf disintegration remains a major potential issue for oil exploration in the Beaufort and Chukchi Seas requiring improved monitoring of the ice islands produced by ice shelf collapse.

4.3.3 Sea bed scouring

Iceberg scour occurs when an iceberg becomes grounded. The length of scour depends on the speed of iceberg movement and the strength of the sea bed. Risk analyses are performed to determine how deep pipelines or communication cables should be buried to avoid breakage.

4.3.4 Impact on ecological systems

The present episode of spectacular ice shelf disintegration in Antarctica has lasted for at least 25 years. During this time several large ice shelves on the Antarctic Peninsula, such as Wordie, and Larsen A and B, have collapsed and larger shelves such as the Ross Ice Shelf also continuously calve large tabular icebergs and ice islands (Figure 4.9). As noted above the largest of these ice masses are monitored by the US National Ice Centre in case they impact shipping or islands in the Southern Ocean. The density of commercial shipping in this area is low (Figure 4.3) and the highest risks are probably run by tourist vessels. The scale of many ice masses in Antarctica is such that they are easily tracked and observed and of limited danger to shipping, but this does not mean that their effects on the broader environment are negligible or benign. Smale *et al.* (2008) have recorded the impact of iceberg grounding and scouring on benthic communities at Adelie Island, Antarctica. Ice scouring was found to be catastrophic with scour assemblages 95% lower in mean macrofaunal abundance and 76% lower in species richness than the undisturbed areas.

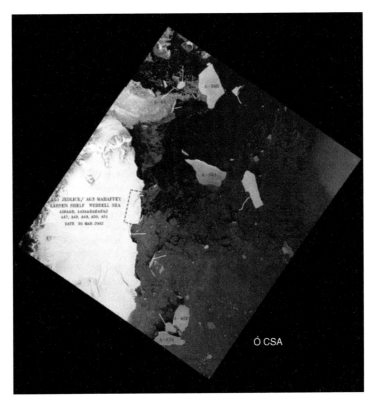

Figure 4.9 Icebergs breaking away from the Antarctic Penisula (Source: NIC, 2009 http://www.natice.noaa.gov/index.htm - accessed 07 March 10 RADARSAT-1 data © CSA 2002.)

It has been suggested (Kooyman *et al.*, 2007) that two enormous icebergs, B-15 and C-19, near Cape Crozier are implicated in the disruption of the breeding cycles of Emperor Penguins. The presence of the icebergs may have forced the penguins to travel further on land to their feeding locations rather than swimming which they prefer. Breeding rates declined and the colony divided into five subgroups.

4.3.5 Impacts on global climate

During the last 'Ice Age' iceberg armadas (Heinrich, 1988) resulting from the collapse of the Laurentide Ice Sheet in the Hudson Bay area transferred masses of fresh water to the North Atlantic which may have been responsible for climate change in Western Europe, if not globally (Lowe and Walker, 1997). Whether similar rapid decay of Greenland could have the same effect is a question for the longer term and not a risk that requires immediate attention.

Similar long-term issues may exist in relation to Antarctica where melting of large icebergs reduces ocean salinity and possibly the supply of high saline waters to the Antarctic bottom water component of the Thermohaline Circulation system (ACECRC, 2009). These are issues to be aware of, but they are not of immediate concern.

4.4 Iceberg mitigation

It is clear from the preceding discussions that icebergs are just one component of a very complex environmental system involving land-based ice sheets and glaciers, shelf ice, sea ice, ocean currents, weather and ocean-bed topography. Some of these components vary significantly from year to year, especially the weather. Weather influences sea surface temperatures, air temperatures, sea ice extent and wind direction, all of which, in turn, affect iceberg survival and trajectory.

The scale of the iceberg hazard in the northern hemisphere and Antarctica seems unlikely to decrease in the near future for a number of reasons. In the north, the present downward trend in Arctic sea ice extent, continued disintegration of Canadian ice shelves on Ellesmere Island and and the increasing flux of Greenland icebergs, are coinciding with the expansion of tourism and commercial shipping activities and the increasing exploitation of polar and sub-polar hydrocarbon resources. These changes must increase the risk of iceberg impacts and a range of strategies will be required to mitigate these iceberg hazards (Eik, 2008).

Mitigation strategies available to risk managers working in areas prone to iceberg drift can be separated into a number of different approaches including:

(1) detection, monitoring, databases and research;
(2) threat evaluation and prediction;
(3) ice management;
(4) avoidance.

Prior to the sinking of RMS *Titanic*, mitigation of the iceberg hazard essentially entailed keeping your eyes open on the ship's bridge and taking any necessary action to avoid a collision. While observation may have sufficed in many cases, the ship–iceberg collision database (Hill, 2004) clearly indicates that it was not a fool-proof strategy, and the massive scale of the *Titanic* disaster, in particular, stimulated an instant response. The first step in the process of iceberg mitigation (1) is to detect where the icebergs are and in which direction they are moving, and then monitor them frequently to keep track of their movement for as long as they remain a threat. Recently much of these data have been assembled in databases which provide a useful historical perspective as well as a source of current information. Over the years, these critical primary data have been supplemented by research (2) into the factors that control the initiation of icebergs and their subsequent behaviour and deterioration. This research and experience has enabled the threat to be evaluated in order to predict (3) potential impacts. Hydrocarbon exploration, in particular, has provided a strong impetus to the development of iceberg predictive models but their success seems to be limited by the considerable variability of the natural systems controlling iceberg incidence. During the middle of the twentieth Century, a number of experiments were conducted with the aim of accelerating iceberg disintegration using gunfire, mines, torpedoes, depth charges, bombing and a material called thermite, which was supposed to explode in ice with an extremely high temperature, but none of these approaches were practicable, successful or cost-effective and were eventually abandoned in favour of monitoring. However, although these destructive options proved to be ineffective, ice management (4) in the form of iceberg control is extensively used, especially by the hydrocarbon industry, to protect static or less mobile drilling platforms and other surface or subsurface installations. Tugs are often used to tow icebergs and deflect them away from their natural path if this is predicted to intersect with the position of an installation. However, some of these installations have been designed to be mobile so that they can mitigate the iceberg hazard by avoidance (5).

4.4.1 Detection, monitoring, databases and research

As discussed in the section on iceberg impact, the available data suggest that prior to the *Titanic* incident in 1912 (Hill, 2004; Table 4.7), little systematic mitigation of iceberg impact risk was in place. Before *Titanic* it was largely the responsibility of individual vessels to avoid the problem. However,

> 'Loss of the *Titanic* gripped the world with a chilling awareness of an iceberg's potential for tragedy. The sheer dimensions of the *Titanic* disaster created sufficient public reaction on both sides of the Atlantic to prod reluctant governments into action, producing the first Safety of Life at Sea (SOLAS) convention in 1914. The degree of international cooperation required to produce such an unprecedented document was truly remarkable and probably could not have been achieved during this period without the catalyst provided by this incident' (USCG-IIP, 2009b).

Today, in accordance with SOLAS, 'ships transiting the region of icebergs guarded by the Ice Patrol during the ice season are *required* (author's italics) to make use of the services provided by the Ice Patrol' (USCG-SOLAS, 2009, p. 124). These services have gradually evolved over the course of a century.

Initially, immediately following the *Titanic* disaster, the US Navy patrolled the area with cutters for the rest of 1912 and the US Revenue (later US Coast Guard) continued the service in 1913. Following an international convention in November 1913, the USA formally undertook to continue an International Ice Patrol (IIP) aided by 18 other nations with an interest in shipping safety in the North Atlantic. From 1912 to 1946 IIP ship reconnaissance was responsible for much of the information with ships patrolling the key areas of the Grand Banks and Labrador Coast. The first aerial reconnaissance using aircraft was undertaken by Canada during the winter of 1927–28. IIP began full-time aircraft visual reconnaissance in 1946 (Figure 4.10) and phased out their reconnaissance vessels by 1973. In 1954 responsibility for 'Ice Services' in Canada was assigned to the Meteorological Branch of the Department of Transport by a joint government committee. The present Canadian Ice Service (CIS) is part of the Meteorological Service of Canada.

Aircraft visual reconnaissance was largely superseded, first by Side-Looking Airborne Radar (SLAR) (Canada 1978; USA 1983) and then, a decade later, by Forward Looking Airborne Radar (FLAR). Tests have shown that FLAR is good at distinguishing ships from icebergs but less effective than SLAR at detecting small and medium icebergs between 50 and 100 m in length. For this reason the two systems have been used by IIP in combination. Targets moving faster than 10 knots, showing a wake and with an intense radar return are likely to be ships, whereas targets with large radar shadows are more likely to be high icebergs. One problem for the film interpreter has been the numerous, small stationary fishing vessels in the Grand Banks which can easily be mistaken for small icebergs (USCG-IIP, 2009c).

Canada launched two satellite reconnaissance platforms: RADARSAT-1 in 1995 and RADARSAT-2 in 2007, both of which use Synthetic Aperture Radar (SAR). Today the IIP system operates with 17 countries participating financially (Belgium, Canada, Denmark, Finland, France, Germany, Greece, Italy, Japan, Netherlands, Norway, Panama, Poland, Spain, Sweden, UK and USA). The US Coast Guard (USCG) is currently responsible for the IIP. Canada, through the CIS, also plays a direct part in monitoring the iceberg hazard adjacent to its coasts, especially in the more northerly areas of the Labrador Sea and Baffin Bay. These are complex systems requiring a great deal of computer power but providing relatively high resolution of the objects. An indication of the breadth of information sources now available to IIP and CIS is provided by the IIP report for 2008, Bulletin 94 (USCG-IIP, 2008). This annual report lists sources of information – generated by land, sea, air and space platforms – which are used in iceberg reports by lighthouses, merchant and Canadian Coast Guard vessels, IIP reconnaissance flights, CIS contracted commercial reconnaissance flights and satellite data processed by the National Ice Centre (NIC) and C-CORE, a commercial satellite reconnaissance provider. There is close collaboration between IIP and CIS using a shared database in the iceBerg Analysis and Prediction System (BAPS).

Figure 4.10 Visual aerial reconnaissance of icebergs from IIP aircraft (Reprinted from http://www.uscg-iip.org/photo_gallery/. Courtesy of US Coast Guard. [accessed 03 April 2009].)

For the IIP in particular, all this reconnaissance activity culminates in the production of a line on a map called the Limit of All Known Ice (LAKI), in effect the southern, eastern, south-eastern, and south-western limit of the iceberg population, and the boundary for ice-free ship navigation. This information satisfies the safety requirements of SOLAS. During the 2009 ice season, IIP began creating and distributing the LAKI to mariners on 16 March when it was tracking 89 icebergs south of 48°N. The majority of IIP's reconnaissance missions focus on this critical boundary (USCG-IIP, 2009b).

The discussion so far has centred on the northwest Atlantic region which, more than any other area, extends across busy shipping lanes. However, a wider distribution of icebergs has already been described, and the risk of iceberg impact in other areas of the Arctic, and adjacent oceans, is likely to increase as access improves. With this in mind, the International Ice Charting Working Group (IICWG) was formed in 1999 to promote cooperation between the world's ice services on all matters concerning sea ice and icebergs and brings together the operational ice services of Canada, Denmark (Greenland), Finland, Iceland, Germany, Norway, the Russian Federation, Sweden, USA and the IIP. These services are charged with establishing procedures, monitoring sea ice and icebergs, and coordinating information for marine safety (NSIDC-IICWG, 2009). The IICWG hold annual meetings to exchange and disseminate information and ideas.

The US National Ice Centre (NIC) has been tracking and archiving all large icebergs in the Antarctic since 1976 (NIC, 2009a). Each iceberg larger than 10 nautical miles (18 km) along its long axis and located south of latitude 60°S is assigned a name composed of a letter, indicating the quadrant in which it was originally sighted:

A longitude 0° to 90° W (Bellingshausen Sea, Weddell Sea)
B longitude 90° W to 180° (Amundsen Sea, Eastern Ross Sea)
C longitude 180° to 90° E (Western Ross Sea, Wilkes Land)
D longitude 90° E to 0° (Amery Ice Shelf, Eastern Weddell Sea)

and a running number (note: the centre of the circle represents the South Pole). When a named iceberg breaks, the individual parts are also numbered by adding a suffix letter. For example, the 15th iceberg found in Quadrant A is named A-15. If it subsequently divides into four parts these are named A-15A, A-15B, A-15C and A-15D, A-15A being the largest fragment of the host iceberg. If the size of an iceberg drops below 10 nm, or visual sighting is lost for 30 consecutive days, of an iceberg north of latitude 60°S, the iceberg will be removed from the database and warnings no longer disseminated to vessels in the region. The iceberg record is transferred to the archive. The database, which includes a list of tracked icebergs, location maps and photographs can be interrogated at the NIC website (NIC, 2009b).

Databases, provide a fund of information that can be analysed to demonstrate existing relationships between variables, longer-term trends and the level of success of monitoring and mitigation procedures. For example PERD (Program of Energy Research and Development), which is an interdepartmental programme operated by National Resources Canada (NRCan) runs a number of databases related to icebergs, including the Iceberg Sighting Database (Verbit, Comfort and Timco, 2006) and the

Comprehensive Iceberg Management Database (Rudkin, 2005) (see below). There is also an Iceberg Shape Database (Canatec Consultants, 1999) and a Ship–Iceberg Collision Database (Hill, 2004). IIP also have their own Iceberg Sighting Database (NSIDC, 1995).

4.4.2 Threat evaluation and prediction

In view of the complex of factors (ocean currents, wind, waves, etc.) that determine iceberg trajectories it is not surprising that predicting their behaviour, for example, in relation to hydrocarbon installations and the shuttle tankers that service them, is a critical problem. The USGS prediction that 25% of remaining hydrocarbon resources is located in the Arctic (Eik, 2009) emphasizes the critical importance of prediction in the mitigation of the iceberg hazard. Marko, Fissel and Miller (1988) differentiated iceberg prediction models according to the time over which they operate and the area they cover. The commonest, short-term, local models forecast iceberg trajectories over periods up to a day and within a range of about 30 km and are most appropriate for iceberg diversion measures such as towing and 'prop-washing', and avoidance (see below). Medium-term or regional models are designed for forecasts up to 14 days and over an area of several hundred kilometres. Such models may encompass groups of icebergs and fixed installations in their forecasts and are appropriate for advisory statements to shipping by allowing time for planning responses and further reconnaissance. Long-term models operating over several months and relating to perhaps thousands of square kilometres have little or no value in terms of day-to-day operations but may, instead, provide predictions of the comparative 'severity' of the iceberg hazard from one year to the next. Within this framework models fall into two categories: 'dynamical' models with input parameters associated with icebergs and their environment, and 'statistical' models based on historical data presented as time-series. These types of models are similar to those employed in avalanche mitigation (Chapter 8).

The area known as 'Iceberg Alley', on the Grand Banks east of Newfoundland, Canada, where the RMS *Titanic* sank and where there are three Canadian oil field installations has probably stimulated most prediction model development. Both the IIP and the CIS have iceberg prediction models in operation. IIP has been running a model since 1979. A more recent predictive model has been developed by the National Research Council Canadian Hydraulics Centre (NRC-CHC) (NRC-CHC, 2007), funded by the Natural Resources Canada's Program of Energy Research and Development (PERD). This new drift model was validated against actual iceberg drift observations around the Hibernia platform. The main improvement is the ability to model complex ocean currents, especially those at depths of about 10 m, which is a key factor in determining iceberg trajectory but one of the most difficult to determine accurately. The Canadian model can also predict iceberg calving and the resulting size distribution and drift of 'bergy bits'.

Elsewhere in the Arctic, Eik (2009) produced a model of iceberg drift in the Barents and Kara Seas but found that more accurate sea current records are required for the model to reproduce iceberg drift accurately. The importance of accurate prediction of iceberg drift cannot be overemphasized as it is the basis of effective ice management.

4.4.3 Ice management

In relation to moving vessels and other mobile objects that can easily adjust their trajectory away from icebergs, iceberg detection, monitoring and drift prediction are usually adequate strategies for mitigating the iceberg hazard. In the case of fixed (e.g. gravity base structure (GBS)) or semipermanent structures (e.g. FPSOs), however, additional mitigation measures are necessary for their long-term survival. When the drilling platform cannot move then it becomes necessary to move (deflect) the iceberg. This strategy which includes several different techniques (Rudkin *et al*., 2005), is referred to as ice management. Eik's (2009) powerpoint presentation to the Fourth Norway–Russia Arctic Offshore Workshop provides a very good outline of the issues involved in the ice management strategy.

Rudkin *et al*. (2005) analysed over 1500 ice management operations, in the seas off eastern Canada, that are contained in the PERD Comprehensive Iceberg Management Database (Rudkin, 2005). This analysis shows (Table 4.9) that towing is by far the most popular (87%) of the techniques employed to divert icebergs and that ramming was discontinued because it was considered to be too dangerous. The success of towing varies with the angle between the trajectory of the iceberg and the direction of the tow. Acute angle diversions are more successful but obviously require starting further from the installation (Figure 4.11). A 45° towing angle has a 50% chance of success. The average time taken to complete an ice management operation successfully was found to vary from 6.9 hours off Labrador to 14.7 hours on the Grand Banks, probably reflecting the more complex current patterns around the shallow Grandbanks. A tow was concluded for a number of reasons: on 61% of occasions because it was successful, on 22% because the tow rope slipped off the iceberg, on 9% because the iceberg rolled and on 1% because of tow-rope failure. A tow was considered successful if 'downtime' (cessation of activity) was avoided. Results of towing improved with experience, up to about five tows, after which there was little change in success rate. The fact that impacts are apparently rare is a measure of the success of the tow technique.

Table 4.9 Types of iceberg management assessed by Rudkin et al., 2005 (source: Rudkin et al., 2005, Table 1)

Total number of individual icebergs	973
Types of management	
Total number of tows	1303
Total number of prop-washings	73
Total number of water cannon management	34
Total number of rammings	5
Total number of two-vessel tows	33
Total number of net tows	45
Total number of other management techniques	8

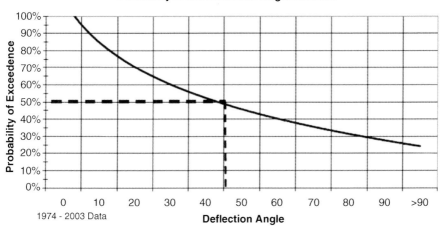

Figure 4.11 Probability of tow deflection angle success (source: Rudkin et al., 2005, Figure 1)

4.4.4 Avoidance

The three drilling sites in the Grand Banks area provide examples of the final iceberg mitigation strategy, avoidance. In deeper water where large platforms are obliged to float, the answer is a floating production storage offloading platform (FPSOP). This is the solution for the White Rose Platform. Given adequate warning this platform can shut down and move out of the drift line of an iceberg. The taps were turned off in April 2008 after an iceberg approached to within 3 km (Calgary Herald, 2008) but a move was not necessary. Wellheads and manifolds are protected in sea bed excavations called 'glory holes' to avoid damage from icebergs that scour the ocean floor (Industry Canada, 2009). The other solution is exemplified by the Hibernia Platform, a huge 1.2 million ton gravity based structure located in relatively shallow (111 m) water and designed to withstand the impact of a 1 million ton iceberg with no significant damage or a

6 million ton iceberg, without harm to workers, the environment or operations (Exxonmobil, no date). The Hibernia platform has space to store 1.3 million barrels of oil and large quantities of ballast to add weight. Being in shallower water the probability of very large, deep-draft icebergs approaching the Hibernia Platform is less than that for the White Rose Platform which is located in deeper water further east (White Rose, 2000).

4.5 Summary

Greenland icebergs rarely feature on global news broadcasts, although the hazard has not disappeared. This reflects the efficiency of iceberg monitoring in the North Atlantic by the CIS and the IIP (who have almost a century of experience and technical innovation behind them), modern on-board equipment available to shipping and the efforts of hydrocarbon exploration and exploitation companies to protect their fixed or movable installations. Today, headlines are more likely to be made in Antarctica by the latest mega-iceberg to be released from ice shelves, with its implications for marine ecology and climate change, or by the sinking of a tourist vessel. It will be interesting to see the extent to which this latter risk is transferred to the Arctic Ocean if current trends in sea-ice loss in that area continue (see Chapter 2).

Box 4.1 The 2008 Iceberg Season on the Grand Banks – from a Report of the IIP, North Atlantic Bulletin No. 94 CG-188-63 (USCG-IIP, 2008)

The IIP conducted aerial reconnaissance from Saint John, Newfoundland, to search for icebergs in the SE, S and SW areas of the Grand Banks, off Newfoundland, Canada. This IIP iceberg data, together with additional reports from other aircraft and ships in the North Atlantic and environmental data was analysed at the Operations Centre in Groton, Connecticut, USA, using the iceBerg Analysis and Prediction System (BAPS) computer model to predict iceberg drift and deterioration. By late March, significant iceberg presence was tracked south of 48°N. Nine icebergs were being tracked and used to define the LAKI boundary. On 28 March weekly iceberg reports were replaced by daily warnings which continued until 15 July, the final day of the 2008 ice season. The 2008 ice season turned out to be in the top 10 most severe based on iceberg populations south of 48°N (Table 4.8), with more than twice the 1900–2007 seasonal average of 472. It was classified as an extreme-season (more than 600 icebergs passing south of 48°N; Trivers, 1994).

Several environmental conditions – temperature, wind speed and direction, storminess and sea ice extent – combined to produce this exceptionally severe iceberg season. December 2007 air temperatures were colder than normal in Labrador, and January 2008 storm tracks, out into the Atlantic south of Greenland, both enhanced vigorous sea-ice formation along the Labrador coast. This, in turn, protected icebergs embedded within the sea ice from wave erosion. (In 2007, in contrast, storms moving northwards into the Davis Strait had destroyed ice or compressed it along the Labrador coastline and icebergs became exposed.) No icebergs passed south of 48°N in January and February, 2008, but the sea ice continued to extend southwards as normal. However, an intense storm on 17–18 March, stalled south of Newfoundland and the resulting strong northeast winds compacted sea ice along the southern coast of Labrador and the north and east coasts of Newfoundland. Instead of beginning its normal yearly retreat, the ice edge, driven southward by the offshore branch of the Labrador Current, extended rapidly eastward and southward in late March and even accelerated in early April until by 8 April it had reached 44°50'N, well beyond the key 48°N line of latitude.

This was bad news for the oil industry that had been drilling in the Grand Banks area since 1997. As the Report dramatically relates: 'Heavy sea-ice and iceberg conditions in the eastern Grand Banks created havoc in early April. Production at the White Rose oilfield east of St. John's was suspended for several days and the exploratory rig, Global Santa Fe Grand Banks, was towed to an ice-free area... The severe sea-ice and iceberg conditions resulted in a flood of iceberg reports into Ice Patrol's Operations Centre during April. PAL flew at least one iceberg reconnaissance flight per day in support of the oil-field activities on the eastern part of the Grand Banks. Two IIP IRDs conducted eight reconnaissance missions and found 585 icebergs. In addition, there were numerous land and ship reports, including numerous iceberg reports from Canadian ice-breakers working to maintain safe domestic marine traffic lanes in the heavy ice conditions... During the last week of April, the IIP Operations Centre began receiving iceberg reports from lighthouses on Newfoundland's south coast, indicating that icebergs were moving past Cape Race... In April, IIP estimated that 712 icebergs passed south of 48°N.'

In May, 173 icebergs passed south of 48°N, but throughout June, sea ice retreated rapidly northwards and the warming ocean severely reduced the Grand Banks icebergs. By mid-June, only 69 icebergs, growlers, and radar targets were still being tracked south of 48°N and during June only 43 icebergs passed south of 48°N. The last daily iceberg warning was broadcast on 15 July, the IIP having verified 'that there were no icebergs threatening the transatlantic shipping lanes'.

5
Glaciers

5.1 Introduction

The global cryosphere is a complex system in which ice occupies several different landscape niches. In contrast to the great ice sheets discussed earlier, glaciers are smaller, usually mountain-based ice masses that are topographically controlled by the valleys in which they are confined. They therefore present a rather different set of hazards from those of the massive, remote ice sheets.

Several features of glaciers, such as crevasses, seracs, melt water and ice avalanches, which are related to their position, structure, motion and temperature, make them inherently hazardous. However, glaciers, being sensitive to climate forcing, also advance and retreat in response to changes in their mass balance. Surging glaciers also respond to internal processes. The nature of glacier hazards varies depending on whether the front of the glacier is advancing (positive net mass balance), retreating (negative net mass balance) or stationary (zero net mass balance). Most glaciers have retreated overall from their Little Ice Age maxima to historically unprecedented positions and this has changed the position of zones of geomorphic activity and hazard distribution. Thus mass balance is a useful criterion for organizing the following discussion of glacier-related hazards and these different categories of hazard, related to advance and retreat, and the inherent properties of a stable glacier will be discussed in the next section (Table 5.1). The difficult issue of their mitigation will then be addressed.

Although some glaciers are located in remote uninhabited regions many occupy the upper reaches of relatively densely populated valleys, such as those in the European Alps where infrastructure is expanding and tourism increasing. Records of glacier hazards in the European Alps can be traced back several centuries (see Tufnell, 1984). Recently, a large inventory called

Cold Region Hazards and Risks, First Edition.   Colin A. Whiteman.
© 2011 John Wiley & Sons, Ltd. Published 2011 by John Wiley & Sons, Ltd.

Table 5.1 Glacier-related hazards and impacts

Inherent glacier hazards	Hazards due to mass balance change		
	Advance	Retreat	Others
Crevasses	Land loss	Valley side instability	Sea-level rise
Seracs	Surging	Proglacial lakes	Albedo change
Ice avalanches	x	Supraglacial lakes	Biodiversity
Complex avalanches	x	Ice avalanches	Skiing
Ponded lakes	x	Loss of glacier ice	x
x	x	Ultimately loss of meltwater	x

GRIDABASE (http://www.nimbus.it/glaciorisk/gridabasemainmenu.asp) of past catastrophic events related to glacial environments in the European Alps and Nordic countries was compiled as part of a collaborative EU Fifth Framework Programme (EVG1 2000 00512), 'Glaciorisk'. The project was designed to facilitate the detection, survey and prevention of future glacial disasters in order to save human lives and to reduce damage costs. However, as Salzmann *et al.* (2004) have pointed out, in the context of present climate change, glaciers are probably experiencing conditions not recorded during the past 2000 years (Haeberli and Holzhauser, 2003; Kääb *et al.*, 2003), and the encroachment of people and their infrastructure into hazardous zones is unprecedented (Haeberli, 1992). In addition, many glacial hazard events happen sporadically in remote areas so that records are often incomplete. Complex chain reactions, involving ice avalanches, rock and snow further complicate analysis. Consequently, glacier hazard and risk assessment, based solely on historical magnitude–frequency analyses, may no longer be entirely valid for current conditions (Zimmermann and Haeberli, 1992; Kääb *et al.*, 2003). This is why considerable effort is being directed towards finding ways of assessing the current risk of the different types of glacier hazard using modern monitoring and modelling technology (e.g. Huggel *et al.*, 2004; Kääb *et al.*, 2005).

5.2 Inherent glacier hazards

5.2.1 Crevasses

Anyone who has travelled even a short distance across the surface of a glacier will be aware of the danger posed by open joints known as crevasses, especially if these are obscured from view by a bridge of snow accumulated during the winter. Crevasses reflect tensional and shear stresses in the ice due to relative movement (Figure 5.1). Complex patterns of crevasses can develop where a glacier accelerates over steepening sections of its valley or spreads out onto lowlands beyond the confines of its valley. Crevasses are also commonly

Figure 5.1 Heavily crevassed surface of a surging glacier: Fridtjovbreen in Van Mijenfjorden in the early 1990s (Photo: M. Sund) (source: SSF NPOLAR, no date)

located near valley sides where friction slows the ice adjacent to its margin relative to the central part of the flow. Interrogation of the Norwegian section of Gridabase reveals that five people died as a result of falling into crevasses in Norwegian glaciers between 1946 and 2002, one every 11.4 years (Table 5.2).

Without data on the number of people who have put themselves at risk during this period by crossing the surface of glaciers, it is not possible to quantify the degree of risk presented by the crevasse hazard. The average of about one death every 12 years suggests that most people are properly aware of the hazard and take appropriate precautions to minimize its impact. Nevertheless the fact that, from time to time, bodies do emerge from the front of glaciers in various parts of the world, is a gruesome reminder of this perennial glacier hazard (see Deem, 2008, for examples).

With regards to mitigation, the best solution is probably to avoid the problem altogether by not walking on glaciers in winter when crevasses are likely to be bridged by snow. Otherwise, safety manuals suggest linking members of a party with ropes, securing members to some sort of anchor and

Table 5.2 Crevasse fatalities on Norwegian glaciers between 1946 and 2002 (source: http://www.nimbus.it/glaciorisk/GlacierList.asp?vista=paese&paese=Norway – Accessed 27 10 09)

Glacier	Date
Sandelvbreen	01 01 1946
Veslgjuvbreen	11 07 1957
Nigardsbreen	14 10 1995
Habardsbreen	27 06 2001
Blåisen	17 09 2002

probing the snow in front as progress is made over the glacier. Even with this advice, and warnings from the authorities, it is unlikely that there will be no further examples of this hazard as Scandinavians, in particular, express their 'allemansrätt' (Swedish – 'freedom to roam').

5.2.2 Seracs

Where a glacier moves from a gradual to a steeply dipping section of the valley it accelerates and tension in the ice creates deep crevasses that pull the ice apart and form large, unstable blocks of ice called seracs. These steep sections of a glacier are usually termed icefalls, one of the most famous, and dangerous, being the Khumbu Icefall which lies on one of the routes to Mount Everest (also known as Chomolungma (Tibetan), Devadhunka and Chingopamari (Nepalese)). Seracs are liable to topple either naturally, as the glacier moves inexorably down-valley, or perhaps when disturbed by earthquakes, or even unfortunate climbers. A number of tragic accidents have occurred to mountaineers forced to traverse this type of terrain to reach their objective. Gillman (1993), in his history of Everest exploration, lists six fatalities (5.2%) due to falling seracs out of a total of 115 fatalities for the period 1922 to 1992, an average of one every 11.8 years. Again, it is difficult to quantify the risk accurately in the absence of detailed figures for the total number of traverses through the icefall, but it is obvious from reading accounts of mountaineering in places like the Himalaya, that serac-rich icefalls are arguably climbers' least favourite places.

The best means of mitigating this hazard is to avoid the problem altogether (not always feasible in a steep-sided valley) or move through the icefall as quickly as possible. It is very difficult to anticipate the fall of a serac. The problem is exacerbated in tectonically active mountains such as the Himalaya, where earthquakes are common and even a small event may trigger collapse.

5.2.3 Ice avalanches

Seracs may also be implicated in the ice avalanche hazard as, like seracs, the breakup of a glacier in a steep topographic position is the basic cause of these dangerous events. In the case of ice avalanches, however, the starting zone is usually at or close to the front of the glacier and at the top of a steep deglaciated slope. Ice avalanches usually involve a much larger mass of ice than a single serac although the volume can vary considerably depending on type (see below). Ice avalanches may represent the normal ablation process of cold-based and/or high-altitude glaciers on steep mountain faces (Salzmann *et al.*, 2004). Unlike most snow avalanches, ice avalanches can occur throughout the year, although winter is the season for the most destructive events as they are likely to release or entrain additional snow masses and therefore travel longer distances (Margreth and Funk, 1999). Unlike crevasses and seracs, which tend to trap only a few people in any single event, particular ice avalanches, like some snow avalanches, have been responsible for multiple fatalities. For instance, in 1965 88 workers were buried in their temporary construction camp near the Mattmarksee dam by ice from the Allalin Glacier, Switzerland (Figure 5.2). This tragic event was made worse by its timing, which coincided with a change of shift so that twice the usual number of personnel was on site when the ice crashed down the valley side (Vivian, 1966, cited in Tufnell, 1984). The Allalin Glacier had previously caused a number of problems, but on those occasions the hazard was flooding, caused by the outburst of lakes dammed by the glacier when it advanced across the Saas-Visp River. Ironically, the workers were in the process of constructing an artificial dam to produce a reservoir for a hydroelectricity scheme when the ice avalanche struck their camp.

In simple terms, ice avalanches occur when part of a steep or 'hanging' glacier breaks away, falls down a steep slope and, if sufficiently large, runs out into a valley where it is likely to have the greatest impact. Ice avalanches may involve anything from a few hundred to millions of cubic metres of ice, falling initially as a large block or several blocks and subsequently as smaller fragments or even powder as the blocks fragment and comminute on impact (note: 1 cubic metre of ice = approximately 0.92 of a metric ton, depending on ice density). Generally, two main types of ice avalanche release are recognized: cliff-edge and ramp depending on the topographic position of the failure (Figure 5.3; Haefeli, 1965; Alean, 1985; Margreth and Funk, 1999; Salzmann *et al.*, 2004). The ramp type can be subdivided on the basis of ice temperature (Salzmann *et al.*, 2004).

Type 1 (cliff-edge) failures occur at a sharp break of slope or at a cliff edge and are referred to as wedge failures (Haefeli, 1965) or frontal block failures (Richardson and Reynolds, 2000). At the edge of the cliff the glacier develops a vertical or even overhanging front. Extending flow giving rise to high tensile stresses and shears within the ice causes crevasses to form resulting in the formation of unstable ice masses. Eventually a stress threshold is crossed and a wedge of ice falls. This process represents the normal means of ablation for hanging glaciers and tends to occur quite frequently under normal conditions. However, ice avalanches from this situation are usually small, with typical empirically-based maximum starting values of 4×10^5 m^3 (Alean, 1985; Huggel *et al.*, 2004) and less often cause serious damage. The Gutz Glacier (see Box 5.1 for details) and the Whymper Glacier in the Swiss Alps are both examples of the cliff-edge type of ice avalanche.

Type 2 (ramp) failures occur at some distance from the front of the glacier and the resulting ice avalanche may be very large, typical empirically-based maximum starting values as much as 5×10^6 m^3 (Alean, 1985; Huggel *et al.*, 2004). The Allalin Glacier (Figure 5.2) in the Valais Alps, and the Altels Glacier in the Bernese Alps, Switzerland, typify the ramp-type of ice avalanche, and this country has experienced some of the worse ice avalanche

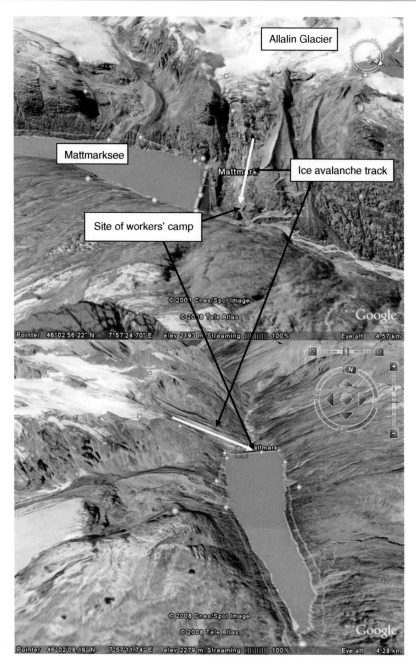

Figure 5.2 Location of the Mattmarksee, Switzerland ice avalanche disaster, 1965

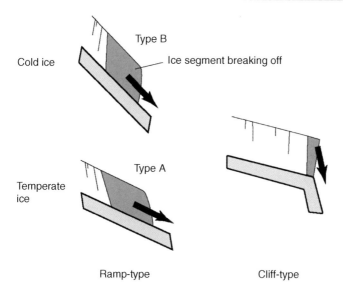

Figure 5.3 Typology of ice avalanche starting positions (Reproduced with permission from Salzmann, N., Kääb, A., Huggel, C., Allgöwer, B. and Haeberli, W. (2004) Assessment of the hazard potential of ice avalanches using remote sensing and GIS-modelling. Norwegian Journal of Geography, 58, 74–84. Fig. 4. © Taylor and Francis.)

Box 5.1 The Gutz Glacier ice avalanche and hazard mitigation

The Gutz Glacier (Gutzgletscher) provides an instructive example, of mitigation procedures applied to the ice avalanche hazard (see Margreth and Funk (1999) and Huggel et al. (2004) for details). This small cirque glacier is situated at 3701 m above sea level on the northwest face of the Wetterhorn, near the famous town of Grindelwald in the Bernese Alps of Switzerland. The 60 m high ice cliff at the end of the glacier is perched on a 60° rock wall above a 1000 m drop (Figure 5.4). Ice avalanches occur in two zones, the Wätterlaui, which is most active, and the Gutzlaui. Small ice avalanches are the normal ablation mechanism of the Gutz Glacier and occur several times a day during more active phases (Huggel et al., 2004). On 5 September 1996 the Gutz Glacier released two ice avalanches from the 'Wätterlaui' zone, at 3 p.m. and 9 p.m. Three people were injured by the avalanche air pressure wave and some hikers were knocked down. The first ice avalanche, with an estimated volume of 80–100 000 m^3 (Margreth and Funk, 1999) reached the Grindelwald to Grosse Scheidegg road. The second, larger event had a volume of 120–130 000 m^3 (Huggel et al., 2004) of ice and covered the road to a maximum depth of 4 m at two locations. In addition to the dense ice, the powder or dust component of the avalanche extended over about 35 000 m^2.

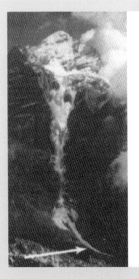 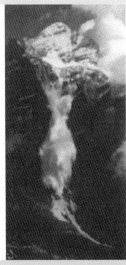 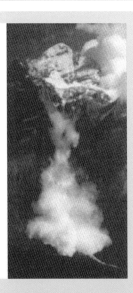

Figure 5.4 Stages in the Gutz Glacier ice avalanche 3 p.m., 5 September 1996 (Reproduced with permission from Huggel, C., Haeberli, W., Kääb, A., Bieri, D. and Richardson, S. (2004) An assessment procedure for glacial hazards in the Swiss Alps. Canadian Geotechnical Journal, 41, 1068–1083. Fig. 10 © NRC Research Press.)

In some cases buildings and forests have been severely damaged and animals killed (Bieri, 1996; Margreth and Funk, 1999) but no other injuries or fatalities were recorded in the Gridabase of the Glaciorisk Project.

These events were analysed retrospectively by the Swiss Federal Institute for Snow and Avalanche Research (SLF) in collaboration with the laboratory for Hydraulics, Hydrology and Glaciology (VAW) (Margreth and Funk, 1999; Huggel et al., 2004). Photogrammetric studies were used to show that between 26 July and 11 September 1996, 220 000 m^3 of ice was lost from the front of the Gutz Glacier. This is close to the probable maximum volume of a single event based on a thickness of 60 m, a length of 180 m and a width of 20 m (216 000 m^3). The maximum travel distance, based on volume and gradient, is 5.9 km, although the abrupt change in direction and the strongly concave longitudinal profile probably reduces the travel distance. In the event the distance was significantly less than this calculated value (Figure 5.5), possibly due to the absence of winter snow which would reduce friction and increase travel distance. Probability of occurrence is more difficult to determine. Nine major events, with an average return period of about 8 years, have been recorded since the first quarter of the twentieth century but they are not evenly spaced (Margreth and Funk, 1999; Figure 5.6). Relevant physical conditions of the glacier such as the presence of melt water and the evolution of the crevasse pattern were not available to the analysis, which makes probability more difficult to assess.

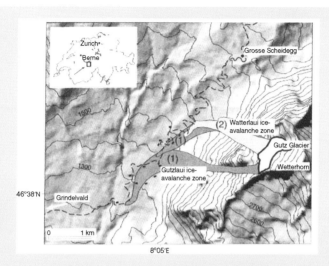

Figure 5.5 Gutz Glacier mapped prior to 1996 event (dark grey) and actual (light grey) run-out areas for the 5 September event: dashed line is the road; black rectangles are buildings (not to scale) in or near the hazard area; contours are in metres (Reproduced with permission from Huggel, C., Haeberli, W., Kääb, A., Bieri, D. and Richardson, S. (2004) An assessment procedure for glacial hazards in the Swiss Alps. Canadian Geotechnical Journal, 41, 1068–1083. DEM25 copyright 2004 swisstopo BA046420 © NRC Research Press.)

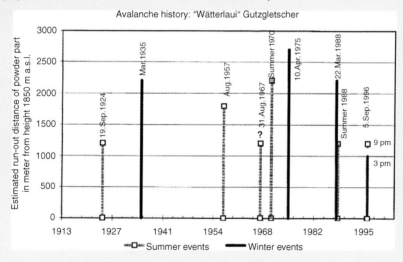

Figure 5.6 Gutz Glacier, 'Wätterlaui' ice avalanche history: winter events, dark grey; summer events, light grey (Reproduced with permission from Margreth, S. and Funk, M. (1999) Hazard mapping for ice and combined snow/ice avalanches – two case studies from the Swiss and Italian Alps Cold Regions Science and Technology, 30, 159–173. Fig. 5 © Elsevier.)

events (Table 5.3). These ramp-type ice avalanches, involving slabs of ice resting on sloping bedrock ramps, occur when adhesion is reduced and a section of glacier shears and slides, either over its bed or along a shear plane close to the bed within the ice. Alean (1985) recognized that, at higher

Table 5.3 Ice avalanches by country according to Gridabase (Glaciorisk, 2003)

Country	Number of glaciers	Ice avalanche events	Events per glacier	Fatalities
Austria	55	11	2.2	0
France	12	36	3.0	5
Italy	14	17	1.2	11*
Norway	1	1	1.0	3
Switzerland	30	119	4.0	243
Total	62	184		262

* NB: Fatalities in Italy due to complex rock/ice avalanches
Three ice avalanches with 88, 81 and 51 fatalities account for 90.5% of Swiss total

altitudes, these ramp-type ice avalanches originate from increasingly steep slopes. Salzmann *et al.* (2004) relate this angular difference to ice temperature: cold ice, frozen to its bed, is stronger and more difficult to move than relatively temperate ice, and so can exist on steeper slopes before avalanching (Figure 5.7). At the lower, warmer sites, the presence of melt water is more likely. This causes an increase in pore water pressure which in turn reduces strength and effective stress.

The type 3 subglacial bedrock failures (Figure 5.8), recognized by Perla (1980), are largely dependent on the structure and strength of the rock rather than of the ice, and are likely to be even more difficult to monitor, analyse and predict.

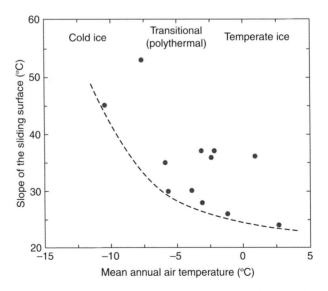

Figure 5.7 Relationship between mean annual air temperature and critical slope for failure on ramp-type glaciers (Reproduced with permission from Huggel, C., Haeberli, W., Kääb, A., Bieri, D. and Richardson, S. (2004) An assessment procedure for glacial hazards in the Swiss Alps. Canadian Geotechnical Journal, 41, 1068–1083. Fig. 4. © NRC Research Press.)

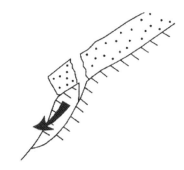

Figure 5.8 Ice–bedrock failure (Perla, 1980)

Most ice avalanches derive from small, steep glaciers with a typical thickness of no more than 30–60 m. It is therefore a reasonable assumption that mean annual air temperature (MAAT) provides an estimate for firn (compacted multiyear snow in the upper part of the glacier) temperature and in turn for glacier bed temperature. Ice temperature is a factor in ice deformation and fracture. Figure 5.7 shows the empirical relationship between MAAT and the gradient of the sliding surface (Alean, 1985; Huggel et al., 2004). This indicates that a minimum of 45° is required to produce an avalanche from a cold-based glacier, but for a warm-based glacier an ice avalanche could be expected from a slope of as little as 25°. Because these small avalanches are relatively thin the surface slope of the glacier offers a useful approximation of this basal slope when remote sensing is being used to obtain data on potentially hazardous glaciers. The starting volume of an ice avalanche is more difficult to determine, especially for the ramp-type. Mechanical conditions of the ice are difficult to measure precisely. It is sometimes possible to deduce the break line from crevasse patterns but these can be misleading. Given the typical thickness of these small cliff-type glaciers (50–60 m) and the fact that they usually break off no more than 10–20 m at a time, Huggel et al. (2004) suggest a simple calculation of maximum starting volume is possible using these two dimensions and the length of the cliff across the glacier front. In the case of the Gutz Glacier (see Box 5.1 on page 107) this is $60 \times 20 \times 180$ m, a total of $216\,000\,\text{m}^3$ which is comfortably within the volume derived empirically by Alean (1985).

Maximum starting volumes for ramp-type ice avalanches are extremely difficult to estimate because of the difficulty of anticipating the position of the break upstream of the glacier front. Consequently, for this type of ice avalanche, it is only possible to gain an estimate of maximum starting volumes from historical observations. This means that these estimates are reliable only in the European Alps where the data were obtained. There have been other exceptionally large events elsewhere (e.g. the Kolka avalanche in Russia and the Huascaran event in Peru) but these were complex events that evolved into debris-laden flows (Kääb *et al.*, 2003) and may not be strictly comparable with those in Europe.

Having estimated the maximum starting volume, or assumed the maximum empirically-derived figure for ramp-type events, the next problem is to work out the likely maximum travel distance and trajectory. Probable maximum travel distances in Switzerland appear to be related to a maximum average slope (given by a straight line between the point of release and the point of maximum run-out) of 17° as no average slopes have been measured as less than this angle in this country. The trajectory of the ice avalanche is controlled largely by the topography below the starting position. Topography will also influence distance through its control of longitudinal gradient. Sharply angled bends reduce run-out distance by increasing friction. Entrainment of snow, for instance, increases the volume and is likely to extend the run-out distance. Exceptional ice avalanches, such as the one mentioned above, produce average slopes of less than 17°. The relationship between average slope and volume (Figure 5.9) produces a regression which fits the equation:

$$\tan \alpha = 1.111 - 0.118 \log(V) \text{ and } r^2 = 0.84$$

where α is the average slope, and V is the volume (m^3) of the ice avalanche (Huggel *et al.* (2004).

The probability of ice avalanche occurrence is difficult to determine due to the shortage of data derived from this relatively rare and complex

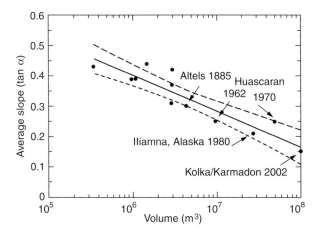

Figure 5.9 Regression (solid line) with 95% confidence intervals (dashed lines) between average slope (tan α) and the volume of large ice avalanches (Reproduced with permission from Huggel, C., Haeberli, W., Kääb, A., Bieri, D. and Richardson, S. (2004) An assessment procedure for glacial hazards in the Swiss Alps. Canadian Geotechnical Journal, 41, 1068–1083. Fig. 5. © NRC Research Press.)

phenomenon. Huggel *et al.* (2004) suggest four potential qualitative indicators of ice avalanche probability: ice avalanche repetition for ramp-type events, precursory smaller ice-fall events from cliff-type sites, excessive inputs of rain or melt water to glacier beds, and crevasse patterns and their temporal evolution. Ice avalanches are not easy to recognize by remote sensing due to snow cover, shadows, irregular topography and the small size of the glaciers.

It is not uncommon for ice avalanches to initiate snow avalanches if they land on large quantities of *fresh* snow during the main snow avalanche season. Often it is the resulting snow avalanche that is the most damaging part of the event because snow avalanches tend to have longer run-out distances. Even more complex avalanche sequences involving large quantities of rock and melting ice have occurred, often with exceptional impacts which do not fit the level of probability described above.

5.2.4 Complex avalanches

Although simple ice avalanches have occasionally caused great loss of life (e.g. the Allalin Glacier at Mattmarksee), their direct impact tends to be restricted because their run-out distances are usually short. However, 'in combination with rock falls, snow avalanches or lakes, ice avalanches have the potential to cause especially far-reaching disasters' with many fatalities (Salzmann *et al.*, 2004, p. 74). These authors list 'such process combinations or chain reactions as:

- the triggering of especially large snow avalanches by ice avalanches in winter (Röthlisberger, 1981, Giani, Silvano and Zanon, 2000);
- the combined break-off of rock and overlying steep glaciers ([Perla (1980), Giani, Silvano and Zanon, 2000, Haeberli *et al.*, 2003);
- the transformation of ice avalanches into mud or debris flows due to friction melting (Kääb *et al.*, 2003);
- lake outbursts triggered by impact waves from ice avalanches (Richardson and Reynolds, 2000) (see Chapter 6).

Amongst the most devastating disasters due to multiple avalanche causes are those which occurred at Huascaran, Cordillera Blanca, western Peru in 1970 (18 000 fatalities; Patzelt 1983) and at Kolka–Karmadon, Northern Ossetia, Russian Caucasus in 2002 (140 fatalities; Kääb *et al.* 2003, 2005).

Kolka–Karmadon avalanche, Russia One of the worst recorded complex ice avalanche-related disasters occurred in a remote valley of the Caucasus Mountains in the late evening of 20 September 2002 (Huggel *et al.*, 2005).

The avalanche began as a slope failure. Rock, and ice from hanging glaciers on the north face of Mt. Dzhimarai-Khokh, fell onto the Kolka Glacier which was disrupted and liquidized. This triggered a massive rock–ice avalanche, containing over $100 \times 10^6 \, m^3$ of ice, snow, rock and mud, which travelled along the Genaldon River valley at speeds estimated to be in excess of $100 \, km \, h^{-1}$. En route this mass of material buried the lower parts of the village of Nizhniy Karmadon and other settlements killing around 140 people. Further along the valley the debris was partially held back at the Karmadon Gorge. The finer sediment continued through the gorge as a mud flow for another 15 km, stopping a few kilometres short of the town of Gisel. The coarse material, held back at the gorge, produced an obstruction which blocked rivers and formed several lakes posing the threat of outburst and catastrophic downstream flooding (Kääb et al., 2003; Haeberli et al., 2003). The mass of debris deposited along the valley destroyed buildings, infrastructure and transport networks. Observations and analysis of this disaster by Huggel et al. (2005) were based mainly on QuickBird satellite images. In view of the extreme acceleration of the avalanche, its high flow velocity, long distance of travel and almost complete erosion of a valley glacier, these authors believe that this is a unique recorded event, at the upper extreme of the average slope/volume relationship (Figure 5.9), and obviously has important implications for subsequent glacier hazard assessment.

Huascaran avalanche, Peru The Cordillera Blanca, a mountain range in the Peruvian Andes has experienced many glacially-related disasters, most associated with avalanches into lakes and the resulting floods (alùviones). However, two particularly severe events were caused more directly by ice–rock avalanches originating from Glaciar 511 on the Nevados Huascarán (Figure 5.10). On 10 January 1962 an ice avalanche, with an estimated starting volume of $10^6 \, m^3$ travelled for 16 km along the valley (Figure 5.11) before destroying the city of Ranrahirca and killing 4000 people. On 31 May 1970, another, catastrophic, rock–ice avalanche, without historical precedence, was triggered by a 7.7 Richter magnitude earthquake. This fractured the partially overhanging granite bedrock beneath a 30 m thick glacier near the summit of Nevados Huascarán. The resulting avalanche, with an estimated starting volume of 5×10^6 to $10 \times 10^6 \, m^3$, travelled along the same valley as the 1962 event but was partially diverted over a ridge towards the city of Yungay. The complete destruction of this city resulted in at least 18 000 fatalities (UNEP, 2007). The avalanche continued for a similar distance almost to Caraz. Geomorphological mapping evidence (Klimeš, Vilímek and Omelka, 2009) suggests that an even larger, prehistorical event may have occurred in this area.

5.2 INHERENT GLACIER HAZARDS

Figure 5.10 The glacierized summit of Nevado Huascarán Norte (6746 m), Peru – note the summit icefield overhanging bedrock that was released by a powerful earthquake (source: Swisseduc, no date; photo: Michael Hambrey)

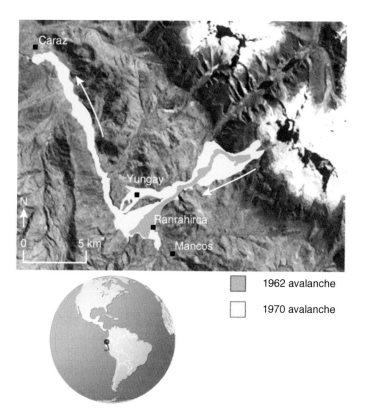

Figure 5.11 Complex avalanches from Nevados Huascarán, Peru (Modified from Google Earth © Google.)

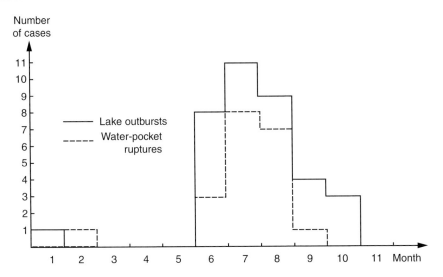

Figure 5.12 Seasonal distribution of historical glacier floods in the Swiss Alps (Reproduced from Haeberli, 1983)

5.2.5 Ponded lakes

Lakes formed between glaciers and valley sides, although less directly glacial than the hazards discussed so far in this section, are nevertheless worth mentioning because glaciers are essential to their existence and they are hazardous. Glacial lakes are dealt with fully in the next chapter so only a brief mention is necessary here. Haeberli (1983) recognized seasonal variations in lake outbursts and water pocket ruptures from Swiss glaciers that produced regular, annual flooding events (Figure 5.12) as the pressure of additional melt season water eventually exceeded the resistance of glacier ice and escaped down-valley. Some irregularity may be superimposed on these expected discharge patterns by the passage of rain-bearing weather systems, but these discharge variations, even during the peak of the early summer melt season (June, July and August), rarely generate floods of destructive magnitude, as mitigation measures have usually been taken in response to known patterns of discharge over long periods of time.

5.3 Glacier mass balance changes

The hazards that have been discussed so far in this chapter reflect the expected characteristics and behaviour of glaciers in mass balance. Other sets of problems arise when the mass balance of a glacier changes positively or negatively and the glacier respectively advances or retreats. Apart from those hazards already mentioned above, which are likely to be present whatever the status of a glacier, advancing glaciers pose little *direct* threat to people, as even the most rapidly surging glacier is no match for the rate of movement of ordinary humans. In contrast, the *indirect* impacts of advancing glaciers can have dire consequences for whole populations if agricultural land, settlements and infrastructure become buried by hundreds of metres of glacier ice, as happened in the more densely populated areas of Europe and elsewhere during the Little Ice Age. Glacier-related flooding was also common at this time as advancing glaciers obstructed the flow of tributary rivers creating lakes that eventually burst out from their glacial trap. In the current climate regime relatively few 'normal' glaciers are advancing. Any threat from glacier advance at the beginning of the twentyfirst century is likely to come from a surging glacier in which the mass balance regime is not related directly to climate but has an inherent glacial cause (Benn and Evans, 1998). However, the temporal pattern of glacier surges tends to be cyclical so that appropriate mitigation measures can usually be put in place once the periodicity of the surge cycle has been established.

5.3.1 Advancing glaciers

It is not easy at the present time to find examples of hazardous, nonsurging, climate-dependent, advancing glaciers, given the global changes in climate that have occurred during the last 30 to 40 years. There are a few exceptions, in the Karakoram Mountains of Asia for instance, but most 'normal' glaciers are responding in an expected way by retreating during the current warming regime. The regular inherent nonclimate-dependent periodicity of surging glaciers, once known, relieves at least some of the uncertainty surrounding their hazardousness. In order to understand the risks associated with 'normal' advancing glaciers, it is necessary to consult historical sources for examples of hazards due to such glaciers.

Many of our prehistoric ancestors must have been forced to retreat in the face of advancing Pleistocene glaciers and ice sheets, but of course there are no records from this time on which to base an analysis of risk. It is not until the significant climatic deterioration of the Little Ice Age (LIA), especially between the sixteenth and nineteenth centuries, that widespread impacts of advancing glaciers in Europe were noted in written records. These textual resources have been extensively analysed. Many useful examples of the impacts of glaciers, both direct and indirect, are presented and discussed in books on Holocene climate change by Le Roy Ladurie (1972) and Lamb (1982, 1995), and Tufnell (1984) has described glacier hazards specifically. Box 5.2 gives selected examples of impacts due to advancing glaciers during the LIA.

> **Box 5.2 Examples of impacts due to glacier advance during the Little Ice Age (from Ladurie, 1972, chapter 4, 'The Problems of the "Little Ice Age"'; see also Tufnell, 1984, chapter 3 for other details)**
>
> - 1589: Allalin Glacier advanced, blocked the Saas-Visp, and formed a lake (the Mattmarksee) which subsequently burst and flooded land.
> - 1595: Gietroz (Switzerland) glacier advanced and dammed the River Dranse, causing flooding of the village of Bagne and 70 deaths.
> - 1600–10: Advances by Chamonix (France) glaciers (Des Bossons, Mer de Glace and Argentière) caused massive floods which destroyed three villages and severely damaged a fourth. One village had stood since the 1200s. In addition, several villages were partially or completely overrun by advancing glaciers. It is interesting to note that between 1560 and 1610 in southern France, the River Rhone froze, olive trees froze and even the sea froze in 1595, indicative of the severity of the climate at this time.
> - 1670–80s: Maximum historical advances by glaciers in eastern Alps, such as the Ruitor and Allalin. Noticeable decline of human population during sixteenth and seventeenth centuries in areas close to glaciers, whereas population in some other parts of Europe rose.
> - 1695–1709: Iceland glaciers, such as Drangajökull and outlet glaciers of Vatnajökull, advanced dramatically, destroying farms.
> - 1710–1735: A glacier in Norway was advancing at a rate of 100 m per year for 25 years.
> - 1741: The farm, Bergseter, at the far end of Krundalen, was seriously damaged by Tuftebreen (note: in Norwegian breen = the (en) glacier (bre), an outlet glacier of the icecap, Jostedalsbreen, southern Norway.
> - 1743: The farm, Nigard, (3 km northeast of Krundalen, was totally destroyed by advance of Nigardsbreen, another outlet glacier of Jostedalsbreen. Grove (1988) reported that farmers applied to the government for tax reductions due to the loss of all or part of their farming area. Norwegian glaciers achieved their historical maximum LIA positions around 1750.

5.3.2 Surging glaciers

Surging glaciers (Benn and Evans, 1998) are a special case within the set of advancing glaciers. While most glaciers move at a normal and continuously fast rate, year on year, others possess a 'surge cycle'; that is they have quiescent phases, when flow is slow, separated by fast surge phases perhaps

Table 5.4 Examples of the regional variation of the surge and quiescent phases of surging glaciers (source: Benn and Evans, 1998)

Region	Surge phase	Quiescent phase
Svalbard	4–10 years at 1.3–16.0 m/day	50–500 years
N.Am, Iceland, Pamirs	1–3 years at up to 50 m/day	20–40 years

10 times the quiescent rate. If the glacier cannot discharge all of the climatic mass input by slow flow alone, the mass may build up until fast flow is triggered and then drain rapidly until slow flow is resumed after the supply of ice is exhausted (Budd, 1975). The trigger may be related to the style of meltwater discharge, varying from discrete channel flow during slow flow to flow distributed across the whole of the glacier bed during fast flow (Fowler, 1987). The scale and cyclicity of surging varies regionally (Table 5.4) and surging glaciers themselves seem to be regionally clustered; in Alaska, Yukon and British Columbia in northwest North America, in Svalbard and Iceland in the North Atlantic, in Ellesmere Island and Axel Heiberg Island in Arctic Canada, in the Caucasus, Tien Shan and Karakoram Mountains in Asia and in parts of the Andes of South America. This suggests that there is a component of climate input to the surge system in addition to changes within the glaciers themselves. Surge-type glaciers often reveal themselves through the presence of looped medial moraines (Figure 5.13) caused by rapid deceleration of the frontal margin or the intrusion of surging tributary glaciers across the path of lateral moraines of the main, nonsurging glacier.

Even though surge glaciers move more rapidly than conventional glaciers, they are unlikely to have direct impacts on humans, not least because most of the surge regions listed above are relatively remote from large

Figure 5.13 Looped medial moraines in a surging glacier (source: Post and Lachapelle, 2000.)

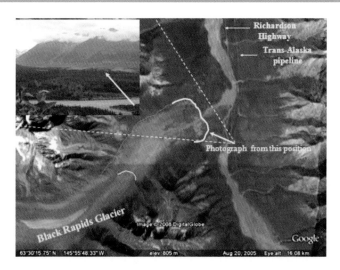

Figure 5.14 Black Rapids Glacier, Alaska, USA, showing advanced post-surge position in the late 1930s close to the present-day Richardson Highway and Trans-Alaskan Pipeline: orange line shows trimline at maximum, thick yellow line is the moraine at maximum position (Modified from Google Earth © Google. Photo: C. Whiteman, August 2008.)

concentrations of people. Additionally, the regularity of surges facilitates prediction and therefore mitigation of the hazard. This does not mean that surging glaciers can be ignored in terms of their hazardousness. One example of a *potential* surging glacier problem is the Black Rapids Glacier (also known as the 'Galloping Glacier') in Alaska. In the winter of 1936–37, Black Rapids Glacier, opposite the Mile 233 post on the Richardson Highway, advanced approximately 4.8 km in only three months and came within half a mile of the highway before ceasing its advance (Figure 5.14). This event became a serious hazard issue for Trans-Alaskan Pipeline engineers when the pipeline was constructed along the route of the Richardson Highway. An assessment of the probability of further surges that could potentially impact the pipeline and the road, based on this event and others inferred from landscape analysis, was undertaken (Heinrichs *et al.*, 1996) and the stated outcome was 'unlikely'.

Secondary effects, such as flooding, may be a greater problem than the actual ice movement, as the termination of the surge is usually accompanied by the rapid release of large quantities of melt water; but again cyclicity offers the possibility of prediction and significant events affecting large numbers of people are difficult to find in the scientific literature. However, to focus exclusively on 'Western' scientific literature is to neglect the rich oral traditions of 'First Nation' people (Cruikshank, 2005). Cruikshank reports that on a visit to Icy Bay, southeastern Alaska in 1888, Harold Topham was informed by the local Tlingit community that a

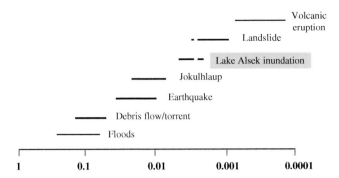

Figure 5.15 Probability of threshold destructive events for natural hazards in the Shakwak Valley, Yukon Territory, Canada (Reproduced from Clague, J.J. (1979) An assessment of some possible flood hazards in Shakwak Valley, Yukon Territory, in Current Research, Part A, Geological Survey of Canada, Paper 79-1B.)

recently advancing glacier had crossed the bay, struck the slope opposite and obstructed the Yahtse River. When the ice dam broke, massive ice blocks destroyed a village as they were carried into the bay. Further east, in Yukon, where the Alsek River traverses the St Elias Mountains, another surging glacier, the Lowell, created problems for the indigenous population there (Cruikshank, 2005). Historical and geological evidence indicates that there were a number of Lowell Glacier surges in the last 3000 years, the last some time between 1848 and 1891, probably in the 1850s (Clague and Rampton, 1982). A similar surge, and lake development today, would flood the upstream town of Haynes Junction on the Alaska Highway. The 1850s surge blocked the Alsek River and created a lake some 100 km in length. A local name for the Lowell Glacier is Nàlùdi or 'fish stop' because it obstructed salmon migration. It is also reported that hundreds of ground squirrels were drowned as the lake rose so quickly. When the glacier dam eventually burst the flood scoured the landscape and drowned Tlingit families. On the basis of previous records, Alsek Lake is unlikely to reform for several hundred years, according to a comparative risk assessment of geological hazards in the Shakwak Valley area of Yukon Territory, carried out for the Canadian Geological Survey (Clague, 1979; Figure 5.15). Given current climatic trends in northern latitudes, this may be a conservative estimate, as the Lowell Glacier, like most others, is currently retreating and thinning. However, as the next section shows, these changes bring their own associated problems.

5.3.3 Retreating glaciers

In contrast to the Little Ice Age, when many glaciers advanced down their valleys and overran useful agricultural land (Tufnell, 1984), the vast majority of the world's glaciers, including surging glaciers like the Lowell, are today experiencing negative net annual mass balance as ablation (melting) exceeds accumulation in response to global climate warming. Consequently, for the last two to three decades at least, most glaciers have not only retreated but also thinned. This means that the position of the frontal margin of a glacier moves up-valley while its upper surface is lowered relative to the adjacent valley side. Both of these changes can initiate new types of hazard for those living or travelling close to the glacier. The actual type of glacier-related hazard is influenced by the extent of the retreat, which varies from a few hundred metres to the total loss of the glacier. It is convenient to consider the hazards of retreating glaciers in relation to different stages of retreat.

(a) Valley-side instability Initially, even a minor amount of glacier retreat will expose glacially 'over-steepened' valley sides, and unconsolidated lateral moraines that have previously been buttressed by the glacier. Consequently, the balance of forces in the hillside changes, features indicative of stress release, such as tension cracks, become apparent and a variety of sediment transfer processes, including rock fall, fluvial erosion, debris flow and landsliding, are initiated. For instance, geological evidence suggests that landsliding was relatively common in formerly glaciated valleys at the end of the last (Devensian) glaciation (Ballantyne, 2002), and since the Little Ice Age (Grove, 1988). The process seems to be accelerating again in alpine regions around the world as climate warming takes its toll of valley glaciers.

This sequence of events – glacier retreat followed by accelerating slope processes – is part of a concept known as para (beyond or after) glaciation. This term was originally applied to the rapid development of alluvial fan sediments and their redistribution in the early Holocene (post-glacial, about 6700 BP) by Ryder (1971a and b). It was subsequently defined by Church and Ryder (1972) as 'nonglacial processes that are directly conditioned by glaciation'. Now, the term is applied to a wider range of post-glacial processes, as well as the timeframe over which these processes operate (see Ballantyne, 2002 for an extensive review). Sediment yields and denudation rates are highest immediately following deglaciation. They then decline through time as the landscape adjusts to the new, nonglacial system, sediment supply becomes exhausted and slopes relax to more stable profiles. Adjustment rates may be very rapid, and the paraglacial period very short where slopes are very steep, bedrock is well-jointed, rates of thaw are rapid, precipitation intensity is high, pore water pressure is high, run-off rates are rapid and the region is tectonically unstable.

These conditions occur most frequently in the world's highest mountain ranges such as the Himalaya, the 'Rockies', the Andes, the European Alps and the Southern Alps of New Zealand. Climate warming after the LIA resulted in widespread glacier ice loss and after a short period of cooling between 1940 and 1970, warming and natural, catastrophic slope processes, which pose hazards to people and development in mountains, has accelerated again. Evans and

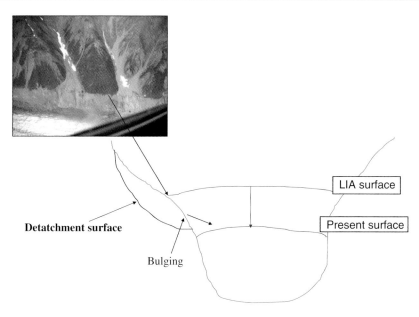

Figure 5.16 Slope instability induced by the thinning of the glacier (Photo: C. Whiteman, 2003.)

Clague (1988 and 1994) describe a number of examples of rock avalanches from the Canadian Cordillera of British Columbia in Western Canada (e.g. Tim Williams Glacier and North Creek, BC). Of 30 large historic rock avalanches in the Canadian Cordillera, 16 have occurred on glacially debuttressed slopes. Field investigations show that detachment surfaces of many of these slides intersect slopes below Little Ice Age trimlines (Figure 5.16). This suggests that the slide is a response to glacier thinning below a critical strength threshold. A similar situation applies to the Mount Fletcher rock avalanche in New Zealand. In some cases (e.g. Melbern and Affliction Creek Glaciers, British Columbia, Canada) cracking, subsidence at the top and bulging at the toe can be seen on slopes where glaciers have thinned and retreated, but catastrophic deformation has not occurred – yet! In Mt Cook National Park, New Zealand, the surface of the Tasman Glacier has lowered by 3.5 ± 0.5 m yr^{-1} (Blair, 1994), although the rate may have increased since the early 1990s in view of the accelerating retreat rate. Lateral moraines and bedrock walls began to undergo slope failure in the 1960s. Tension cracks were followed by bedrock debris slides, topples and debris avalanches, especially in response to spring thaws and wet storms. Dry moraine walls appear to adjust by slope failure when vertical relief between the moraine crest and the surface of the glacier exceeds 129 ± 10 m. Knowledge of this threshold is a useful means of predicting future problems as the glacier continues to retreat up its valley. However, the typical complexities of bed rock distribution and slope properties in fold mountains, means that this threshold figure cannot be used universally as a marker for monitoring purposes. Some roads, trails and huts, prime tourist and recreations assets adjacent to the glacier (Hay and Elliott, 2008), have had to be closed and access diverted to less risky locations.

Since Blair's (1994) work further significant landscape change has occurred. A recent study of the whole Mt Cook region (Allen, Schneider and Owens, 2009) provides an excellent summary of glacially-related hazards (ice avalanches, debris flows and lakes) in this area, set in the context of new conceptual and technical (e.g. monitoring) developments of the last couple of decades. One of the most dramatic changes has been the formation, since 1973, of a 7 km long lake in front of the Tasman Glacier, which has become another glacier-related hazard.

(b) Lakes Glacier retreat inside large moraines often produces proglacial lakes, such as that in front of the Franz Josef Glacier in New Zealand. (Boating on this lake was popular until a large ice block held down by debris, rose rapidly to the surface near a boating party.) This process is especially active where glaciers are retreating within easily erodible or tectonically active valleys, typical of those in New Zealand and the Himalaya. Where the moraines are even partially ice-cored, the potential for a disasterous escape of the melt water is enhanced unless the ice remains securely insulated by overlying sediment. This can no longer be guaranteed in the context of current climate warming. Similar hazards have arisen where melt water is trapped on the surfaces of melting glaciers by large accumulations of surface debris in their terminal zones. In recent decades, in the Himalaya, numerous small glacier surface lakes have expanded and coalesced to form very large lakes constituting major potential hazards. Outburst floods from these glacial lakes are now commonly referred to as GLOFs (glacial lake outburst floods). These and related glacial flood hazards will be considered in the next chapter.

(c) Ice avalanches Continuing glacier retreat may enhance the extent of possible slope failure, lake formation and flooding, but the nature of the hazards is unlikely to change until the glacier terminus retreats to another significantly different position. This is often the lip of a cirque (corrie) at the top of a steep valley side. Here, the front of the glacier is 'hanging' and in the perfect position to release ice avalanches into the valley below. Ice avalanches have already been discussed in some detail, but it is worth emphasizing that it is frequently in this 'hanging' position that a glacier presents its greatest potential hazard judging from past disasters. As glaciers retreat in response to climate change more valley glaciers will come to occupy this particularly hazardous landscape position, although, the balance may be redressed to some extent by the total loss of other small glaciers already in precarious positions high on mountain slopes.

(d) Loss of glacier ice Concern about the total loss of glaciers was expressed in several paragraphs of the IPCC 2007 report (see Box 5.3 for details). It is anticipated that equatorial glaciers outside South America, such as Mount Kilimanjaro in East Africa, will not persist for more than a couple of decades (Thompson *et al.,* 2002), and those in the Andes have shrunk alarmingly in recent years, or even melted completely (e.g. Cotacachi, Ecuador). The situation is similar in temperate regions (e.g. southern Rockies, European Alps) where some glaciers are projected to be completely lost by the end of this century: the Shephard Glacier appears already to have succumbed to the warming climate (Figure 5.17).

Box 5.3   Extracts from IPCC, 2007 concerned with impacts of glacier loss around the world

'Climate change is expected to exacerbate current stresses on water resources from population growth and economic and land-use change, including urbanization. On a regional scale, mountain snow pack, glaciers and small ice caps play a crucial role in freshwater availability. Widespread mass losses from glaciers and reductions in snow cover over recent decades are projected to accelerate throughout the twentyfirst century, reducing water availability, hydropower potential and changing seasonality of flows in regions supplied by melt water from major mountain ranges (e.g. Hindu-Kush, Himalaya, Andes), where more than one-sixth of the world population currently lives.' (Source: IPCC, 2007, SYR, 3.3.1).

'Climate change-related melting of glaciers could seriously affect half a billion people in the Himalayan-Hindu-Kush region and a quarter of a billion people in China who depend on glacial melt for their water supplies (Stern, 2007). As glaciers melt, river run-off will initially increase in winter or spring but eventually will decrease as a result of loss of ice resources. Consequences for downstream agriculture, which relies on this water for irrigation, will be likely unfavourable in most countries of South Asia.' (Source: IPCC, 2007, AR4, WGII, 10.4.2.1).

'Himalayan glaciers cover about three million hectares or 17% of the mountain area as compared to 2.2% in the Swiss Alps. They form the largest body of ice outside the polar caps and are the source of water for the innumerable rivers that flow across the Indo-Gangetic plains. Himalayan glacial snowfields store about 12 000 km^3 of fresh water. About 15 000 Himalayan glaciers form a unique reservoir which supports perennial rivers such as the Indus, Ganga and Brahmaputra which, in turn, are the lifeline of millions of people in South Asian countries (Pakistan, Nepal, Bhutan, India and Bangladesh). The Gangetic basin alone is home to 500 million people, about 10% of the total human population in the region.' (Source: IPCC, 2007, AR4, WGII, 10.6.2).

'Vulnerability studies foresee the ongoing reductions in glaciers. A highly stressed condition is projected between 2015 and 2025 in the water availability in Colombia, affecting water supply and ecosytem functioning in the páramos (IDEAM, 2004), and very probably impacting on the availability of water supply for 60% of the population of Peru (Vásquez, 2004). The projected glacier retreat would also affect hydroelectricity generation in some countries, such as Colombia (IDEAM, 2004) and Peru; one of the more affected rivers would be the Mantaro, where a hydroelectric plant generates 40% of Peru's electricity and provides the energy supply for 70% of the country's industries, concentrated in Lima (UNMSM, 2004). (Source: IPCC, 2007, AR4, WGII, 13.4.3).

'During the twentieth century, the areal extent of Mount Kilimanjaro's ice fields decreased by about 80% .... It has been suggested that if current climatological conditions persist, the remaining ice fields are likely to disappear between 2015 and 2020 (Thompson *et al.*, 2002'. (Source: IPCC, 2007, AR4, WGII, 9.2.1.4. Box 9.1).

It might be thought that total loss of glaciers removes many, if not all, significant hazards from mountain regions, and, indeed this could be true. However, other substantial impacts – economic, social, political and cultural – will be felt by many people who have become used to the presence of glaciers in their vicinity. The risk is, in the case of total glacier loss, that without the glacial meltwater resource, populations will be deprived of access to vital water for

Figure 5.17 Shephard Glacier, Montana, USA, in 1913 (left) and 2005 (right) (Reproduced with permission from USGS Repeat Photography Project, nrmsc.usgs.gov/repeatphoto/ Fig. 2, 1913 photo W.C. Alden, GNP Archives, 2005 photo Blasé Reardon, USGS.)

domestic, industrial and agricultural use, not to mention a source of power in the form of hydroelectricity. Maximum impact is likely to be felt in relatively arid regions, where glacier melt water balances periods of seasonal drought. Although this particular glacial hazard impacts *indirectly*, the effects of such changes pose serious threats to substantial numbers of people (Stern, 2007), not just within the glaciated mountain areas themselves, but beyond, in lowland regions such as the Ganges valley of India, where mountain glaciers function as 'water towers', storing vital supplies not readily available locally. Where glaciers and the consumers of their melt water are located in one country the problems are severe enough: where the glacial supply is nationally divorced from some of its consumers, as in central Asia (Kyrgyzstan, Tajikistan, Uzbekistan, Kazakhstan), then the problems are even more acute and could become the source of conflict (Weinthal, 2006; Luterbacher *et al.*, 2008; Orlove, 2009).

Predictably, more vulnerable, less-developed regions are of greatest concern and have received most attention, but richer nations are not immune to the effects of glacier loss. It was a surprise to this author to learn from a Guardian newspaper headline that a 'plastic sheet save[d] Swiss glacier from meltdown' (The Guardian, 2005). This referred to the Gurschen glacier, above the village of Andermatt, Switzerland, where $2500\,m^2$ of reflective plastic sheeting costing €62 000 was wrapped over ice between the cable car and the ski slopes to save having to build an artificial snow ramp each season. Of course, artificial snow has been around for many years but the concept of protecting glaciers from the effects of summer solar radiation to maintain the economic value of ski tourism is novel and demonstrates the substantial value of this activity to the economies of some European Alpine countries. According to a video article published by swissinfo.ch (2009), 10 Swiss ski resorts in the 1970s were open for summer skiing on glaciers and the snow quality was good. Now, only two resorts, Zermatt and Saas-Fee provide a summer service. At Graubünden what snow remains from the winter fall is covered with reflective foil, as illustrated in Figure 5.18, so that the winter skiing season is not also lost. Even more remarkable, an Italian ski resort on the Vedretta Piana Glacier is reported (Smiraglia *et al.*, 2008) to shift firn from the upper to the lower part of the glacier to facilitate skiing in the lower part, although this procedure actually accelerates glacier shrinkage by removing material from the accumulation zone to the ablation zone!

A final issue related to lost glaciers is one that is rarely if ever discussed in scientific literature but deserves a mention: that is, the cultural importance of glaciers. It was alluded to in the section on surging glaciers and is a major theme in a recent publication, *Darkening Peaks: Glacier Retreat, Science, and Society* (Orlove, Wiegandt and Luckman, 2008). For instance, white glaciated peaks in the Andes have cultural significance as spiritual homes for many indigenous

Figure 5.18 Workers cover the ski slopes on the Pitztal Glacier in Austria with white fleece in an effort to protect the mountain from glacier melting and preserve the snow for skiing (Reproduced with permission from Andy Eckardt, 2005. http://www.msnbc.msn.com/id/8432120/.)

Andeans, and some distress is felt as white summits become dark (Bolin, 2001, and Regalado, 2005, quoted in Orlove, 2009).

(e) Other impacts Finally, a number of more indirect impacts of retreating glaciers should be mentioned. First, assuming that not all the glacial melt water released by retreating glaciers is abstracted for the use of people downstream, some will reach the oceans and contribute to sea level rise, amounting to about 0.7 mm per year according to the last IPCC (2007) report. Second, there will be a change in surface albedo, as light, reflective ice is replaced by dark, absorptive rock. The effect of surface albedo change on climate systems and weather is even more difficult to predict than rising sea levels (IPCC, 2007). Third, the retreat of glaciers will expose bedrock and a range of glacial sediments of doubtful stability. How soon these relatively unstable sediments will be colonized by flora and fauna and to what extent biodiversity will be altered is not easy to predict. Fourth as noted above, there have already been imaginative attempts to maintain the viability of the skiing industry in some places. A longer-term issue is the question of how tourists will react if or when iconic glaciated mountains become ice-free.

Clearly, a wide range of glacial hazards has to be confronted by many different populations worldwide. Numerous researchers have commented upon the complexity of the glacial environment and the hazards that it produces, but in the context of climate change the problems are magnified, and it is not surprising that mitigation of glacial hazards is rarely easy in spite of the enormous amount of recent research directed to this end.

5.4 Mitigation measures

There was a time, before science began to influence glacial studies significantly in the first half of the nineteenth century (Haeberli, 2008), when glacier-related hazardous events were seen as an 'act of God' or 'divine retribution' and risk perception was fate-based (Orlove, Wiegandt and Luckman, 2008). To a degree, culture still has a bearing on the way that different people view hazards and their mitigation (e.g. Carey, 2005, 2008). Early, well-meaning attempts at mitigation, based on *some* understanding of the situation, have not always succeeded. Before the Giéto Glacier dam broke in 1818, a channel had been cut through the glacier to evacuate water and relieve the pressure. Unfortunately, friction from the flow of water melted the ice further and the dam collapsed, the resulting flood killing 40 people and causing considerable damage (Wiegandt and Lugon, 2008). However, not surprisingly, during nearly 200 years since the Giéto Glacier disaster, knowledge of processes has advanced enormously, although it is still not perfect. Today, with potential impacts of climate change as the primary driving force, rapid advances in technical expertise (e.g. remote sensing, geoinformatics, GIS, GPS and database construction) are beginning to provide the data on which to base more successful mitigation strategies for glacier hazards (for details, see, for instance, Haeberli, 2008; Kääb *et al.*, 2005; Kääb, 2008; Paul, Kaab and Haeberli, 2007; Paul *et al.,* 2009; Quincey *et al.,* 2005).

However, it should be obvious from the wide range of hazards associated with glaciers, that they form a very complex natural system in their own right. When rapid climate change is superimposed on a complex situation, it is not difficult to appreciate that solutions to the problems of glacier hazards will not be obtained easily (Haeberli and Beniston, 1998). For instance, while there are legal requirements on Swiss cantons to provide hazard maps for snow avalanches, the same does not apply to glacier hazards because they are still perceived, for several reasons, to involve too much uncertainty (Wiegandt and Lugon, 2008). (Budget allocations for glacier hazard monitoring are very small in comparison with those for other hazards in the Valais Canton, Switzerland (Charly Wuilloud, personal communications, 2005, quoted in Wiegandt and Lugon, 2008)). There are other difficulties to overcome. Events do not always happen in the same place, high magnitude but low frequency events often generate too little data to be statistically significant (Wiegandt and Lugon, 2008), and complex factors involving both glaciers *and* permafrost have only recently been widely recognised and addressed (Kääb *et al.*, 2005). Glacial environments (and socio-economic and environmental systems too) may just be too dynamic and complex for successful local prediction and mitigation in all cases at the present time.

The 'Gridabase' database, an inventory of glacier-related hazard events produced by the trans-national EU Glaciorisk Project in 2003, is an interesting historical resource but coverage across the six participating countries is not even. In the context of rapid climate change the value of historical precedence as the basis of mapping, prediction and zoning has been questioned (e.g. Haeberli *et al.*, 2009). Decisions about zones can be subjective, in the interests of attracting industry and tourists (Orlove, Wiegandt and Luckman, 2008). The enforcement of formal restriction on movement and limits on construction can be difficult. There are also cultural challenges and different perceptions of risk, and response to hazards varies with experience as Carey (2008, pp. 229–230) found in his study of the residents of Yungay, the Peruvian town destroyed in the Huascaran disaster of 1970:

> 'Hazard zones and hazard zoning held widely divergent meanings for scientists, government officials, and local residents. To scientists, hazard zones represented paths that avalanches and outburst floods could follow. Hazard zoning was seen as the most prudent way to avoid future glacier disasters, and because the 1970 earthquake and avalanche had destroyed most of the region it did not entail moving structures or communities; rather, it required shifting reconstruction to new areas. To the avalanche survivors, however, the hazard zones were historically produced spaces with cultural, economic, social, and political meanings. Relocating to a safe place meant major compromises and significant risks to their livelihoods, connection with ancestors, material well-being, social status, and political power. Consequently, their decisions about whether to remain or to relocate involved the ranking of risks. For those who rejected hazard zoning, there were cultural, social, economic, and political risks associated with leaving that outweighed the risk of unknown and unpredictable glacier disasters. In other words, while scientists focused on a single risk, residents contended with a host of them. By analyzing the rationality of Yungay residents' risk perception–the historical forces informing that perception, the multiple meanings they assigned to the hazard zone, and their reasons for resisting relocation–it is possible to understand why experts, policy makers, and local residents clashed over the 1970s disaster mitigation policies.'

5.5 Summary

Few cold region hazards can be as complex as those associated with glaciers. Besides the hazards inherent in glaciers – crevasses, seracs, ice avalanches – there are a host of others introduced by changes in mass balance which cause glaciers to advance and retreat. Retreat is *the* current problem as climate warms and adds the hazards of excess water (but eventually maybe the lack of it) to those of ice itself. At the present time landscape destabilization appears to be an increasing problem. More uncertainty surrounds the indirect hazards of sea level rise, albedo changes, biodiversity and tourism. There have been advances in understanding processes, and new ways of mapping and monitoring the hazards have been developed in the last few decades, especially remote sensing. National and international systems are in place to facilitate the rapid transfer of knowledge. However, while there has been substantial progress in science and its organizations, a more comprehensive solution to the problems posed by glacier hazards, in particular the question of water supply, will require careful integration of politics, economics and sociology with the science in order to resolve some of the more intractable difficulties of hazard mitigation.

6
Glacial Lake Outburst Floods (GLOFs)

6.1 Introduction

The previous chapter dealt with the hazardous effects of glaciers as they advance or retreat across the landscape, including loss of vital water supplies if glaciers completely melt from an inhabited catchment. In contrast, this chapter focuses on the risk of excessive amounts of melt water released from impounded lakes. In an historical review of glacier hazards Tufnell (1984) showed that the release of stored water quickly and without warning has caused serious damage to property and loss of life on numerous occasions in areas such as the Alps. Richardson and Reynolds (2000) speculated that the destruction by glacial floodwater of a nearly completed, $US4 million, Nepalese hydroelectric power plant in 1985 could eventually cost $US500 million by setting back development in the area for a generation. Although there are records of earlier Himalayan glacier floods (see for instance Ives, 1986), the costly 1985 glacial lake outburst flood (GLOF) event clearly raised international and local awareness of the risk of glacier flood hazard in this remote, relatively undeveloped, high mountainous region (Kattelmann, 2003). Subsequently, groups from several developed countries (e.g. Austria, Japan, The Netherlands and the UK) as well as the United Nations became involved, either directly or in an advisory capacity, in the mapping, monitoring and mitigation of the GLOF hazard in the Himalaya. Although GLOFS have been recorded from other glaciated mountain ranges (e.g. Alps, Andes, Rockies), this chapter will concentrate on the Himalayan context because it not only illustrates many of the issues surrounding the GLOF hazard per se, but also highlights the difficulties faced by remote communities in developing countries in responding and adapting to this hazard.

Cold Region Hazards and Risks, First Edition.   Colin A. Whiteman.
© 2011 John Wiley & Sons, Ltd. Published 2011 by John Wiley & Sons, Ltd.

6.2 The glacial meltwater system

The melting of glaciers produces large volumes of water at different locations within the glacier system (Benn and Evans, 1998). Water produced at the glacier surface by warm air, radiation and rainfall may be transferred to the front of the glacier along meandering sub-aerial channels, or it may flow through the glacier along sub-vertical channels known as *moulins*. At the base of the glacier, more water is produced by geothermal heat, pressure melting and frictional heat generated by ice deformation and sliding. Much of this water travels as major meltwater rivers along sub-glacial tunnels (Rotlisberger channels) and leaves the glacier system through large portals at the front of the glacier. Water evacuated in this way forms a 'normal' proglacial river system downstream from the glacier. Proglacial rivers typically show frequent, significant variations in discharge, as glaciers respond to daily and annual temperature cycles. However, these 'normal' discharge variations, even during the peak of the early summer melt season, rarely generate floods of destructive magnitude, as mitigation measures have usually been taken in response to *known* patterns of discharge over long periods of time. This also applies to situations where glacially-impounded lakes build up seasonally and evacuate annually once lake water pressure exceeds the resistance (weight) of the blocking glacier, usually in late summer or autumn. This pressure threshold may be one of the reasons why some Alpine lakes empty at regular times each year (Haeberli, 1983), though some variation can be expected following mass balance changes and the resulting fluctuations in the position and thickness of the glacier. Additional irregularity is also superimposed on these expected patterns by the passage of weather systems, such as the monsoon, which may bring high rainfall events in summer and early autumn, facilitate snow and ice ablation, and add rainfall directly to catchment run-off. Iceland provides a different context for periodic glacial flooding events. Here, volcanic activity beneath the Vatnajökull icecap melts ice until the crater lake of Grimsvötn is full and the lake empties catastrophically beneath the ice (termed a jökulhlaup in Icelandic). While this is not an annual event, there does appear to be a broad cyclicity of several years controlled by volcanic activity. The inherent cyclicity of some glacial and volcanic systems facilitates hazard mitigation. The last big jökulhlaup from Vatnajökull in 1996 did considerable infrastructural damage but there were no human casualties because the time lapse between the initial volcanic activity and the actual flood allowed precautions to be taken. In the absence of known cyclicity or prior warning, mitigation becomes much more difficult, and destructive GLOF events more costly in terms of land, infrastructure and lives lost.

6.3 GLOFs

The most destructive glacier flood events tend to be those associated with sudden, unexpected releases of *stored* glacier melt water and surface run-off.

Water storage associated with glacier systems depends partly on the density of snow and ice forming the glacier. Fresh, dry snow has a low density and melt water is able to pass freely through it. However, once the interconnected air passages between ice crystals have been sealed off by pressure from accumulating snow above, any surviving air is restricted to isolated bubbles and ice permeability is sufficiently reduced to create a barrier to water. Some water may be stored in small sub-glacial or englacial cavities, but the largest bodies of water usually accumulate either on the surface of glaciers, or in different positions around their margins (Cooke and Doornkamp, 1990). The advance and retreat of glacier margins tends to produce complex relief systems which facilitate water storage (Figure 6.1).

For instance, kettle holes (depressions left by melting blocks of ice) provide basins in which melt water can accumulate. Landslides, initiated by removal of the buttressing effect of a glacier following ice retreat, can block melt water

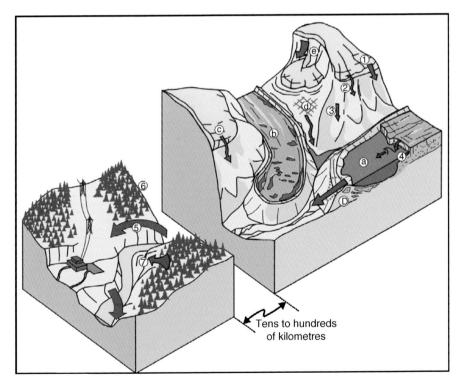

Figure 6.1 Model of GLOF terrain showing source area and downstream impact locations (Reproduced with permission from http://geomarineuk.com/mountain_hazards_group/pdf/Chapters_1_4.pdf © Reynolds Geo-Sciences Ltd (2003).). Key: 1–3, sources of ice and debris; 4, calving glacier and displacement wave; 5, downstream flood; 6–7, slope run-off; a, proglacial lake; b, debris-covered glacier; c, ice avalanche from hanging glacier; d, melting permafrost; D, ice-cored terminal moraine and flood breakout location; e, rock avalanche site

run-off and produce large lakes. Local melting of ice by volcanic activity may cause ponding of water in a crater obscured from view. Lakes may form between adjacent glaciers, or between a glacier and the valley side. Ponds and lakes also develop *on* the downstream surface of glaciers, or *immediately in front of* glaciers, behind large, ice-cored terminal moraines. Lakes in surface and proglacial positions form easily in some parts of the Himalaya because large quantities of surface and marginal debris accumulate from the long, steep, unstable slopes typical of the region. Also, large moraines, some distance in front of glacier margins, are common in alpine environments because many glaciers retreated significantly after the Little Ice Age (LIA) ended in the nineteenth century. In the Himalaya glaciers have generally retreated by about a kilometre since the LIA (Mool, Bajracharya and Joshi, 2001a).

6.4 Trigger mechanisms

Potentially hazardous stores of water can therefore accumulate in a many different situations. However, the flood hazard posed by this wide range of water storage sites is further complicated by the variety of flood-release trigger mechanisms (Figure 6.2), some inherently glacial, others of external origin related to atmospheric or tectonic events. Where glacier ice provides the barrier holding back the water, the trigger for a flood is likely to be related to a pressure threshold. Once the weight of water in the lake exceeds ice pressure the ice will float and release the water as a flood. Such pressure thresholds are generally measurable and easily predictable. In contrast, where dams are created by ice-cored terminal moraines or supra-glacial debris, the strength of the barrier is much less predictable and consequently more hazardous. It is not surprising, therefore, that moraine-dammed lakes are usually responsible for the GLOF hazard in the Himalaya and elsewhere.

One common type of trigger is the avalanche displacement wave (Figure 6.2). This mechanism may account for over 60% of 26 Himalayan GLOFs for which failure mechanisms and timing are known (Richardson and Reynolds, 2000), including the Dig Tsho GLOF that destroyed the Namche Bazaar hydroelectricity plant (see Box 6.1 for details). In many cases, however, these triggers operate in sequence before eventually setting the GLOF on its potentially destructive course. An example of these complex events is provided by the 1998 Sabai Tsho GLOF (Figure 6.3) in the Hinku Valley, 20 km south southeast of Chomolungma (Mount Everest).

A warm, wet monsoon earlier in 1998 may have topped up the lake and slightly destabilized the hanging Sabai Tsho glacier. Three local earthquakes on 2 and 3 September, 1998, culminating in a 3.77 Richter Scale event at 5.50 on the second day, probably tipped the glacier over its stress threshold causing a mass of ice, about 300 m by 100 m by 10 m thick to crash into the head of the

6.4 TRIGGER MECHANISMS

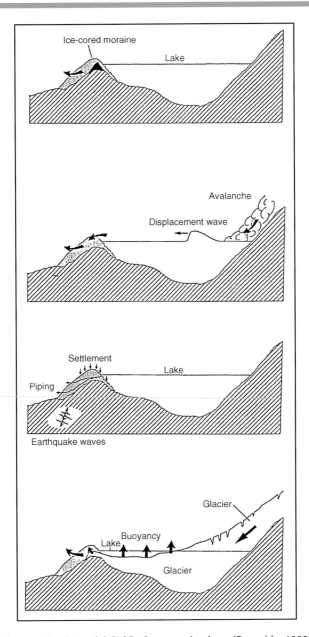

Figure 6.2 Potential GLOF trigger mechanisms (Reynolds, 1992)

lake. This was the trigger for a displacement wave which would have passed down the lake and turned the trickle through the notch in the moraine dam into an erosive torrent. This would have lowered the lake exit until the magnitude of the flow was sufficient to cause catastrophic collapse of the moraine and the initiation of the GLOF. The initial massive flood lasted 5–10 minutes, emptied

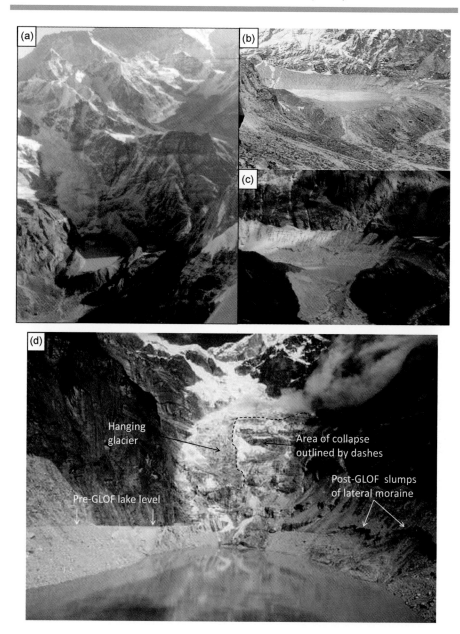

Figure 6.3 Sabai Tsho GLOF, Khumbu Himal, Nepal, 3 September 1998: (a) topographical setting of Sabai Tsho Glacier and Sabai Tsho (lake); (b) and (c) views of moraine-dammed Sabai Tsho before and after the GLOF, respectively; (d) view looking up Sabai Tsho towards the Sabai hanging glacier, showing area of glacier fall, former lake level and post GLOF slumps of lateral moraine sediment; (e) location of Tangnag village adjacent to the debris fan deposited by the GLOF below the eroded moraine (Reprinted from http://p6.hostingprod.com/@treks.org/lakeng98.htm. Courtesy of Jean van Berkel.)

Figure 6.3 *(Continued)*

the lake of about 25 million m³ of water and lowered its surface by some 50 m. Smaller intermittent floods continued for the next 20 hours, probably as other pieces of ice fell into the lake or piles of boulders and ice blocks near the lake exit collapsed or melted under the flow of water.

Box 6.1 Dig Tsho GLOF, Khumbu Himal, Nepal, August 1985 (Ives, 1986)

The GLOF
On 4 August 1985, 6–10 000 000 m³ of water with a possible peak discharge of 2000 m³ s⁻¹ (average 500 m³ s⁻¹ over 4 hours) rushed down the Bhote Kosi and Dudh Kosi valleys, transporting 3 000 000 m³ of debris within a distance of less than 40 km (Ives, 1986; Vuichard and Zimmermann, 1986, 1987). The catastrophic, 10–15 m high flood surge (Sherpa: chhugyümha) originated from Dig Tsho (Tsho = lake), a moraine-dammed lake in the Khumbu Himal of eastern Nepal. The lake was about 50 ha in area with a maximum depth of 18 to 20 m. During the summer melt season the lake overtopped the moraine at its lower end and drained through a steep channel down the outer slope of the moraine forming a debris cone at its base. In 1985 the lake was again full to the rim of the moraine at the time of the burst following a particularly warm summer. Seepage was also evident at the foot of the moraine indicating some subterranean drainage, though this was not the ultimate cause of the GLOF. The upper end of Dig Tsho was partly

covered by a layer of 'dead' ice, the detached remnant of the small Langmoche Glacier which had retreated upslope into a 'hanging' position on steep bedrock (Figure 6.4).

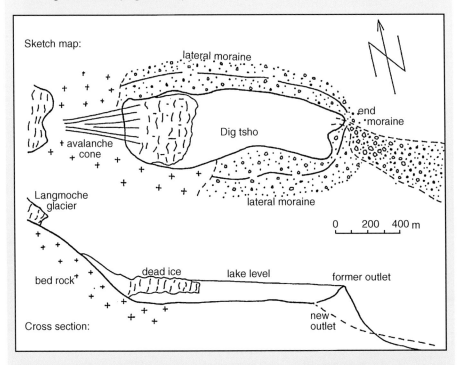

Figure 6.4 Sketch map and cross section of the Dig Tsho GLOF terrain showing ice avalanche trigger and lake at maximum level behind terminal moraine (Source: Vuichard and Zimmermann, 1987)

The trigger for the GLOF certainly came from further up the valley though there was some dispute about its precise details. Either a large mass of ice on the rock wall above the Langmoche Glacier became detached in warm weather, or a massive rockfall occurred following heavy monsoon rain and snow. Either way, a fall of ice and/or rock debris dislodged the toe of the hanging glacier which then avalanched down the steep slope towards the lake. The ice avalanche may have crashed onto the floating mass of ice below, depressing it and producing a static lifting of the water of about 40 cm accompanied by a superposed wave of unknown size. Alternatively, the dead-ice may already have been broken into blocks by melting so that the ice avalanche produced an impulse wave which travelled across the lake surface before overtopping the moraine. Impulse wave calculations suggest a wave height of 5 m giving it considerable capacity to erode a notch and release progressively more water from the lake. According to eyewitness accounts reported by Ives (1986, p. 27), 'the surge front appeared to move down-valley rather slowly as a huge black mass of water full of debris. The movement was of a rolling type, splashing from one river bank to the other.... Waves overtopped the river banks in places. Trees and large boulders were dragged along or bounced around; some of the trees were in upright positions. The surge emitted a loud noise, "like many helicopters", and a foul mud smell. The valley bottom was wreathed in misty clouds of water vapour, the river banks trembled, houses shook and the sky was cloudless.'

Impacts
In spite of these dramatic descriptions, fatalities from the Dig Tsho GLOF were not exceptional by the standards of some global hazards – tsunami and hurricanes, for instance. Only four or five deaths were reported. Fortunately, many Himalayan villages are high above valley bottoms and many Sherpas were in their villages for a festival. At other times of the year, for instance during the height of the trekking period in October, the casualty list might have been much greater. However, given the context in which the GLOF occurred, a relatively poor developing country, and the HEP impact in particular, this event assumed a much greater significance. It provided a tipping point in terms of the level of international interest in the GLOF problem in the Himalaya.

All the bridges (14) for 42 km below Langmoche were destroyed, including four new high suspension bridges. Dozens of houses were also lost, either by direct erosion of river banks or by destabilization of river terraces and gradual collapse, a process that continued for several days after the GLOF. Even surge vibrations caused the collapse of some less robust properties. The main trail was undermined for long stretches, cutting off access to Namche Bazar market for several weeks and reducing trekking. Immediate attempts to replace the vital trail after the GLOF were not always successful and there were at least two fatal accidents during October 1985. Cultivatable level land, an important part of the subsistence base, was lost by erosion. Other land was buried under the coarse bouldery debris. Sediment loads in the river remained high for several years. Undercut valley sides similarly remained susceptible to subsequent monsoon rains. However, the most devastating aspect of the GLOF disaster was the virtual destruction of the Namche Bazar HEP station, which was nearing completion after nine years of negotiation, planning and construction. Unfortunately the planning does not seem to have included an adequate risk assessment – no investigation of catchment hydro-glaciology was made prior to site selection – and no remedial measures were undertaken once seepage from the moraine and overtopping had been noticed in August 1984. The equivalent of about $US 4 million had been spent on materials and labour, as well as expensive helicopter transport of turbines and generators, but the overall economic costs including the loss of future benefits for instance are more difficult

> to determine. 'It is even more difficult to estimate the psychological impact of the indefinite postponement of hydroelectric power supply to the main Sherpa villages after such a long wait and high expectation.' (Ives, 1986, p. 31).

Displacement waves in glacial environments may be generated in several ways: by glacier calving and the collapse of hanging glaciers, which together may account for 53% of displacement wave failures, or by valley side rock falls and debris flows. Calving is especially dangerous in the case of floating ice tongues but even grounded glacier margins collapse due to thermal erosion of ice at the water level and the exploitation of crevasses. Rock falls become increasingly likely as valley sides destabilize when they lose the support of glaciers as these thin and retreat in the warming climate. Whichever displacement wave mechanism is responsible, the sudden entrance of masses of falling ice or rock into the lake abruptly displaces some of the water, producing a wave which moves down the lake. If the wave has sufficient amplitude it will overtop and erode the morainic barrier, especially at the drainage notch, releasing more water capable of further erosion and even more water release. A similar outcome will occur if excess water enters the lake abruptly as a result of catastrophic glacial drainage from sub- or en-glacial channels or other supra-glacial lakes. Exceptional weather events, such as rainstorms and snowmelt may also overwhelm lake storage capacity. It is not surprising that most recorded Himalayan GLOFs have occurred at the height of the summer monsoon season from July to September (Yamada, 1998) when lakes usually attain their highest level.

Alternatively, the capacity of lakes to store water may be reduced, for instance by lowering the level of the moraine dam. There are several possible causes for such lowering. Moraines are often ice-cored so by melting this core the level of the moraine will be reduced, though Richardson and Reynolds (2000) reckoned that only one GLOF in their study was due to this cause. However, this process of thermokarst, similar to the retrogressive thaw slumps of permafrost regions, might become more likely during the current period of climate warming. Settlement may also occur following piping within the moraine dam due to progressive seepage. Seismic shaking can rearrange the moraine sediment so that it occupies less space. Unfortunately, remedial engineering works have also been the cause of GLOFs, such as the incidents in Peru, South America, recorded by Llibourty *et al.* (1977).

6.5 Risk

Given such a complex set of environmental preconditions for GLOFs, it is not surprising that the assessment of GLOF risk is correspondingly difficult. Key parameters for assessing degree of risk are lake volume, moraine dam structure, presence of buried stagnant ice, amount of freeboard (difference in height between lake level and moraine crest) and the nature of the surrounding topography. Maximum risk will apply when a large lake, with limited freeboard, contained behind a high, narrow, ice-cored moraine, is surrounded by steep slopes susceptible to ice avalanche, landslip or debris flow. In these circumstances, only a small change in any one of a number of environmental parameters will suffice to exceed the failure threshold and trigger a potentially catastrophic GLOF.

GLOFS are not a new phenomenon in Himalayan valleys. The climate change record of the Quaternary Period suggests that conditions favourable for GLOF formation would have occurred on many occasions during the last 2.5 million years, though much of the evidence has probably been removed by successive glacier advances. Of more immediate interest is the current period of rapid climate change from about 1970 when global temperatures have risen dramatically, not least in the Himalayan area. Huge LIA terminal moraines and large volumes of supra-glacial debris obstruct the outflow of melt water in many Himalayan valleys. Instead of a regular outflow being maintained at the level of the valley bottom, melt water is released over a shallow notch at the crest of the moraine and the rest of the height of the moraine is available to pond back lakes of considerable depth. The insulating properties of supra-glacial and morainic debris ensure that many moraines and supra-glacial debris accumulations retain ice at their core. This is not a problem as long as the ice remains covered but any slippage of the covering debris is likely to expose the ice to melting and the resulting water reduces the strength of the moraine. The current environment of the earlytwentyfirst century could hardly be more conducive to the production of GLOFs. Global temperature increases, reflected in glacier recession rates, have been recorded in Nepal, with the most pronounced changes at high altitudes (Shrestha *et al.*, 1999). With more precipitation falling as rain the albedo effect of snow is reduced making radiative melting of glaciers even more likely (Mool, Bajracharya and Joshi, 2001a).

Historical records indicate that even during the four decades up to 1970, several GLOFs occurred in Nepal, though a GLOF in 1977 in the Khumbu Himal seems to have been the first to have received significant scientific study (Kattelmann, 2003). At least seven other GLOFs occurred between 1977 and 2000. A compilation of Himalayan GLOFs since the 1930s suggests that GLOF frequency is increasing (Richardson and Reynolds, 2000, Figure 6.4). In the eastern Himalaya, where the problem is most serious, damaging GLOFs appear to occur with a *mean* recurrence interval of three to four years, though the regularity of this pattern is not high. There is a marked contrast between the regularity of ice-dammed lakes subject to an annual pressure threshold and others, especially moraine-dammed lakes, which are generally less predictable (Ives, 1986). The hydrological system of glacierized basins often changes suddenly and discontinuously as drainage conduits are blocked off or new ones opened by ice flow. Furthermore, while annual monsoon flood events generate a recognizable temporal pattern that can be easily quantified, GLOFs from particular moraine-dammed lakes may be very rare or even 'one-off' occurrences. In effect, they may be a delayed response to glacier retreat from a Little Ice Age maximum, or a more immediate response to the special case of current climate warming.

The increasing risk of GLOFs, highlighted by Richardson and Reynolds' (2000) data, is graphically illustrated by the growth of supra-glacial lakes near a number of glacier margins. Over recent decades the terminal areas of some glaciers have supported increasing numbers of small ponds which gradually expand and merge to produce a single large lake. One example that attracted considerable attention during the 1990s, but has since become less significant, is located on the lower Imja Glacier in Nepal. According to photographic and anecdotal evidence assembled by Watanabe, Ives and Hammond (1994), the debris-covered surface of this glacier supported five small ponds, no more than about 100×250 m in size, in the period 1956 to 1963 (Figure 6.5(a)). This situation applied until at least 1971 but by 1973 a larger lake had developed extending across the full width of the glacier. Subsequent air photography in 1975, 1978 and 1984 showed the outline of the lake becoming less irregular, presumably as it grew and deepened. At the time, these photographs recorded an alarming rate of change. Rates of decay of stagnant ice in front of Imja Glacier Lake varied from 0.1 to 2.7 m per year between 1989 and 1994, with a maximum rate in excess of 5.0 m per year where the ice surface was submerged by the lake. If this rapid melt rate had been sustained, it is thought that the western shoreline of the lake could have migrated to the edge of the terminal moraine in about 7 years causing it to be undermined and leading to its catastrophic collapse. However, instead, the lake level fell from a height of 5022 m in 1984 to 5017 m in 1989 and to 5007 m in 1994, apparently due to the melting of debris-covered ice along the line of the outflow stream in front of the main glacier. Thus, the immediate danger of lateral moraine collapse appeared to have lessened. No flood events have occurred during the subsequent 15 years, though Imja Glacier Lake remains on the list of at-risk glacial lakes subject to monitoring in Nepal.

Tsho Rolpa, another hazardous Nepalese glacier lake, experienced a similar expansion to the extent that it became the largest moraine-dammed proglacial lake in Nepal. Figure 6.5(b) illustrates the gradual enlargement of this lake for some 40 years after the late 1950s during which time the area of the lake increased more than sevenfold from $0.23 \mathrm{km}^2$ to $1.65 \mathrm{km}^2$. In this case, concern was so great that remedial mitigating measures were undertaken to lessen the risk of a GLOF (see Box 6.2; Figures 6.6 and 6.7).

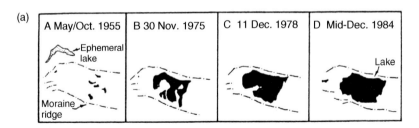

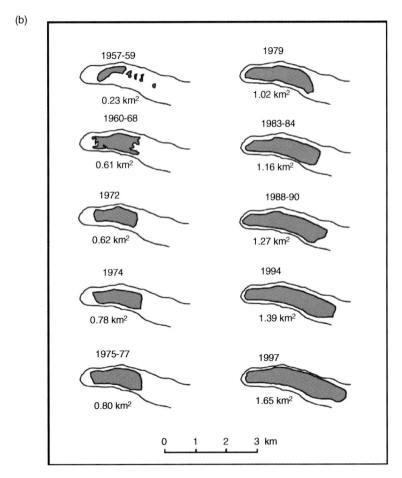

Figure 6.5 Expansion of glacier lakes in the Himalaya: (a) Imja Tsho, Nepal (Watanabe, Ives and Hammond, 1994); (b) Rolpa Tsho, Nepal (Rana et al., 2000).

Figure 6.6 Hard mitigation measures at Tsho Rolpa, Nepal: (a) view of terrain; (b) upper end of channel leading from lake; (c) view of gate looking upstream; (d) view of channel and gate looking downstream; (e) channel stream on the outer edge of the moraine (source: Rolwaling, no date).

Box 6.2 Mitigation at Tsho Rolpa, Rowaling Valley, Nepal (Reynolds, 1999; Rana *et al.*, 2000; Kattelmann, 2003)

Tsho Rolpa is the largest proglacial moraine-dammed lake in Nepal. The lake has grown alarmingly since the late 1950s. The lake level is close to the moraine crest. The moraine itself is partly ice-cored. There is a high level of risk of a GLOF from this lake. A Tsho Rolpa GLOF threatens the Khimti 60MW HEP complex located about 80 km downstream from the lake. Rebuilding would cost $US22 million plus losses incurred from the resulting lack of electricity. It was the 1991 Chubung outburst, from a lake less than 1 km from Tsho Rolpa, that alarmed many inhabitants of nearby Na and Beding villages (Reynolds, 1999). In May 1992, in a first step towards hazard mitigation, many Beding villagers petitioned the embassies of Nepal's major donor nations to assist with the Tsho Rolpa hazard, potentially a much larger problem than nearby Chubung (Kattelmann, 2003). The British Emergency Aid Department of the Overseas Development Administration financed visits by a British geologist familiar with the area to assess the Tsho Rolpa hazard (Reynolds, 1999). The Summit Trekking company of Katmandu, with an obvious interest in the safety of the growing number of tourists, publicized warnings of the Tsho Rolpa hazard (Kattelmann, 2003). Still on the international theme, residents of the Rowaling Valley solicited help through the Netherlands–Nepal Friendship Association, resulting in the financing and installation of a siphon system. At the same time a joint Nepal–World Bank project installed an early warning system of 19 automatic sirens in 17 villages to alert inhabitants to seek refuge on higher ground in the event of a GLOF (Kattelmann, 2003). The sirens were linked to sensors placed to detect the passage of a flood wave.

The project at Tsho Rolpa, in the Rolwaling Himal, about 110 km north east of the Nepalese capital, Katmandu, involved the installation of siphons, the excavation and lining of an open channel in the moraine crest, and a controlling gateway. Rana et al. (2000) provide a useful summary of the Tsho Rolpa problem. During the second half of the twentieth century, Tsho Rolpa had been growing at a rapid rate as the Trakarding Glacier terminus retreated, partly through calving into the lake, a possible GLOF trigger. By the early 1990s the lake measured 3.3×0.5 km and covered an area of 1.65 km^2. With a mean depth of 55 m (maximum depth 132 m), Tsho Rolpa contained some 100 million m^3 of water. Even more alarming, the 150 m high terminal moraine dam was showing signs of distress. Geophysical observations had confirmed the presence of ice-cored areas in the moraine and melt water from this buried ice was causing slumps which reduced dam strength and exposed fresh ice for further melting. Reynolds and Pokhrel (2001) recorded subsidence on the slopes of the moraine in ice-cored areas of 2–3 m/year with exceptional rates of 22 m/year. In summary, the key factors responsible for the high level of GLOF hazard at Tsho Rolpa were: accelerated growth of lake volume, melting of the moraine ice core, seepage from the end moraine, slumping of moraine slopes and rapid ice calving from the glacier terminus. If the dam were to burst an estimated 30 million m^3 of water, at least three times as great as the destructive 1985 Dig Tsho GLOF, would be released. In the worst scenario modelled by the Nepalese Department of Hydrology and Meteorology (DHM) peak discharge could be 7000 m^3/s with peak flood heights up to 17 m at valley constrictions and a maximum downstream extent of 100 km. It was predicted that a Tsho Rolpa GLOF could threaten 10 000 lives and thousands of livestock, inundate agricultural land, destroy bridges and other infrastructure and damage the Khimti HEP station, under construction 80 km downstream and expected to employ 1500–2000 workers during the peak of the construction effort. With such a potent combination of possible triggers, coupled with the list of potential impacts, it is not surprising that a decision was eventually taken to carry out emergency remedial action.

The first (unpublished) report on Tsho Rolpa (Damen, 1992) made several recommendations: the use of siphons to lower the lake level by several metres, the monitoring of the lake level and the discharge of the outlet river, the Rolwaling Khola and flood mitigation works. However, these were by no means easy tasks. The lake is at an altitude of 4580 m and its surface is frozen during the winter months. Construction is limited to the period from April to October. There is no road access and the nearest significant settlement is Dolakha, seven days *walk* away. Nepal is one of the world's poorest states and does not have its own resources to transport personnel and large quantities of heavy equipment to high altitudes as well as provide for their welfare over an extended construction period. Financial constraints initially inhibited large-scale construction work. Test siphons, donated by WAVIN Overseas BV, were first installed in May 1995 with the aid of the Nepal–Netherlands friendship associations.

Meanwhile, additional – largely unpublished – investigations were still being carried out to determine the growth pattern and bathymetry of the lake and, using electrical resistivity techniques, the distribution of buried ice in the moraine. A geotechnical hazard analysis concluded that the dam was stable on an engineering timescale of 50–100 years *providing no large-scale events like earthquakes occur*, but that dam failure was eventually bound to happen without some mitigation measures being undertaken. This appears to have convinced the Nepalese government to become committed to dealing with the Tsho Rolpa GLOF hazard and an additional five locally manufactured siphons were installed in 1997, supported by the Natural Disaster Fund of the Ministry of Home Affairs in the absence of dedicated funds. Although the siphons worked satisfactorily, with some maintenance, in the freezing conditions, the scale of the enterprise required to make a significant difference to lake level, and the need for constant maintenance, led to the abandonment of siphons in favour of a larger and more permanent solution. In 1997, a DHM team recommended lowering the lake by 3 m by means of an open, lined channel in the southwest part of the terminal moraine, where no buried ice had been recorded to at least a depth of 48 m. Fortunately, the Netherlands provided a grant of about $US3 million to fund the Tsho Rolpa GLOF Risk Reduction Project in 1998. The Nepalese government contribution, for ongoing monitoring and other support, amounted to $US115 414 reflecting the relative wealth of the nations. An automatic weather station was established in June 1999 and a ground-penetrating radar survey in the construction area was undertaken before the work commenced. The trapezoidally-cross-sectioned channel, 6.4 m wide at its base, utilizes a depression crossing the moraine ridge. The channel contains a gate structure to regulate the flow, and is designed to accommodate a flood flow of 14.6 m^3/s with a maximum capacity of 35 m^3/s, capable of lowering the lake surface level by at least 3 m, which is greater than the height of the largest displacement wave so far recorded at Tsho Rolpa. The remediation work was completed in July 2000, the first construction of its type in the Himalaya. Nevertheless, it was viewed by the so-called formulation team in 1997 as an interim measure. The ideal solution, funds permitting (probably another $US5 million), would be enacted through a Tsho Rolpa Permanent Remediation Project. This would lower the lake surface level by a further 17 m sufficient apparently to eliminate the Tsho Rolpa hazard permanently.

The complexity of risk management is illustrated by a model proposed by Geomarineuk UK (2003; Figure 6.8). Survey design and specification, hazard assessment, vulnerability assessment and risk assessment are all involved in managing the GLOF risk.

Figure 6.7 Some of the sensors and audio warning installations of the Tsho Rolpa GLOF warning system (source: Bell *et al.*, 1999)

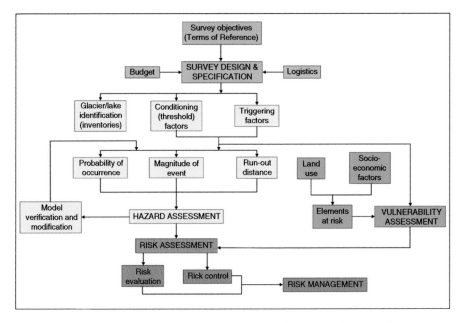

Figure 6.8 Model of GLOF risk management (Geomarineuk, 2003) accessed 22 02 08

6.6 Mitigation

During the current period of rapid climate change the risk of GLOF disasters has been increasing and demands a response from appropriate authorities if significant human and economic losses are to be avoided in the future. A range of measures are now available, designed to mitigate the risk of GLOFs. It is convenient to separate them into two types: 'hard' and 'soft'.

6.6.1 Hard measures (engineering)

These methods are often employed as an immediate or emergency response. At the site:

- drainage by syphon (supplement low season flows for fish and agriculture) or pump (expensive in remote areas);
- cut drainage channels (some benefit is possible from micro-hydroelectricity);
- tunnelling through bedrock (expensive and time-consuming);
- early-warning sensor system at the GLOF site and along the downstream river channel.

Downstream:

- early-warning audio system adjacent to population centres;
- flood control structures;
- building reinforcement.

6.6.2 Soft measures

These methods include:

- mapping of hazards and vulnerabilities;
- computer modelling of potential GLOFs;
- geophysical analysis of lake dams to determine structure and ice content;
- observation and forecasting systems;
- ground-penetrating radar;
- increase in community awareness and participation;
- promotion of afforestation and conservation;
- National Adaptation Programme of Action (NAPA);
- planning.

Hard measures are essentially those which require some degree of engineering. They are generally applied at the lake and its dam. Tsho Rolpa, the largest glacial lake in Nepal, is a good example of a situation where hard remediation measures were carried out in response to a perceived emergency (see Box 6.2). The task here included lowering the lake level by siphoning and channelling through the moraine (Figure 6.6), and the installation of an early warning system (Figure 6.7) with sensors at the dam and along the river, and audio signals close to settlements to provide inhabitants with some time to move to higher ground. This is the first example of such measures being used in the Himalaya. Other hard mitigation measures can be applied at downstream sites in the form of flood control systems and the reinforcement of buildings, but their precise location and scale can be difficult to determine in the variable terrain of Himalayan mountain valleys. There are also financial constraints in developing countries like Nepal and Bhutan.

The hard mitigation measures at Tsho Rolpa were put in place around the year 2000. It is likely that the Tsho Rolpa measures were taken because of a growing awareness of the GLOF problem in the Himalaya. Prior to the 1980s, general awareness of the Himalayan GLOF hazard, as reflected in the western literature and press, was very limited (Kattelmann, 2003). Then in 1985 a new, almost-completed hydroelectricity facility worth $US4 million was destroyed by a GLOF from the Langmoche Glacier lake in Nepal. Suddenly these flood events were no longer a purely local issue to be faced and overcome by remote and impoverished people, but posed a serious threat to significant economic development through loss of essential power supplies, and involved foreign aid donors. The circumstances surrounding the Dig Tsho GLOF disaster (Figure 6.4; see Box 6.1) provided dramatic evidence of the urgent need for a detailed assessment of the GLOF risk in the Himalaya. This incident, in particular, seems to have been the catalyst for a significant increase in awareness of the GLOF hazard in the Himalayan region, and there has since been a great deal of activity directed towards a reduction of GLOF risk (e.g. DPNET Nepal, no date). Destruction of the power station provided a clearly measurable level of loss, substantial in the context of a small developing country, against which costly mitigation measures could be balanced. With similar types of investments in the region likely to become larger and more frequent, even greater potential losses from future GLOFS would justify an enhanced level of response. In its detailed review of the Langmoche Disaster (Ives, 1986), the International Centre for Integrated Mountain Development (ICIMOD), set out its conclusions and recommendations as follows:

(1) GLOFs (referred to in the report by the Icelandic term, jökulhlaup) *will* continue to occur with sufficient frequency and magnitude to cause significant loss of life and property.
(2) As infrastructure develops and tourism increases the scale of potential impacts will rise.
(3) Many potential GLOF sites (sub-aerial lakes) can be inexpensively located and mapped.
(4) Enhanced assessment of the dynamics of glaciers perceived to be associated with potential GLOFs, would aid GLOF predictability.
(5) There should be a legal requirement for all development projects, especially HEP projects, to carry out an appropriate hazard impact assessment.
(6) The Nepalese government should formalize its response to the GLOF hazard, including a physical analysis of the extent of the problem using remote sensing, and a consideration of human issues through an examination of the behaviour and attitudes of local communities in response to GLOFs.
(7) Public, government and development agency awareness should be raised.

In contrast to site-specific, 'hard', engineered projects, such as the one carried out at Tsho Rolpa, these recommendations are focused firmly on 'soft' GLOF mitigation measures. Members of ICIMOD clearly recognized a growing Himalayan problem and, in particular, that more information about its extent and magnitude was urgently required. Little could be achieved in terms of 'hard' mitigation until the full scale and distribution of the problem was known. A number of preliminary studies (cited in Kattelmann, 2003), were carried out soon after these disasters. The first priority was an inventory of glaciers to determine the most vulnerable sites. An early joint Nepalese/ Japanese attempt to produce an inventory of dangerous lakes in eastern and central Nepal was undertaken, on a reconnaissance basis, in 1991 (Yamada and Sharma, 1993), using light aircraft to overcome accessibility problems. However, although many lakes and their 'mother' glaciers were recorded photographically, difficult flying conditions, including cloud cover and turbulence, meant that not all river basins were covered and not all glacial lakes within some basins were observed and photographed. In view of the relative inaccessibility of much of the region, remote sensing was the obvious next step, followed by a closer field inspection of those sites deemed to be in the most critical condition. However, although remote sensing platforms solved the accessibility issue, they still needed to avoid local weather conditions. An important development in remote sensing technology, especially applicable to high mountain terrains which attract a disproportionate amount of cloud, is cloud-penetrating radar such as Synthetic Aperture Radar (SAR). More comprehensive inventories for Nepal and Bhutan, derived from analyses of topographic maps, aerial photographs and satellite imagery, were conducted by ICIMOD and UNEP/RRC-AP and eventually completed in 2000 (Mool, Bajracharya and Joshi, 2001a; Mool, Wangda and Bajracharya, 2001b). Although they have been criticized as relying too heavily on out-of-date topographical maps and not being amenable to GIS analysis (Reynolds and Taylor, 2004) they do constitute a useful first step. The final inventories listed 2315 glacial lakes in Nepal and 2674 in Bhutan. Of these, 26 in Nepal were classified as 'potentially dangerous' and 24 in Bhutan. Although these figures may seem small as a proportion of the total number of lakes present in these countries, they present a significant hazard in the context of remote communities in relatively poor developing countries. This gathering of information to provide a baseline inventory of potential GLOF lakes is the preliminary step in developing 'soft' mitigation measures against the GLOF hazard. Subsequent measures include hazard mapping, vulnerability zonation studies, participation and awareness-raising initiatives, international (donor) friendship associations, organization of early warning systems, monitoring and forecasting and the development of adaptation strategies are some of the longer-term mitigation responses that fall into the 'soft' mitigation category.

The Tsho Rolpa and Rowaling Valley situation (see Box 6.2 for details) is an example of the apparent success of both 'hard' and 'soft' mitigation in response to the GLOF hazard. However, this is not always the case. The successful Tsho Rolpa warning system contrasts with that in neighbouring Solu Khumbu where an attempt by a local nongovernmental organization to establish a warning system for the Imja valley has so far failed for want of funds and political agreement. The situation highlights the breadth of opinion amongst the local Sherpa inhabitants, with views ranging from anxiety, especially amongst those who had lost property during the Lamoche flood, to relative complacency. Others believed that human intervention would exacerbate the problem, especially if carried out by non-Sherpa people. They cite the relative impact of the 1985 flood which destroyed the foreign-built HEP station but impacted relatively few local residents. Although a detailed hazard map has been prepared for the valley below the Imja Glacier lake (Braun and Fiener, 1995), discussions have continued without a decision for more than a decade. These different responses to the GLOF problem highlight difficulties that may arise when the different belief systems, experiences and expertise of indigenous and outsider communities come into contact with each other. To what extent can hazard mitigation be left to local decision-making rather than being imposed centrally? How far should external governments and their agencies control mitigation procedures and the installations? How can the relative risks of action and inaction be portrayed to central governments and local communities? Kattelmann (2003) advocated careful evaluation of the Tsho Rolpa mitigation system, being the first of its kind in the Himalaya, and also an assessment of its general applicability to other valleys. He implied that many residents of the Rowaling valley below Tsho Rolpa 'were led to believe that a catastrophe was imminent' (p.152). He added that 'technical advisors have a responsibility for communicating ... knowledge without creating ... panic' and that 'there are lessons to be learned from the social aspects of the Tsho Rolpa experience as well'.

6.7 Summary

This discussion of approaches to GLOF mitigation brings out the overall complexity of the GLOF hazard. Although its basic physical characteristics may be reasonably well known in outline, detailed, site-specific technicalities are often much more difficult to determine to a high level of probability. In part, this is due to the remoteness and high relief of many sites, but mainly it stems from the complicated and dangerous nature of ice-marginal environments and problems of access to critical internal structures of glaciers and their moraines. The variety of possible trigger mechanisms only serves to exacerbate the physical difficulties associated with the hazard and complicate risk assessment. To these difficult physical problems must be added social, economic and political issues that may, in some cases, be equally intractable. The poverty of some GLOF-susceptible countries, such as Nepal and Bhutan, means that financial constraints restrict easy access to remedial measures. Poverty is also likely to be implicated in the difficulties of communicating complex technical issues to parties with limited appropriate technical education. Without external aid little progress would have been made in understanding the risks and implementing solutions. However, cultural and political contrasts between recipient societies and external donors require tact and careful management in order to avoid misunderstandings which could inhibit progress towards solving the problem. There can be no doubt that GLOFs in the Himalayan region are a serious hazard in the context of today's climate change. What is equally clear is that solutions to the GLOF problem, in the Himalaya in particular, while reasonably obvious on paper, will require a high level of communication skills and involve some difficult physical, economic, social and political decisions.

7
Permafrost

7.1 Introduction

Nearly 25% of the land of the northern hemisphere, amounting to millions of square kilometres, is underlain by permafrost (perennially frozen ground), especially in the area of the Former Soviet Union (FSU). Around 49% of the FSU, 50% of Canada, 22% of China and 80% of Alaska (USA) are influenced by permafrost and the phenomenon also occurs in Greenland, Antarctica and Scandinavia as well as many high mountain ranges around the world (French, 2007). This may be a surprise to many people and the old adage, 'out of sight, out of mind', aptly describes the hazards associated with permafrost. Early colonizers of permafrost terrain were either unaware of its presence or did not understand how this invisible environmental system works. It is not easy to find examples of fatalities unequivocally due to permafrost processes (but see Nelson, Anisimov and Shiklomanov, 2002, p. 212 for Noril'sk, 20 fatalities), but numerous examples of distorted buildings and disrupted roads and railways (Figure 7.1) are scattered across the permafrost of North America and Eurasia (e.g. Ferrians, Kachadoorian and Green, 1969). Natural landscapes show long-term scars due to nothing more than the passage of vehicles or the removal of a shallow layer of tundra vegetation (Figure 7.2). Perhaps because of the paucity of direct human fatalities, permafrost hazards rarely receive the same level of media coverage as that devoted to more dramatic events such as earthquakes and floods and, until very recently, were neglected in the areas of social science and policy (Nelson, Anisimov and Shiklomanov, 2002). However, the economic cost of the permafrost hazard is considerable and is likely to increase in the context of climate change. Authorities are becoming more aware of the potential consequences of neglecting this complex, yet unobtrusive phenomenon.

Cold Region Hazards and Risks, First Edition.　Colin A. Whiteman.
© 2011 John Wiley & Sons, Ltd. Published 2011 by John Wiley & Sons, Ltd.

Figure 7.1 Buildings and infrastructure distorted by permafrost thaw: (a) buildings on Third Avenue, Dawson City, Yukon, Canada; (b) railway near Strelna, Alaska (from Ferrians, Kachadoorian and Green, 1969, Fig. 20; photo: L.A. Yehle); (c) road near Umiat airstrip, Alaska (from Ferrians, Kachadoorian and Green, 1969, Fig. 25); (d) roadhouse, Mile 278.5, Richardson Highway, Alaska (from Ferrians, Kachadoorian and Green, 1969, Fig. 33; photo: T.L.Péwé)

With over 11 million km² of permafrost terrain and the relatively early colonization of Siberia, it is not surprising that Russia led the way in permafrost studies, though the parlous state of some of their northern towns and cities suggests that good practice in relation to combating permafrost hazards has often been neglected (cf. Grebenets, 2003). In North America, it was probably the Second World War (1939–1945) that focused attention on northern permafrost with the need to construct the Alaskan Highway and the Canol pipeline as part of the war effort and military engineers were quickly made aware of the inadequacies of traditional construction methods (Muller, 1943, quoted in French, 2007). (Today, the CRREL (Cold Regions Research and Engineering Laboratory), a leading investigator of permafrost problems in the USA, remains under the auspices of the US Army Corps of Engineers.) Although hard lessons have been learned and modern engineering practice is greatly improved, past mistakes will impose high financial, if not direct human, costs as climate change inevitably disturbs permafrost systems and renders existing engineering design criteria obsolete.

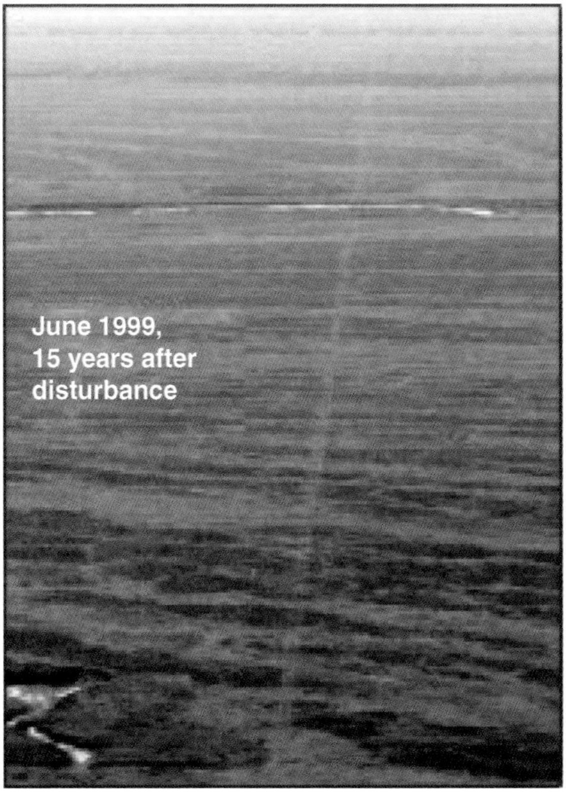

Figure 7.2 Seismic line scar, North Slope, Alaska (Reproduced from http://arctic.fws.gov/issues1.htm. Courtesy of U.S. Fish and Wildlife Service.)

7.2 Permafrost distribution and characteristics

Permafrost is essentially climate dependent. It forms where the air temperature is sufficiently cold to maintain soil and rock at or below 0 °C for at least two consecutive years (Figure 7.3). Persistent cold temperatures will maintain a negative heat balance at the ground surface which means that heat will be released from the ground and the permafrost will thicken. However, at the same time as heat is lost at the surface it is being gained at depth, at the rate of about 1 °C per 30–60 m (the geothermal gradient; Lachenbruch, 1968) due to the flux of interior geothermal heat upwards. The thickness of permafrost should therefore reflect a balance between the surface air temperature and the geothermal gradient. However, other factors, such as large water bodies (lakes and the ocean) may intervene. If lakes do not freeze to their bed, cooling of the ground beneath is reduced or inhibited. Permafrost thickness will either be less than expected, or the ground will remain unfrozen as a talik (from the Russian verb, tait, to melt) within the surrounding permafrost. In contrast, the presence

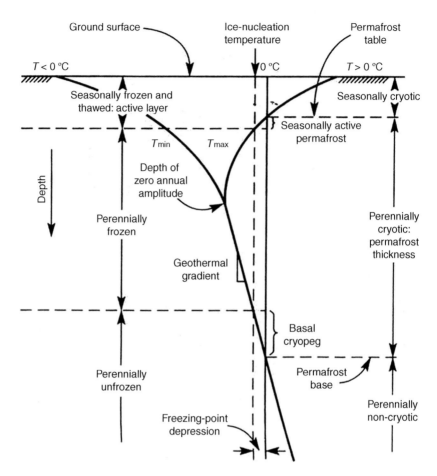

Figure 7.3 Permafrost ground-thermal regime (Reprinted with permission from French, H.M. (2007) The Periglacial Environment (3rd Edition). Wiley, Chichester, UK. 458 pp. Fig. 5.1.)

of areas of exceptional permafrost thickness (e.g. 1600 m in parts of Siberia) suggests that permafrost has progressively aggraded on several occasions during Pleistocene glaciations and has not been completely removed during short warm interglacial periods. Such permafrost is probably relict and not directly related to the present climate. These variables, and other complicating factors considered below, mean that permafrost cannot be simply mapped theoretically, a fact that enhances its hazardousness.

Figure 7.4 illustrates the general lateral distribution of permafrost in the northern hemisphere, where most of it occurs. It reveals a broad differentiation into high latitude and high altitude regions which, incidentally, provides a relevant basis for differentiating permafrost hazards. Three key

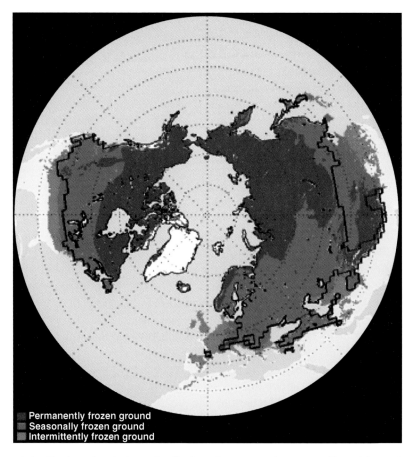

Figure 7.4 Northern hemisphere distribution of permanently, seasonally and intermittently frozen ground (source: http://en.wikipedia.org/wiki/File:Frozenground.gif [accessed 17 July 2010]

scale-related factors influence the distribution of permafrost (Gruber, 2005). Global and continental climate patterns, and associated atmospheric and oceanic circulations, account for permafrost in polar regions and its extension southwards in the centre of the North American and Eurasian continents. Regionally and locally, topography becomes the dominant factor, controlling air temperature and influencing the amount and type of precipitation (rain, snow, etc.).

High latitude permafrost is subdivided into continuous (90–100% of area underlain by permafrost), discontinuous (50%–90% of area underlain by permafrost), sporadic (10–50% of area underlain by permafrost) and isolated (0–10%) zones according to the International Permafrost Association (IPA) Circum-Arctic Map of Permafrost and Ground Ice (Brown *et al.*, 1997). Traditionally the term high altitude or 'alpine' permafrost referred to permafrost only on middle and low-latitude mountains. However, mountains also occur in high-latitude regions where it becomes difficult to distinguish between the two types. As all rugged mountains (as distinguished from high plateaux) possess distinctive features, especially steep slopes with different aspects which strongly influence permafrost distribution, the term increasingly being used to refer to permafrost in all high-altitude areas (except the Tibetan Plateau) is 'mountain', rather than 'alpine', permafrost (French, 1996). Mountain permafrost can also be subdivided into continuous (below the snow limit), discontinuous, sporadic and isolated zones but in this case reflecting altitudinal rather than latitudinal effects on climate.

Figure 7.3 illustrates the fact that permafrost does not usually extend up to the ground surface. This is because *seasonal* changes in temperature can cause the layer of ground immediately below the surface to thaw and freeze on an annual cycle. The temperature of the ground fluctuates seasonally only down to the depth of zero annual amplitude. The ground thaws only above the level where the temperature rises above zero during the summer. This seasonally thawed layer, known as the 'active layer', changes broadly with latitude from several metres in the south to no more than a couple of decimetres in the far north, reflecting the maximum mean annual depth of the $0\,°C$ isotherm. However, its actual thickness depends on local factors, such as vegetation type, rock and soil properties, water content, gradient and aspect, and varies from year to year, due to annual differences in weather conditions, especially snow thickness. The thickness of the active layer is important because it affects the heave ('jacking') potential of the ground during freeze-back in the cold season. In general a thicker active layer imparts greater heave stresses on surface structures and their supporting piles.

As permafrost exists beneath the ground surface, it is only revealed by excavation or, these days, by some form of geophysical remote sensing. Even when exposed, its presence may not be obvious, unless it is ice-rich, when rock joints and sediment pores may be seen to be filled or coated with ice. Ice content of permafrost is a key to the degree of hazardousness of permafrost. Sometimes, larger lenses of ice, called segregation ice, may be present, especially close to the permafrost table (Figure 7.5(c)). In suitable ground, especially in unconsolidated sediments, wedges of ice (Figure 7.5(a)) form a distinctive interlocking pattern (ice wedge polygons (Figure 7.5(b)) over large parts of the Arctic landscape. In some areas, such as the Tuktoyaktuk Lowlands in northwest Canada or the Yamal Peninsula in northern Russia, thick, extensive masses of buried glacier ice from the last glaciation, which have survived in the cold polar temperatures, may still be found. The presence of large ice bodies like these is a major factor determining the hazardousness of permafrost (see next section). This is because of the volumetric difference (about 9%) between a body of ice and its equivalent amount of water after melting. The growth of segregation ice lenses therefore causes the ground to

Figure 7.5 (a) Ice wedge, Ellesmere Island, Canada; (b) ice wedge polygons, NWT, Canada; (c) segregation ice, Ellesmere Island, Canada (photos: C. Whiteman)

heave, while melting of these ice lenses results in subsidence of the ground surface.

7.3 Permafrost hazardousness

Permafrost is not necessarily inherently dangerous. In some situations, if undisturbed, it can enhance natural landscape stability and provide a stable foundation for buildings and infrastructure. However, it can possess properties – or exist in circumstances – that make it potentially hazardous. The following factors, in particular, enhance permafrost hazardousness:

- irregular permafrost distribution (discontinuous lowland permafrost and lower level mountain permafrost) complicates mapping;
- a high permafrost temperature ($0\,°C$ to $-3\,°C$), close to the $0\,°C$ threshold, means that permafrost is more sensitive to warming;
- a high ground-ice content allows greater settlement of the surface on thawing;
- a thick, moist active layer increases potential heave pressure during freeze-back;

- steep slopes increase gravitational stress in mountain permafrost contexts.

From the human perspective permafrost hazardousness is increased in situations where:

- a high population density and/or vulnerability increases the likelihood of impacts on people, as well as human impact on the permafrost;
- valuable infrastructure is located on permafrost;
- engineering regulation and procedures are inadequate or not carried out effectively.

Given its defining property – a sub-zero temperature – permafrost is obviously sensitive to any natural (e.g. climate change) or human-induced disturbance that causes its temperature to increase towards, or move above, the critical threshold of 0 °C. Thus, the present episode of climate warming is having a significant impact on permafrost and is a major cause of concern. Temperature increases in the Arctic average about 5 °C (IPCC, 2007). Increases in permafrost temperatures at various depths (Table 7.1); see IPCC, 2007, Table 4.5 for a more extensive list) reflect this, but other factors, such as changes in vegetation and snow cover, must also be taken into account. There is considerable regional variation in temperature change, and even negative changes have been recorded in eastern Canada from the late 1980s to mid-1990s (Allard, Wang and Pilon 1995), although subsequent measurements in the same region suggest that this negative trend may have been reversed (DesJarlais, 2004). Air thawing (ATI) and freezing (AFI) indices, expressed as degree–days above or below 0 °C respectively, have been compiled and show rising and falling trends, respectively, in line with temperature changes. The ATI is a measure of thaw season magnitude and can be used to calculate the key parameters of active layer thickness and permafrost temperatures (ACIA,

Table 7.1 Examples of permafrost temperature changes (Data compiled from French, H.M. (2007) The Periglacial Environment (3rd Edition). Wiley, Chichester, UK. 458 pp., with permission from John Wiley & Sons, Ltd.)

Region	Depth of recording (m)	Period of record	Permafrost temperature change in degrees celsius
N. Alaska	20	1983–2003	2–3
C. Mackenzie Valley	10–20	Mid 80s–2003	0.5
N. Quebec	10	Late 80s–Mid 90s	<-1
N. Quebec	10	1996–2001	1.0
E. Siberia	1.6–3.2	1960–2002	~1.3
N. European Russia	6	1970–1995	1.2–2.8
Svalbard	~2	Past 60–80 years	1.2
Tibetan Plateau	~10	1070s–1990s	0.2–0.5

2005). The engineering design of infrastructures built on permafrost must take into account the seasonal thaw (active layer) depths and permafrost temperatures expected during the lifetime of the structure, not just those applicable during construction, as these parameters control key cryogenic processes, such as creep, thaw settlement, adfreeze bonding, frost heave, and frost jacking (see Glossary for definition) (Esch and Osterkamp, 1990, quoted in ACIA, 2005). For instance warm permafrost enhances creep and a thicker active layer increases the potential for adfreeze bonding, frost heave and frost jacking.

The ground materials in which permafrost exists possess different sensitivities to climate change and other types of disturbance. Massive, nonjointed rock is the least sensitive to change because of its inherent strength. Sensitivity increases through gravels and sands, silts, clays, organic soils and peat to pure ground ice, the material most sensitive to changes in temperature (ACIA, 2005). Given this high sensitivity to temperature change, ground ice is a major cause of permafrost hazards for several reasons. First, ice occupies approximately 9% more volume than the equivalent amount of water after melting. Consequently the melting of ice, and expulsion of water during consolidation, leads to subsidence of the ground surface, or thaw settlement, while the growth of segregation ice lenses during freezing causes the ground to heave. Second, there is a substantial difference in strength between crystalline ice and liquid water, respectively below and above the 0 °C threshold. In unconsolidated sediments, melting ice is likely to raise pore water pressures, thereby reducing soil stability and leading to mass movements such as shallow 'skin' flows within the active layer, a common feature in the Canadian Arctic during the warm summer of 1998 (Lewkowicz and Harris, 2005). Ice bonding can be stronger than rock-to-rock bonding at low temperatures (Harris *et al.*, 2001), but the main influence of ice on material stability may be to exclude water from the site or inhibit its movement through sediment or along rock joints and thereby reduce its lubricating effects. On the other hand, if some melting occurs in rock joints lubrication may be enhanced. Thirdly, even if ice just warms its shear strength decreases and it becomes more susceptible to creep or failure. Creep is the main cause of rock glaciers (Figure 7.6) in which any overlying sediment, or structure such as a mountain hut, is moved as the ice deforms. If the strength to stress ratio ('factor of safety') becomes less than zero, the bond will fail and the bedrock will fall or slide. Davies, Hamza and Harris (2003) found that failure may occur at just sub-zero temperatures and did not require complete melting. Ice was observed on joint surfaces following rock falls in the Alps in the very hot summer of 2003 (Figure 7.7).

Although there is a long history of permafrost study, considerable theoretical understanding (French, 2007), and a growing awareness of the impact of climate change, permafrost terrain continues to present substantial problems with regard to its hazard management because, in practice, it is difficult to determine exactly where it exists and what its properties are in any particular location. Permafrost distribution is especially hazardous wherever it is patchy,

Figure 7.6 Muragl rock glacier, Upper Engadine, Swiss Alps (Reproduced with permission from Harris, C. and 21 others (2009) Permafrost and climate in Europe: Monitoring and modelling thermal, geomorphological and technical responses. Earth-Science Reviews, 92, 117-171. Fig. 32. © Elsevier.)

in the discontinuous or sporadic zones. Such irregular distribution requires very detailed mapping on a large scale to determine the distribution of the different geotechnical responses needed for hazard mitigation.

The rest of this chapter is divided into two sections: lowland permafrost hazards and mountain permafrost hazards. Although permafrost retains the same essential temperature requirements wherever it exists, a topographic differentiation is convenient to highlight variations in the balance of processes and hazards associated with these two contrasting types of terrain. In lowland regions, with relatively uniform, horizontal or low gradient topography with, commonly, unconsolidated sediments, heave and subsidence are the dominant processes, characteristically leading to ground instability and the deformation of buildings and infrastructure if the state of the permafrost is disturbed. Disturbance can occur naturally, through climate change, or due to human interference with the environment. In contrast, permafrost in steep, complex mountain terrain with variable aspect and shading effects commonly has a less regular distribution and is more difficult to locate precisely. The commonest hazards associated with mountain permafrost are rockfalls, landslides, debris flows and deforming rock glaciers, reflecting the height and steepness of alpine terrain and the availability of coarse sediment produced largely by frost weathering and glacial erosion.

Figure 7.7 Ice-covered detachment surface (left-hand side of photograph) exposed by release of a rock fall in 2003 on the Matterhorn Lion Ridge (Reproduced with permission from Harris, C. and 21 others (2009) Permafrost and climate in Europe: Monitoring and modelling thermal, geomorphological and technical responses. Earth-Science Reviews, 92, 117-171. Fig. 8. © Elsevier.)

7.4 Lowland permafrost hazards

In 1994, J. Demek published a 'catastrophic scenario' for global warming and permafrost in Eurasia, in which widespread subsidence of the Siberian floodplains due to permafrost melting allows the transgression of cold Arctic ocean

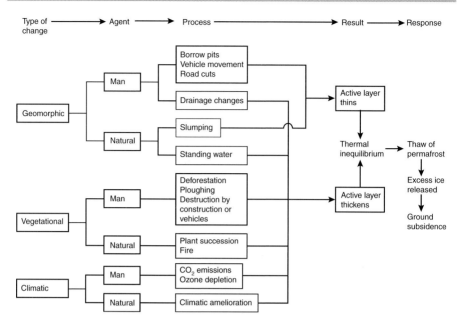

Figure 7.8 Factors involved in permafrost degradation (Reprinted with permission from French, H.M. (2007) The Periglacial Environment (3rd Edition). Wiley, Chichester, UK. 458 pp. Fig. 8.2.)

water into the lower reaches of Siberian rivers, and warming permafrost reduces the strength and bearing capacity of the ground. Probably not everyone would accept this degree of pessimism for the future of Arctic Eurasia, but during the latter half of the twentieth century the circum-Arctic permafrost regions of the high northern latitudes have become a focus for extensive human development (Nelson, Anisimov and Shiklomanov, 2002). As already discussed, permafrost is delicately balanced and can be very sensitive to environmental change. Unfortunately, there are many different ways to disturb permafrost (French, 1996; Brown et al., 1997; Figure 7.8), both naturally and through human activity.

Lowland permafrost areas possess a number of characteristics that make them potentially hazardous. First, being low-lying, they are relatively warm, > -5 to $-3\,°C$, especially in the more southerly discontinuous permafrost zone where they may be within one or two degrees of zero, and therefore especially sensitive to warming. Second, lowland permafrost areas are frequently large river floodplains and terraces with fine-grained frost-susceptible alluvial sediments, favouring high ground-ice concentrations in the form of wedges, lenses or veins. Some areas, such as the Yamal Peninsula, Russia, and the Tuktoyaktuk Lowlands, northwest Canada, contain extensive buried glacier ice in addition to periglacial ground ice. Third, these warmer lowlands usually support more people with the potential to disturb the ground surface,

7.4 LOWLAND PERMAFROST HAZARDS

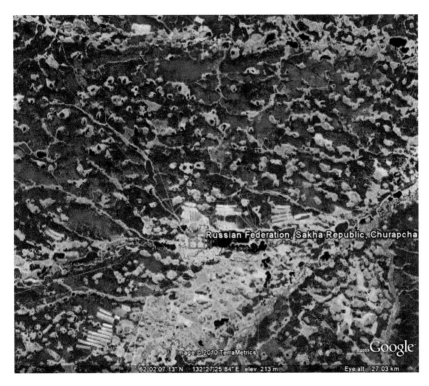

Figure 7.9 Thermokarst depressions (alases) around Churapcha, Central Yakutia, Sakha Republic, Russian Federation – note, some of the alases contain lakes (dark); compare this image with Demek, 1994, Fig. 15

for instance by clearing woodland, removing or compressing tundra vegetation by the passage of vehicles or by the digging of borrow pits for construction materials (French, 2007).

Demek's (1994) article was mainly about thermokarst terrain, that is land which has become uneven due to subsidence resulting from the melting of ice-rich permafrost. The irregular surface can take many forms from small hummocks and hollows a few metres across (Demek, 1994) to flat-bottomed, steep-sided basins, termed alases, which may extend to several square kilometres in area (Figure 7.9). Some of these alases may contain thaw lakes. The Lena and Aldan valleys in Siberia are classic areas for this type of terrain but thermokarst is also widespread in the northern lowlands of Siberia, in the Mackenzie Delta region of northwestern Canada and in parts of Alaska. A satellite study (Smith *et al.*, 2005) of lakes in the area between the rivers Ob and Yenisey in western Siberia discovered significant variations in their presence during the period 1973 to 1997. In the continuous permafrost zone there had been a 12% increase in lake area and a 4% increase in the number of lakes as ground-ice melted. In the discontinuous permafrost zone lake area had declined by 9% and there were 13% fewer lakes, presumably due to drainage through taliks developed in the thinner permafrost. In the continuous zone where, presumably, the permafrost was thicker the melt water was retained at the surface in thermokarst lakes. These changes may disrupt local water supplies and fishing economies. Also, within the study area is an extensive, developing pipeline system servicing large new onshore and offshore oil and gas fields. However, whether the degree of change outlined above continues at the same rate is not certain as the effects of an increase in mean annual temperature are complex (French, 2007). If warmer temperatures increase precipitation and cloud cover, this might lower the summer soil temperature and decrease the thaw depth rather than increasing it (Demek, 1994; French, 2007). The future of thermokarst development and hydrological systems in these lowland permafrost areas may be uncertain, but the widespread development of thermokarst is not without precedence. There is evidence for a number of episodes of extensive thermokarst development across up to 50% of the Yakutia lowlands and eastern Siberia (Czudek and Demek, 1970; Demek, 1994). A thaw unconformity recording significant warming during the early Holocene Period is widespread in the northwest Canadian Arctic. It is significant that both of these regions are located in areas of moderate to high risk on hazard zonation maps (Nelson, Anisimov and Shiklomanov, 2002; see mitigation section below).

Even without complete melting of ground ice in permafrost, ground strength and bearing capacity are both reduced as permafrost warms. At the same time the increase in temperature thickens the active layer and potentially increases heave stresses. These effects are of particular relevance for the population of urban centres with the capacity to modify permafrost artificially. Overall, the permafrost zone is not densely populated. There may be no more than eight million people living on permafrost terrain in the Arctic (French, 2007), but some, especially in Russia, are concentrated in substantial industrial (e.g. Norilsk, Vortuka) or administrative (e.g. Yakutsk and Tyumen) cities, with populations around 100 000 or more. This scale of urbanization has the capacity to modify climate significantly through the classic 'heat island' effect. Prior to development the mean annual ground temperature in Norilsk was $-3\,°C$ (Grebenets, 2003). In the centre of Norilsk ground temperatures in a borehole at a depth of 20–60 m increased $0.5\,°C$ during the 30 years up to the mid-1990s, while records from a borehole on the outskirts of the city showed a rise from $-3.5/-4.0\,°C$ to $-1.5/-2.0\,°C$ during the previous 50 years (Grebenets, Fedoseev and Lolaev 1994; quoted in Grebenets, 2003). These increases in ground temperatures occurred even though air temperatures recorded around Norilsk apparently did not reflect the broader pattern of climate warming in the Arctic.

Unfortunately, the negative effects of these human-induced permafrost temperature changes in Norilsk have been exacerbated, according to Grebenets (2003), by other adverse consequences of industrial development, such as imperfect foundation and construction design, heat flow into the ground during piling, inadequate ventilation and cooling systems, and the mechanised redistribution of the snow cover. Beneath Norilsk is a lattice-shaped heat-extracting subway system located in the permafrost at a depth of 2–6 m, usually along street axes. Taliks developed round these heat collectors are extremely unstable with frost heave or ground subsidence occurring depending on the season. More than 60 km of the network (20%) is severely damaged and needs replacement, while 70% requires some repair. Freeze–thaw processes rapidly deform roads and the underground system rarely lasts more than 5–8 years before requiring maintenance. Where these heat exchangers enter buildings an additional hazard occurs as they often pass heat to foundations. Consequently, as many as 80% of the apartment buildings of Norilsk have suffered from severe deformation. In May 1976 at Kayerkan, 15 km west of Norilsk, 'dozens of people were killed' when a building collapsed catastrophically (Grebenets, 2003, p. 306). The situation is often made worse during snow clearance: piling up snow in courtyards or over the heat exchange subways reduces winter ground cooling while blocking ventilated cellars with snow causes overheating. Another permafrost overheating problem, perhaps affecting 60% of deformed buildings, is caused by pipes in cold cellars leaking water, which reduces the bearing capacity of foundations.

In addition to its apparent inability to control the thermal regime of the permafrost, the mining and industrial complex at Norilsk is notorious for its output of pollution. Flooding and salinization of near surface soils thickens the active layer and increases thermal conductivity (Grebenets, Kerimov and Savtchenko, 1997). Annually, 2 million tons of sulphur dioxide, 60 000 tons of chlorine, 10 000 tons of dust and other pollutants are discharged and enter open water systems and the active layer with the help of rain and melting snow. Flooding damages the ice-rich layers of the permafrost causing thermokarst and thaw subsidence. Salinization lowers the freezing temperature, which helps to

Figure 7.10 Decayed ferro-concrete pile, Dudina, Russia (from Grebenets, 2003, Fig. 2)

Figure 7.11 Concrete piles for buildings in Yakutsk, Russia, 1973 (Reprinted with permission from French, H.M. (2007) The Periglacial Environment (3rd Edition). Wiley, Chichester, UK. 458 pp. Fig. 14.4.)

thicken the active layer, reduce bearing capacity and ultimately to deform buildings. The chemical effects of polluted groundwater aggressively attack the ferroconcrete piles (Grebenets, 2003; Figure 7.10) originally favoured in Siberian buildings (French, 2007; Figure 7.11). Around 8500 Norilsk foundations may be subject to such active destruction. By the end of 1998, about 250 large structures in the Norilsk industrial area were deformed and about 100 in a critical condition (Makarov *et al.*, 2000; Weller and Lange, 1999). Thirty four five to nine storey apartment blocks erected between 1960 and 1980 had to be demolished.

Table 7.2 Percentage of buildings in Yakutsk, Russian federation, that have failed or are expected to between 1990 and 2030 according to the year of commission (Reprinted with permission from French, H.M. (2007) The Periglacial Environment (3rd Edition). Wiley, Chichester, UK. 458 pp. Table 14.1 (based on Khrustalev, 2000).)

Construction year	Year of predicted failure				
	1990	2000	2010	2020	2030
1950	8	28	72	94	100
1960	6	27	72	94	100
1970	4	25	71	94	100
1980	2	24	71	94	100
1990	0	22	70	94	100

Yakutsk is another city with severe permafrost-related problems. Here, a substantial number of buildings, erected between 1950 and 1990 have already technically failed (Table 7.2; Khrustalev, 2000), and *all* are predicted to have similarly failed by 2030, due to a decrease in the bearing capacity of the permafrost around the pile foundations as the mean annual air temperature rises, unless mitigation procedures are successful. Numerous other Russian Arctic cities (e.g. Dudinka, Igarka, Khatanga, Talnakh, Tiksi and Vorkuta) have experienced similar impacts due largely to inadequate management of the permafrost hazard during the Soviet era.

The effects of lowland permafrost hazards such as cracked and distorted buildings and thermokarst tend to be obvious even to the casual observer. In contrast, the flux of carbon dioxide and methane, the two best known 'greenhouse gases', only becomes apparent using appropriate instrumentation. However, with some 12–16% (121–191 Gt) of global carbon stored in the northern circumpolar arctic tundra (Tarnocai, Kimble and Broll, 2003; Christensen, Friborg and Johansson, 2008), the area perhaps most vulnerable to climate warming, it is not difficult to appreciate the potential magnitude and significance of this hazard. Because ice-rich permafrost beneath subdued topography inhibits drainage, Arctic soils are often wet and anoxic, and slowly accumulate carbon as peat before releasing methane (CH_4). During the present interglacial (last 11 500 years), most Arctic tundra has been a net carbon sink though this does vary, with drier areas exporting carbon to the atmosphere. As the lake study (Smith *et al.*, 2005) illustrated, the outcome of permafrost melting is complex: it may lead to wetter or to drier conditions. The relatively few observations of annual carbon budgets currently available (Christensen *et al.*, 2008) indicate considerable variability in Arctic terrestrial ecosystems. This suggests that more long-term investigations are urgently required before firm predictions can be made regarding the contribution of the Arctic carbon store to climate warming. On balance, however, Arctic permafrost areas are considered (ACIA, 2005; Christensen, Friborg and Johansson 2008) to be a net source of both CO_2 and CH_4 to the atmosphere.

7.4.1 Mitigation measures

It is clear from the foregoing section that permafrost presents a range of difficulties on different scales, some induced naturally, others self-inflicted by humans. Perhaps the simplest solution to these permafrost-related problems would be to 'let sleeping dogs lie'! (This is an English idiom meaning 'do not disturb a situation as it would result in trouble or complications' (http://www.usingenglish.com/reference/idioms/let + sleeping + dogs + lie.html– accessed 30 11 09).). Unfortunately, this is very unlikely to be a viable option in view of the enormous economic and political pressures that already exist in Arctic and sub-Arctic regions, and the fact that global climate change is one of the critical factors impacting upon permafrost. Attempts to mitigate the problems will correspondingly involve a global response; nothing less than the modification of global climate or large-scale human migration will be required if some of the worst effects are realized (Demek, 1994). At regional and local levels, mitigation can probably be achieved through planning regulations supported by technology. However, there is plenty of evidence to suggest that available geotechnological and engineering capabilities have not always been adequately employed for the task of counteracting the adverse effects of permafrost and active layer processes, and that economic and political imperatives have sometimes been allowed to override the requirements of safe practice. In the last 20–30 years, however, cold-regions engineering and geotechnology have undoubtedly 'raised their games' in an effort to avoid the problems of the past and adequately service the domestic and industrial demands currently being placed upon them. Although cold regions do not support large numbers of people, they contain substantial natural resources, such as hydrocarbon reserves and minerals, which attract industry to the northern regions of Canada, Russia, the USA (Alaska) and elsewhere. Support for existing populations and new economic developments, demands cryogenic expertise across a wide range of activities including protection of municipal services (e.g. water supply and sewage disposal), construction of buildings, provision of transport networks (road, rail, air and sea), exploration facilities, mining and associated waste disposal and the installation of pipelines. The importance of these activities is reflected in specialist symposia and workshops held in Russia (Kamensky, 1998), Norway (Svalbard) (Senneset, 2000) and China (Cas.cn, 2009), for instance. CRREL, based in the USA, continues its response to permafrost problems and many other countries support national organizations concerned with geocryological engineering. There are also numerous commercial companies that have developed specific expertise in permafrost engineering to mitigate the impacts of this particular landscape system. In the next few sections, different approaches to the mitigation of permafrost hazards will be examined.

7.4.2 Climate change

As climate is the primary influence on permafrost, it follows that permafrost will remain stable if the climate is stable. However, the evidence is now overwhelming that climate is changing rapidly and polar temperatures in particular are increasing, due largely to the anthropogenic input of greenhouse gases into the atmosphere. Consequently one obvious way to mitigate widespread decay of permafrost is to reverse the trend in greenhouse gas emissions. This is an easy statement to make but the solution is proving extremely difficult to achieve politically (e.g. the Copenhagen conference on climate change, December 2009). Obviously, controlling and reducing CO_2 emissions is not likely to happen quickly. The CO_2 increase that has already occurred will ensure (unless there is a significant natural reversal of global temperature) that further lowering of the permafrost table and deepening of the active layer will take place. Consequently, methane and CO_2 will continue to escape into the atmosphere at the continental scale for some time. At regional and local scales, the population will be forced to respond to the effects of climate-driven permafrost decay. However, whereas pioneer settlers on permafrost may have failed to recognize its problems or lacked the knowledge to be able to respond to them effectively, nowadays cryotechnological solutions are available and effective, providing that they are carried out satisfactorily. One of the most important initial steps towards achieving mitigation is the accurate mapping of ground conditions.

7.4.3 Mapping

At first, empirical records of permafrost presence enabled a broad picture of permafrost zones, with their different susceptibility to permafrost hazards, to be established. Discontinuous permafrost generally creates more problems than continuous permafrost. This is partly due to the difficulty of mapping discontinuous permafrost precisely, but also because the more southerly discontinuous zone is generally warmer and therefore more sensitive to melting and ground subsidence. However, given that permafrost hazards are not just temperature dependent, other factors, especially ice content, must also be considered (Nelson et al., 2002). In recognition of this complexity, a digital, colour-coded, 1:10,000,000, circum-Arctic map relating permafrost zonation to ground ice conditions was prepared for the International Permafrost Association (IPA)(Brown et al., 1997). Using this permafrost and ground ice information, together with a 'settlement index' ($I_s = \Delta_{al} \cdot V_{ice}$ where Δ_{al} is the percentage increase in active layer thickness and V_{ice} is the volumetric proportion of near surface soil occupied by ground ice), general circulation models (ECHAM1-A and UKTR) and a mathematical solution for active layer thickness, Nelson, Anisimov and Shiklomanov, (2002) created permafrost hazard zonation maps, in effect, thaw-sensitivity maps (ACIA, 2005), for northern North America and Eurasia (note: Anisimov and Lavrov (2004) have subsequently suggested the addition of a coefficient of soil salinity, K_s to take account of the effect of salt on the freezing temperature). The hazard zonation maps (Nelson, Anisimov and Shiklomanov, 2002, Figures 6 and 7; Figure 7.12 (a) and (b)) were generated using a GIS to differentiate stable, low, moderate and high susceptibility to permafrost degradation and thermokarst development under warming conditions. Initial results predicted a significant northward movement of the southern boundary of permafrost over the next few decades and an increase in the thickness of the active layer especially along the northern (coastal) and southern margins of the permafrost (Nelson, Anisimov and Shiklomanov, 2002, Figures 4 and 5). Both of these changes are likely to

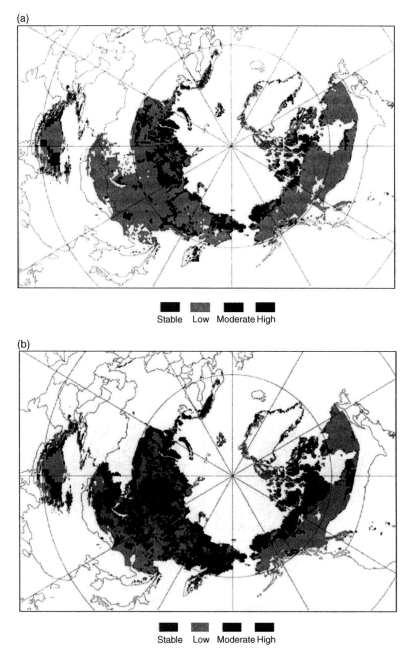

Figure 7.12 Hazard zonation maps obtained using (a) ECHAM1-A and (b) UKTR climate scenarios. Shading density indicates susceptibility to disturbance under conditions of warming climate (Reproduced with permission from Nelson, F.E., Anisimov, O.A. and Shiklomanov, N.I. (2002) Climate change and hazard zonation in the circum-Arctic permafrost regions. Natural Hazards, 26, 203-225. Figs 6 and 7. http://www.springerlink.com/content/cy73mhwccl1jubeb/?p=0c21e4b67ab04d70bb78256bf287438c&pi=0.)

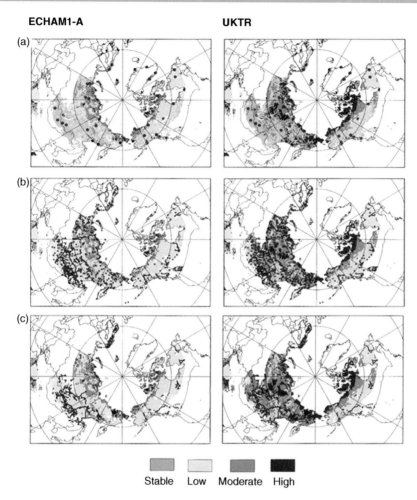

Figure 7.13 Hazard potential maps of Figures 7.12(a) and 7.12(b) with the distribution of contemporary settlements and infrastructure (US National Imagery and Mapping Agency, 1997) superimposed: (a) population centres (red) and settlements (pink); (b) roads including winter trails (yellow), railways (blue) and airfields (red); (c) electrical transmission lines (blue) and pipelines (yellow). Only settlement/infrastructure features within the contemporary permafrost zones (as defined by Brown *et al.*, 1997) are shown (Reproduced with permission from Nelson, F.E., Anisimov, O.A. and Shiklomanov, N.I. (2002) Climate change and hazard zonation in the circum-Arctic permafrost regions. Natural Hazards, 26, 203-225. Figs 8. http://www.springerlink.com/content/cy73mhwccl1jubeb/?p=0c21e4b67ab04d70bb78256bf287438c&pi=0.

cause some subsidence and thermokarst, depending on the amount of ground ice present. The final stage in this procedure was to superimpose the distribution of human infrastructure (population centres, settlements, utility lines, pipelines, airfields, transport networks and the Russian Bilibino nuclear plant) onto the hazard *zonation* maps to produce hazard *potential* maps (Nelson,

Anisimov and Shiklomanov, 2002, Figure 8; Figure 7.13 (a)–(c)). These maps of potential hazard indicate that mitigation measures against the permafrost hazard should be focused on central and northern Alaska, northwestern Canada, western Siberia and the Sakha Republic in Siberia where thaw-induced disruptions to settlements and industry have already been reported (e.g. Grebenets, 2003). According to Nelson, Anisimov and Shiklomanov, (2002), major settlements located in areas of moderate or high hazard potential are Barrow (Alaska), Inuvik (NW Canada) and Vorkuta and Yakutsk, both in Siberia. Transport facilities, such as the Dalton Highway in Alaska, the Dempster Highway in Yukon, Canada, and the extensive road and rail network in central Siberia, are located in the high risk category, while the Trans-Siberian, Baikal-Amur Mainline, Hudson Bay and Alaska railways generally cross areas of lesser hazard potential. The Trans-Alaska Pipeline is particularly vulnerable in two areas and the Bilibino nuclear power station and its grid is similarly located in high-risk areas. The oil and gas fields of the West Siberian Plain are particularly vulnerable as this permafrost is ice rich.

Obviously, mapping at a scale of 1:10 000 000 will not solve the detail of permafrost hazards, but it is a useful way of focusing the attention of hazard scientists, policy makers, planners and cryogenic engineers on areas of greatest risk. These small-scale hazard potential maps should facilitate the development of 'strategies to mitigate detrimental impacts of warming and [assist] adaptation of the economy and social life to the changing environment of northern lands' (Nelson, Anisimov and Shiklomanov, 2002, p. 219). More detailed mapping should then be undertaken regionally, at the local municipal level and, as required, on a site-by-site basis. Wolfe's (1998) account of Yellowknife, capital of the North West Territories (NWT), Canada, is a good illustration of work at the scale of a single community. Here, glacially moulded bedrock is interspersed with glacial outwash gravels and sand and fine-grained, ice-rich, lacustrine deposits. Onto this geological base is superimposed discontinuous permafrost forming, overall, a very complex mosaic of permafrost hazards. Both highways and buildings in Yellowknife have required specially engineered solutions and remedial actions (addressed in the next section) to mitigate the detrimental effects of permafrost.

7.4.4 Geotechnical engineering

Numerous geotechnical and engineering solutions have been devised over the years to mitigate the main lowland permafrost hazards of heave and subsidence. Chapter 16 of the ACIA report is a useful source for this section (ACIA, 2005). The range of approaches to the problem of lowland permafrost hazards can be summarized as follows:

1. maintain thermal equilibrium by the use of thaw-stable pads, usually of gravel or similar aggregate;
2. remove frost-susceptible sediments, if possible and economically feasible;
3. maintain thermal equilibrium by keeping the ground cold;
4. maintain building stability by the use of piles driven into stable permafrost;
5. insulate piping which carries warm liquids or gases either above or below ground;
6. introduce and effectively enforce adequate regulations to control ground disturbance.

Thaw-stable foundation pad Road travel across any permafrost region with frost-susceptible sediments is always along a pad of gravel, or similar material, that raises the road above general terrain level by a few metres. Traditionally, many buildings are also located above their surroundings on gravel pads (e.g. skating/ice hockey rink in Inuvik, NWT, Canada). In the case of the road, vegetation would be removed and/or the active layer sediment compressed resulting in warming the permafrost and inducing settlement. In the case of the building, warmth from the structure would similarly melt ice in the permafrost which would cause settlement leading to deformation. With an appropriate thickness of thaw-stable material between the structure, road or building, and the ground, the thermal stability of the underlying permafrost is maintained and a potential hazard mitigated. However, the construction of an embankment often alters the surface microclimate and increases the temperature, and the complex relationships between the embankment, natural vegetation and maintenance operations can make prediction difficult (ACIA, 2005).

Removal of frost-susceptible sediments It is obvious that the complete removal of frost-susceptible sediments, or their replacement by frost-stable material, such as gravel, is impractical on a large scale, but this approach has been used locally to solve particularly difficult permafrost problems. In Dawson City (Yukon Territory, Canada), 5–7 m of ice-rich silty sediments beneath some historic buildings have been replaced by thaw-stable gravels in order to maintain the stability of the buildings *and* their original level in relation to the streets. Use of the tradition gravel-pad technique in this case was ruled out by the Canadian Federal Governments agency, Parks Canada, as it would have reduced the heritage status of the buildings (French, 2007).

Maintenance of thermal equilibrium by refrigeration Shifting large quantities of sediment around is expensive and not every location has easy access to appropriate aggregates. Consequently, a frequently-used approach now is to keep the ground cold artificially in order to maintain its thermal equilibrium and stability. Ice-bonded sediments usually possess substantial bearing capacity and can actually contribute to the design of engineering solutions

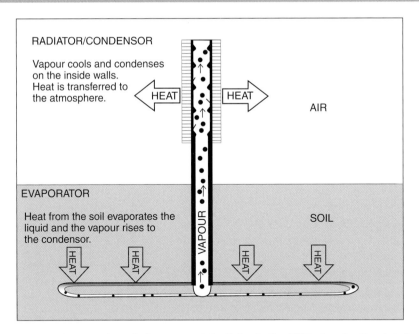

Figure 7.14 Thermosyphon design as used in Yellowknife, NWT, Canada (Reprinted with permission from French, H.M. (2007) The Periglacial Environment (3rd Edition). Wiley, Chichester, UK. 458 pp. Fig. 14.6A.

for permafrost hazards. The task, therefore, is to remove any excess heat from the sediments underlying and immediately surrounding a structure. One of the commonest solutions to this problem, which has been available since being pioneered on the trans-Alaska pipeline in the late 1960s, is the thermosyphon. Thermosyphons (or thermoprobes) are passive, self-powered refrigeration devices designed to transfer heat from the ground to the atmosphere (Arctic Foundations.com, accessed 07 12 09). The thermosyphon is essentially a closed tube containing a two-phase working fluid (Figure 7.14). Beneath the ground the liquid is evaporated using heat drawn into the tube from the soil that is thereby cooled. The resulting vapour rises into the upper, exposed part of the tube which is normally covered with thousands of small fins (heat exchangers) on its outer surface. The cold, ambient air, passing across the fins, causes heat to radiate from the vapour in the tube to the atmosphere. This condenses the vapour to a liquid, which returns to the underground section where the cycle begins again. Thermosyphons have been used extensively throughout the northern regions, to maintain the thermal equilibrium of permafrost in a variety of settings including buildings (houses, schools, factories, hospitals, aircraft hangers and fuel storage tanks), highways and earth dams (Figure 7.15).

Figure 7.15 Thermosyphons installed in an earthen dam, BHP Ekati Diamond Mine, NWT, Canada (Reprinted with permission from French, H.M. (2007) The Periglacial Environment (3rd Edition). Wiley, Chichester, UK. 458 pp. Fig. 14.6B.

In situations where the temperature gradient is insufficient to maintain a thermosyphon circulation system naturally, for instance where heat is to be transferred from the underlying permafrost into a building, rather than into the atmosphere, a heat pump may be required to drive the heat from the system. The extra power of a pump is especially applicable where heat is escaping into the warm permafrost foundations of a building and needs to be returned quickly back into the building. This technology was employed in the construction, in 1974, of a school for the Ross River community in Yukon Territory, Canada, where very sensitive ($-0.5\,°C$), discontinuous permafrost presented difficulties (Baker and Goodrich, 1990; Goodrich and Plunkett, 1990;

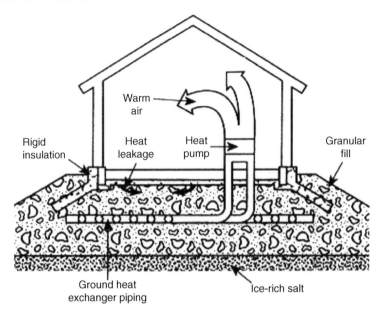

Figure 7.16 Thermosyphon system modified to include a heat pump as employed at Ross River School, Yukon, Canada (Reprinted with permission from French, H.M. (2007) The Periglacial Environment (3rd Edition). Wiley, Chichester, UK. 458 pp. Fig. 14.3A.)

Figure 7.16). However, the building has continued to settle. First, a plumbing breakage directed water onto the permafrost and caused melting, then the ammonia gas refrigerant reacted with the piping and, thirdly, the 3 m spacing of the sloping evaporator pipes was probably too wide to cope with the air temperature warming, given the sensitivity of the permafrost (Holubec, Jardine and Watt, 2008). Unfortunately, even a replacement building (2002) with a flat loop thermosyphon system has since shown signs of cracking and settlement as the air temperature continues to warm. The University of Alaska, Fairbanks, has included the Ross River School in its outreach 'Permafrost Health' programme for IPY in 2009, which involved installing a permafrost temperature monitoring system at the school (Permafrost outreach, 2009).

In a different context, a modified version of the thermosyphon, termed a 'hairpin thermosyphon' for obvious reasons (Figure 7.17(a)), has been installed beneath Thompson Drive, a new highway in Fairbanks, Alaska, which crosses ice-rich permafrost. In this case the carbon dioxide refrigerant in the lower (evaporator) (Figure 7.17(b))part of the system boils by removing a large amount of heat from the ground, in effect, supercooling the permafrost during the winter so that it can survive through the following summer. The boiled refrigerant moves to the upper (condenser) section where it cools and condenses as heat is radiated out to the cold atmosphere above the road. The

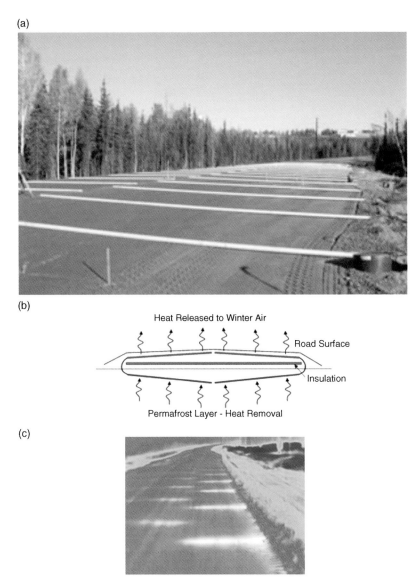

Figure 7.17 (a) 'Hairpin thermosyphons' placed in a roadbed along Thompson Drive, Fairbanks, Alaska; (b) diagram showing the arrangement of the installation; (c) photograph of a thermal image of the roadbed with thermoysyphons installed (source: http://www.alaska.edu/uaf/cem/me/news/thompson_drive_thermosyphons.xml, accessed 24 01 10)

system is completed with a layer of insulating material between the condenser and the evaporator to assist permafrost survival during the summer. The success of the hairpin thermosyphon is monitored using infrared technology which highlights in orange the heat being given off by the condensers

Figure 7.18 ACE ventilated shoulder being constructed on Thompson Drive, Fairbanks, Alaska, USA, September 2003 (Article by Douglas J. Goering (C) 2004, Alaska Business Publishing Company)

(Figure 7.17(c)). This project was a collaboration between the Alaska Department of Transportation, University of Alaska, Fairbanks academics and Arctic Foundations, Inc. of Anchorage, Alaska (UAF Thermosyphons, no date).

The hairpin thermosyphon technology used in the Thompson Drive scheme is supplemented by another passive cooling technique, the air convection embankment (ACE), which relies for its effect on natural air convection through a highly porous, well-sorted embankment material (Goering, 2004; Figure 7.18). At Thompson Drive the porous material is used either just on the shoulder of the embankment (a 'ventilated shoulder') or across its full width. In the latter case a fabric separator is installed beneath the finer material of the roadbed foundation to maintain the porosity of the ACE, which is essential for effective air circulation. Given sufficient winter cooling of the embankment surface, relative to its interior, an unstable pore–air density gradient develops which causes a convectional circulation with cold dense upper air sinking and warmer, less dense air at the base rising. This air movement in the embankment can transfer heat out of the embankment faster than conductive cooling. Proof of convection was observed in January 2004 when vapour plumes were seen to rise through holes in the surface snow layer, produced by the rising warm moist air. In summer, a stable pore–air density gradient develops and convection ceases. The net effect of the process is to reduce mean ground temperatures, raise the level of the permafrost table and increase highway stability. (Incidentally, to avoid introducing heat to the permafrost beneath the road, some of the finer foundation layers of the embankment were emplaced during the winter and allowed to freeze before the next layer was applied.) An instrumentation system has been installed in the ACE to monitor temperature. The

Figure 7.19 Thermosyphons in a tailings dam, Kubaka Gold Mine, eastern Russia (source: Arctic Foundations, 2009)

ACE stabilization technique is also appropriate for rail and airport embankments, containment dams and mining waste heaps, providing always that the embankment material is sufficiently porous, and winter temperatures are low enough to generate an unstable pore–air density gradient to produce the convective circulation.

Waste disposal ponds and reservoir dams are a third setting in which thermosyphons have proved effective in maintaining structural integrity. At the BHP Ekati Diamond Mine at Lac de Gras, NWT, Canada, a long row of thermosyphons maintains the frozen state of an earthen tailings dam so that the dam remains impermeable to seepage and resistant to the thawing effects of unfrozen lake water (French, 2007; Figure 7.15). In eastern Russia, thermosyphons were installed in a tailings dam at Kubaka Gold Mine after settlement had occurred due to disturbance of peaty and silty permafrost during construction (Edlund, Gordon and Robinson, 1998; Figure 7.19).

Piling As noted earlier, ice-bonded permafrost sediments can possess substantial load-bearing capacity and this characteristic can be used in the protection of the permafrost itself, especially where substantial quantities of heat may be lost from a building to the underlying permafrost, or where suitable aggregate is in short supply. Under these circumstances, the preferred solution for stability is to place structures on wood, concrete or metal piles, even though this is usually a more expensive option. Piles need to be secured within the permafrost in order to resist the heave stresses of the active layer. There are three main piling methods: (a) the freeze-back method,(b) the back-fill method and (c) the driven method (Patentstorm, no date). With the freeze-back method, a hole is augured

Figure 7.20 Thermopile supporting an antenna tower at the HAARP facility, Gakona, Alaska (source: http://www.arcticfoundations.com/index.php?option=com_content&task=view&id=28&Itemid=56, accessed 07 12 2009)

just larger than the pile, and wet slurry is back-filled around the pile and allowed to freeze back and adhere to the pile, hopefully not introducing too much warmth and wetness to the existing permafrost. Back-filling requires removal of a large amount of permafrost, the availability of a similar amount of nonfrost-susceptible aggregate and, again, disturbance of the existing permafrost regime. Driven piles, the increasingly preferred

Figure 7.21 Traditional (a) and modern (b) utilidor system in Inuvik, NWT, Canada, used for transporting water, sewage, etc. between buildings and the central plant, above the permafrost (source: http://www.pws.gov.nt.ca/pdf/GEP/02-Piping-Apr04.pdf, accessed 12 12 2009)

method, are hammered into the permafrost, as long as it is not too cold and hard, with minimal disturbance and no back-filling problems.

Piles support structures above ground level, leaving a 'crawl space' to allow warm air beneath the building to be dispersed by air circulation or cold air to enter the space in winter to keep the permafrost frozen. The ventilation space can be seasonally regulated using flaps around the edge. However, there are some circumstances in which ordinary piling is not a reliable solution: for instance, in a frost-susceptible active layer where heave and pile jacking are excessive, in warm permafrost where pile creep rates are high, in saline permafrost where substantial quantities of unfrozen water may be present in the soil and where development has warmed the permafrost (Arctic Foundations, 2009). Temperature increases due to climate change may accelerate settlement of shallow pile systems over the design life of a building (about 20 years) (ACIA, 2005). Under these conditions it may be preferable to employ load-bearing thermosyphons, or thermopiles, which both support structures (such as the antenna tower at the High-frequency Active Auroral Research Program facility at Gakona, Alaska; Figure 7.20), *and* cool the ground. Ground cooling reduces heave pressures and pile creep and, in saline locations, freezes water around piles while pushing the salts away.

Pipe insulation   The transfer of liquids and gases through pipes is a hazardous undertaking in permafrost terrain. Either the liquid cools or freezes completely, thereby interrupting the supply, or warmth introduced to counteract this problem causes melting of the permafrost, deformation and cracking of the pipe, loss of the liquid or gas and pollution of the environment. This is an issue not only for the transport of oil and gas on an industrial scale, but for the supply of water to, and removal of sewage from, domestic properties. The answer to this problem is a fine balance between fluid temperature and the amount of insulation; between keeping the heat in and the cold out. An expensive solution, applicable only to larger settlements, such as Inuvik in the Mackenzie Delta, NWT, Canada, is the utilidor (French, 2007). In Inuvik, this was originally a system of rigid continuously-insulated aluminium boxes supported close to the ground on piles (Figure 7.21(a)). Modern replacements (Figure 7.21(b)), still supported overground, are circular-sectioned pipes, chained at the supports to allow a limited amount of seasonal adjustment which avoids distortion of the system. Careful planning of the layout, including the provision of numerous small bridges and tunnels, is needed to link all the buildings to a central system, and avoid traffic. These systems require frequent maintenance and are easily damaged as the following statement from the Inuvik authorities at the town hall suggests:

> 'The utilidors are a dangerous place for children to play. Please instruct your children not to play on the Utilidors.... Walking on the utilidor is considered trespassing.... During the winter months do not push any snow on or near the utilidors. The system must be clear at all times. If you are caught shoveling (sic) or pushing snow on or near the utilidors, you will be responsible for the removal of it at your expense.' (Inuvik, no date)

At Resolute, Canada, the population is resistant to replacement of their utilidors by a water trucking system favoured by the Government of

Nunavut, as renewal of the utilidors is anticipated to cost $Can29.5 million (Nunatsiaqonline, 2009).

Regulations Most countries with permafrost possess a set of regulations or guidelines for the use of planners and civil engineers and others operating in this type of terrain. For instance, in the past 30 years the Alaska Department of Natural Resources has reduced the time allowed for working on the tundra from over 200 days down to 100 days as the cold season has shortened and potential damage to the thickening active layer increased. Consequently there has been about a 50% reduction in the time for oil and gas and mining exploration in northern tundra areas. This is a severe restraint on development and there is a possibility that the regulations, based on tundra hardness and snow conditions, might be relaxed.

The evidence of severe permafrost impacts on buildings and infrastructure, already discussed, suggests that regulations for the control and mitigation of disturbance generally work better in North America than in Russia. While both areas are subject to rising Arctic temperatures, it would appear that inadequate engineering and maintenance was practiced in the FSU. Rail lines have been deforming, airport runways in several cities are in an emergency state, oil and gas pipelines have broken and polluted their surroundings, open pit walls and mining tailings are all at risk, especially in the discontinuous permafrost zone. The weight of buildings may also be relevant to the problem: contrast the common multi-storey buildings in northern Russia where severe deformation problems occur with generally lighter-weight buildings in North America where damage over a similar timescale appears to be less extensive (French, 2007).

7.5 Mountain permafrost hazards

The permafrost concept has already been defined in relation to lowland permafrost. This definition remains valid for mountain permafrost as it simply refers to the thermal state of soil and rock beneath the ground surface. However, while lowlands are, by definition, low relief areas usually with simple relief patterns and shallow gradients, most alpine mountain ranges, such as the European Alps and Asia's Himalaya, are regions of high, complex relief patterns and steep gradients. Regionally and locally, topography becomes the dominant factor controlling air temperature and influencing the amount and type of precipitation (rain, snow, etc.). Solar radiation is modified by the insolation angle, the angle at which the radiation strikes the surface, and by shading. Temperature declines with increased altitude at the dry adiabatic lapse rate of approximately 1 °C per 100 m, so that permafrost is most extensive at higher altitudes. Broadly, the lower limit of mountain permafrost in Norway is around the $-3\,°C$ to $-4\,°C$ isotherm and, in Iceland, about $-3\,°C$. In Scandinavia the lower limit of permafrost rises rapidly from east to west

Table 7.3 Potential hazards related to permafrost degradation (from Harris et al., 2001)

Slope class (degrees)	Bedrock			Sediment	
	Noncompetent lithologies (shales, soft mudstones, etc.)	Competent well-jointed lithologies	Competent massive lithologies	Fine-grained (silts, clays, some tills)	Coarse-grained (screes, gravels, sands)
>75	Rockfall	Rockfall	Occasional rock fall	—	—
30–74	Debris flows and landslides (including deep-seated failures)	Rockslides, debris flows	—	Debris flows	Debris flows
15–29	Landslides, thaw subsidence	Rockslides	—	Landslide/mudflow	Accelerated permafrost creep (rock glaciers)
<15	Thaw settlement	—	—	Thaw subsidence, solifluction, mudslides on steeper slopes	Accelerated permafrost creep
0	Thaw settlement	—	—	Thaw settlement	—

towards the Atlantic coast under the influence of oceanic warming, and from north to south. In Iceland there is also a clear north to south rise in the altitude of the permafrost limit in response to maritime conditions and increased snow cover. Locally, the properties of the surface and subsurface, including vegetation, snow cover (an effective insulator above 50 cm thickness), rock texture, and air and water movement, significantly modify the basic temperature signal and therefore the detailed pattern of permafrost distribution.

Steep, high relief not only increases shear stress, but also imposes strong aspect and shadow effects on an already heterogeneous land surface of exposed bedrock, bouldery talus slopes and moraines, gravelly river beds and the fine sediments in the basins of former lakes (Gruber, Hoelzle and Haeberli, 2004), each with different properties of thermal conductivity and water (or ice) retention. In steep terrain, large rockfalls, landslides and debris flows are the main types of hazard associated with mountain permafrost, but solifluction, permafrost creep and subsidence, though less dramatic, can also destabilize structures, especially where ground ice content is high (Table 7.3).

Rock falls are a characteristic hazard of high mountains, especially where seasonal freezing or permafrost occurs. Traditionally rock falls have been attributed to volumetric expansion (9%) during phase change, but ice segregation (the progressive growth of ice lenses in jointed rocks as unfrozen pore water migrates to a freezing front) may be even more important as it does not require high rock saturation but may receive water from the melting of the active layer (Harris *et al.*, 2009). Anyone who has spent time in the vicinity of a high alpine rock face will almost certainly have been aware of the frequent clatter of rock fragments onto the scree below, the rate of debris production depending on the frequency of the freeze–thaw cycle, snow cover and moisture availability (Sass, 2005a, b; quoted in Harris *et al.*, 2009). Larger-scale boulder-sized falls are likely to be seasonal or episodic in frequency, and tend to occur in summer during seasonal thaw or refreezing (Stoffel *et al.*, 2005; quoted in Harris *et al.*, 2009). This type of rockfall activity, while potentially hazardous (rock climbers have been seriously injured and even killed by such events) does not constitute a threat to large numbers of people or extensive property. What *has* stimulated considerable recent research is the perceived potential for the release of much larger rock masses, either as falls or slides, as climate change impacts on mountain permafrost. Noetzli, Hoelzle and Haeberli (2003) analysed 20 major twentieth century alpine rockfall events and concluded that as many as 18 could have a permafrost-related cause, several originating in potentially sensitive warm permafrost areas. It is possible that other factors contributed – loss of slope stability through glacier retreat or bedrock hydrology, for instance – but the greatly increased rockfall activity of 2003 in the Alps, coinciding with the hottest June, July and August on record and a much deeper active layer, implicates permafrost as a major contributory factor. Given that a large proportion of the permafrost in high, steep mountains is in bedrock slopes, ice-bonded joints and faults are likely to be a key factor in rock face stability (Haeberli and Gruber, 2008). The detachment surfaces of several rockfalls, such as the Matterhorn Lion ridge (Figure 7.7), revealed the presence of ice that had presumably helped to secure the rock in place. Under a warming climate regime permafrost ice is *the* factor that is likely to respond most rapidly to climate change. Gruber and Haeberli (2007, p. 144) listed six factors that in their view underline the importance of permafrost in bedrock destabilization. These are:

(a) the high proportion of rock fall events originating in permafrost areas;
(b) ice on fresh detachment surfaces;
(c) presence of wide ice-filled fissures in bedrock;
(d) the potential temperature-dependent loss of stability in permafrost;
(e) recent research demonstrating physical processes that actively widen frozen rock joints;
(f) observed warming in both atmosphere and rocks.

Unlike Arctic regions, already discussed above, European mountain permafrost terrain does not directly support large, permanent settlements (Harris *et al.*, 2009, Section 11.1). However, this does not mean that mountain permafrost does not constitute a hazard, as densely populated valleys and their transport networks are potentially vulnerable to the types of processes associated with steep permafrost slopes-rockfall, landslide, debris flow – that are capable of travelling considerable distances in high relief terrain. Tourist infrastructure, hydroelectric power facilities and communication networks, of high economic and social significance, are often located directly on permafrost. To some extent, mitigation responses to these two hazard contexts, permafrost and nonpermafrost areas, will be different. The lower valley areas require careful land management solutions, beginning with the production of up-to-date hazard zonation maps, and appropriate engineering solutions, such as retention dams, to restrict or divert the movement of rock and debris away from vulnerable areas. Within the permafrost area itself, these mitigation strategies are obviously also applicable, but the key requirement here is to maintain the thermal stability of the permafrost in order to reduce the risk of structure or slope failure. In lowland permafrost areas, passive cooling systems (thermosyphons, thermoprobes, air-duct cooling systems and gravity-driven air convection) are well established but, according to Harris *et al.* (2009, Section 11.5), require more testing in mountain environments. Construction processes often disturb the permafrost regime, but risks can be restricted by using insulation materials or chilled-air flushing rather than liquids, during drilling. Slow permafrost creep, as demonstrated by rock glaciers, or ground settlement can sometimes be accommodated using 'adaptable systems' that can be corrected as required. At Grächen, Switzerland, the midway station of the new chairlift is supported on three point bearings that can slide horizontally to maintain its optimal position (Phillips *et al.*, 2007).

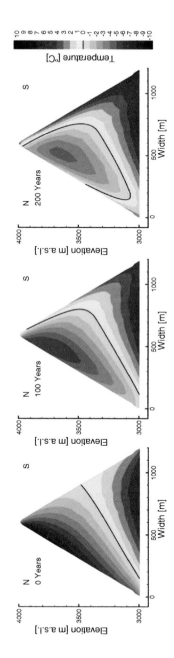

Figure 7.22 Modelled evolution of subsurface temperatures in a simplified ridge with a gradient of 60° for steady-state (0 years) and after time periods of 100 and 200 years; the warming at the surface was set to +3.5 °C for north slopes, +2.5 °C for south slopes and +3 °C for east and west slopes over a time period of 100 years and the black line corresponds to the 0 °C (Reproduced with permission from Harris, C. and 21 others (2009) Permafrost and climate in Europe: Monitoring and modelling thermal, geomorphological and technical responses. Earth-Science Reviews, 92, 117-171. Fig. 30. © Elsevier.)

As with all icy hazards, climate change will impact mountain permafrost, and is perhaps *the* major catalyst for the wealth of research currently being undertaken into the subject of mountain permafrost. The modelling of climate in the Swiss Alps for 2050, published in a report for the Swiss Federal Government (OcCC, 2007), indicates likely temperature increases of 2 °C for winter and 3 °C for summer by 2050. The boundary at the lower level of permafrost in the Swiss Alps has been rising since 1850 at a modelled, average, annual vertical rate of 1–2 m per year and is now close to 2 m per year. At first this was most marked on steep south-facing slopes of mountains like the Matterhorn but now the pattern is being repeated on slopes with a northern aspect (Figure 7.22). Many rockfalls and landslides were reported during the exceptionally warm summer of 2003, the year of the massive Matterhorn collapse and the fall on the Mont Blanc path. Similar, recent events appear to have been concentrated around the lower margin of the permafrost (Harris *et al.*, 2009) suggesting a link between the incidence of rockfalls and landslides and the location of decaying permafrost as its lower boundary moves gradually up-slope under the influence of climate change. Deep warming of permafrost in mountain peaks is leading to decreased slope stability and the increasing probability of major rockfalls, as well as to more complex impacts (Haeberli and Hohmann, 2008), such as rockfalls into lakes forming on the surface of melting glaciers which may produce severe down-valley flooding. In the densely populated European Alps it is not surprising that a great deal of recent research has been carried out in the hope of mitigating the risks of the mountain permafrost hazards. Consequently, much of the following discussion of mountain permafrost will be focussed on Europe, where analysis of permafrost hazard and risk is probably most advanced. For instance, a noteworthy European Union-sponsored project, the Permafrost and Climate in Europe (PACE) Project (Harris *et al.*, 2001) (see Box 7.1 for details of the EU PACE Project) grew out of awareness of the growing permafrost hazard problem, and has been instrumental in generating further studies in this field. Much of this research has been comprehensively reviewed in a paper written by 22 authors and containing around 500 references (Harris *et al.*, 2009) and this will serve as an important source of reference for the rest of this chapter.

Many miles of chair-lift and cable-car infrastructure are supported by structures based in permafrost. Numerous mountaineering huts and other tourist facilities have their foundations in permafrost. As long as the permafrost remains stable there is little cause for concern and the risk of collapse is negligible. However, in the Alps as elsewhere, the mean annual air temperature trend is upwards, permafrost is warming and some is melting. It is often erroneously said that icy permafrost is the 'glue' that holds high mountains together. In fact the real function of permafrost ice in the rock joints of mountains is to exclude rainwater which might lubricate the joint and induce collapse in the form of a landslide or rockfall. However, complete melting may not be necessary for the rock to fail. As permafrost warms its shear strength decreases and it may begin to creep. In their engineering laboratory simulation of mountain permafrost, Davies, Hamz and Harris, (2001) showed that rock walls were most susceptible to failure at a temperature of about $-0.5\,°C$, below the zero threshold, zero when creep was sufficiently strong to allow failure. Once the ice has completely melted the friction between the rock masses on each side of the joint *may* be sufficient to support the rock mass, though in practice the ingress of moisture to the joint plane may overcome the friction and release the slide or fall.

Clearly, the topography of alpine terrain presents substantial problems for those with responsibility for mountain permafrost hazard mitigation. The fact that mountain permafrost shows complicated patterns of distribution and is invisible from above makes accurate mapping a primary requirement of hazard mitigation, yet very difficult to achieve.

7.5.1 Mitigation strategies

Modelling and mapping As permafrost is largely invisible from the ground surface it is particularly difficult to map to a level of detail required in alpine-type terrain. Boreholes are a very effective method for determining the presence of permafrost but they are limited in number and extrapolation of conditions away from the borehole site is unreliable in variable mountain terrain. Consequently, Harris *et al.* (2009) suggested that the best approach to the problem of locating permafrost is to model distribution on the basis of a thorough understanding of processes. A range of techniques has been developed, especially since the PACE initiative (Harris *et al.*, 2001; Riseborough *et al.*, 2008), to tackle mapping at various scales. Two broad types of modelling are currently employed to predict permafrost distribution (Harris *et al.* (2009). Relatively simple empirical-statistical models relate documented permafrost occurrences to easily-measured topographic factors (altitude, slope and aspect, mean air temperature, solar radiation). More complex process-oriented models require more detailed understanding of energy fluxes between the atmosphere and the permafrost and a large amount of precisely measured or computed data. The PACE project developed a distributed energy-balance model linking ground snow and atmosphere interactions, but snow cover estimates and links between coarse debris and the atmosphere require greater accuracy for the model to be effective. So, while statistical-empirical models, based on (a) the bottom temperature of snow and topo-climatic factors such as altitude, (b) potential incoming radiation and (c) potential topographic wetness, do not explain complex heat transfer processes, they do provide an *easily* obtainable map of permafrost distribution which has some applied value in focusing the attention of hazard managers and permafrost engineers onto the most likely hazardous areas.

In Norway and other Nordic countries including Iceland, a regional map of permafrost distribution related broadly to Mean Annual Air Temperature (MAAT) has been produced (Harris *et al.*, 2009). This map provides a useful

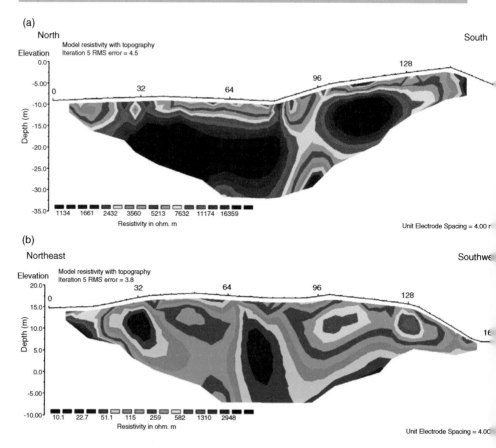

Figure 7.23 Electrical resistivity tomograms through contrasting pingo types in Svalbard: (a) Innerhytte pingo, Adventdalen, above the marine limit; (b) Longyear pingo, Adventdalen, below the marine limit (from Harris et al., 2009, Fig. 17)

first approximation for many parts of Norway where the high ground is more plateau-like and topography is less influential in controlling details of permafrost distribution. In Switzerland, a regional map of permafrost distribution, produced by the Swiss Federal Office for the Environment is based on a digital elevation model, ground surface characteristics (coarse debris, bedrock and glaciers and water) and a permafrost index related to topographic parameters. Again, it provides a useful *estimate* of permafrost distribution, but here complex topographical factors mean that there is still considerable uncertainty about its location at the local level.

Three-dimensional geophysical methods (see Harris et al., 2009; Hauck and Vonder Mühll (2003); Hauck and Kneisel (2008) and Kneisel et al. (2008) for more details) offer cost-effective solutions to this problem. Parameters such as electrical resistivity, dielectric permittivity and seismic compressional- and shear-wave velocities are geophysical properties that vary essentially with the

7.5 MOUNTAIN PERMAFROST HAZARDS

Box 7.1 The PACE (Permafrost and Climate in Europe) Project (Harris et al., 2001, 2009)

The PACE Project, subtitled 'climate change, mountain permafrost degradation and geotechnical hazard' was an EU-sponsored, multinational (Norway, Sweden, Switzerland, Germany, Spain, Italy and the UK), three-year research project initiated in December 1997. It was the 'first coordinated European programme of mountain permafrost monitoring and measurement' (Harris *et al.*, 2001, p. 4). The core aim of the project was to measure ground temperature and monitor permafrost change. The basis of the PACE project is a series of seven 100 m deep boreholes, extending along a north–south transect from the polar latitudes of Svalbard (78°N) to Mediterranean southern Spain (37°N) (Figure 7.24). Except for the Spanish site, which was shown not to contain permafrost, each borehole contains a thermistor string to record temperatures at 30 different depths down to 100 m. Measurements are taken at least once every 24 hours. At each site 'a meteorological station...records wind speed and direction, air temperature and relative humidity, net radiation and snow height, for use in energy flux studies' (Harris *et al.*, 2001, p. 8). Alongside this basic monitoring network, a series of other work packages covering 'geophysical surveys (Vonder Mühll *et al.*, 2001; Hauck *et al.*, 2001), microclimate investigations (Hoelzle *et al.*, 2001; Gruber and Hoelzle, 2001), numerical modelling of permafrost distribution (Etzelmüller *et al.*, 2001), and physical modelling of permafrost-related slope instability' (Harris *et al.*, 2001; Davies, Hamza and Harris, 2001) are being undertaken to 'improve assessment of potential permafrost hazards in the context of land-use planning and geotechnical engineering' (Harris *et al.*, 2001, p. 4). The PACE21 (Permafrost and climate in the twentyfirst century) programme (2003–2006), supported by the European Science Foundation was a follow-up to the original PACE project. The objectives of PACE21 were to coordinate and integrate the European data collection and regional synthesis on the impact of climate change in permafrost regions, the provision of a European forum for knowledge transfer and information dissemination to the larger international permafrost community and establishment of an expert group capable of providing clear, accurate and unbiased information to educationalists, agencies, the press and the public. An extensive review of the PACE project and subsequent research was published in 2009 in the journal *Earth-Science Reviews* (Harris *et al.*, 2009), and is an essential source for this permafrost hazard chapter.

phase change of water between a frozen or nonfrozen state (Harris *et al.*, 2009, Fig. 17; Figure 7.23). These authors suggest that Electrical Resistivity Tomography (ERT) is most effective for mapping ice content, the most critical

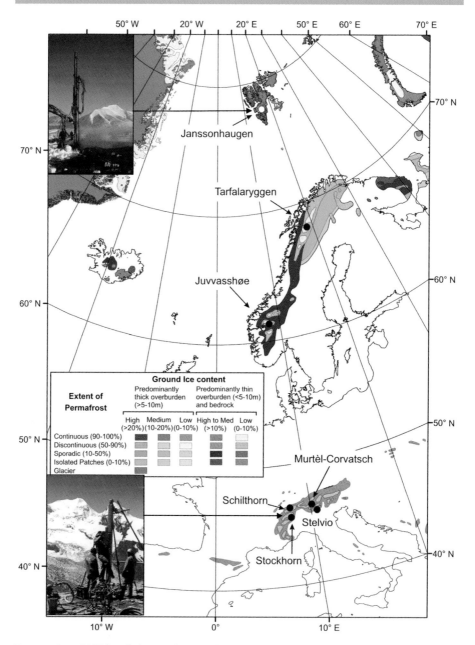

Figure 7.24 PACE boreholes on the IPA Circum-Polar Map of Permafrost (Brown *et al.*, 1997) showing the distribution of permafrost in the European sector (from Harris *et al.*, 2009)

hazard factor in permafrost, while the active layer is best measured using Ground Penetrating Radar (GPR).

Monitoring Having located the permafrost, at least to some useful degree of accuracy, the next step in mitigation is to monitor its behaviour in order to recognize and predict any change in risk, especially an increase. An example of a monitoring system that has recently been established is PERMOS (Permafrost Monitoring in Switzerland; Vonder Mühll, Nötzli and Roer, 2008). Potential changes in the hazardousness of the alpine permafrost regime, due to both climate change (temperature and snowfall) and increasing population and tourist pressures, have persuaded the Swiss authorities to formalize a long-term system of permafrost monitoring in Switzerland in an effort to reduce the risk of serious impacts resulting from avoidable events. Investigations began with the drilling of a borehole through the Murtèl–Corvatch rock glacier in 1987 (Vonder Mühll and Haeberli, 1990). The PACE Project (see Box 7.1 for details) extended the borehole network between 1998 and 2001 and a more formal, government-sponsored pilot phase ran from 2000 to 2006 (Vonder Mühll *et al.*, 2001, 2004). The PERMOS network currently includes 16 drill sites with 27 boreholes, three GST (ground surface temperature) sites and five kinematic sites, recording surface displacement (creep) in rock glaciers (PERMOS, 2009). Aerial photographs, taken regularly at 12 sites, also allow assessment of movement. At the drill sites sensors in boreholes record temperatures down to 100 m, while equipment measures ground surface and air temperatures and records snow thickness. In addition to these temperature, snow and displacement readings that imply permafrost presence, the PERMOS network of monitoring sites now have long-term geophysical monitoring facilities, such as Electrical Resistivity Tomography (ERT) to record the presence of ground ice and unfrozen water. It is likely that this type of monitoring programme will be more widely used, especially where warming of sensitive ice-rich permafrost sites may initiate debris flows or ground settlement beneath structures.

PERMOS has recently concluded its implementation phase (2007–2010) during which it was evaluated and became part of federal monitoring structures in Switzerland. PERMOS is coordinated by the PERMOS Office and supervised by the Cryospheric Commission (CC) of the Swiss Academy of Sciences (SCNAT). In Switzerland PERMOS complements the Glacier Monitoring Network, and at a larger scale forms part of the Global Terrestrial Network for Permafrost (GTN-P), previously discussed in relation to lowland permafrost. It is clear that, at least in Europe, sophisticated modelling, mapping and monitoring are increasingly being developed and put into operation as authorities become more aware of the threats to permafrost-based structures and communication systems posed by a warming climate.

However, once the location and state of the permafrost is known, responsibility then passes to the engineers to find appropriate methods of ensuring that mountain infrastructure is capable of withstanding the stresses put upon it by the behaviour of permafrost.

Engineering Essentially, the engineer's function is to assess ground conditions to determine the thermal and geotechnical properties of the permafrost soil, and then find technical solutions capable of maintaining permafrost stability in the face of the current challenges of climate change and population pressure. Warming permafrost and melting ground ice are particular challenges, and maintaining thermal stability of the permafrost is the first priority. Drilling with the aid of chilled-air flushing rather than liquids will minimize thermal disturbance (Thalparpan, 2000, quoted in Harris *et al.*, 2009). Passive cooling systems, such as those discussed in the Lowland Permafrost section are also being used, although Harris *et al.* (2009, p. 157) maintain that 'their efficiency in alpine environments is still to be tested'. Harris *et al.* (2009, p156) also suggest that 'in contrast to other [nonpermafrost] regions, the design life of a structure built in a permafrost environment should be planned to be 30 to 50 years, rather than 100 years'. This recognizes the fact that engineering properties associated with the current state of mountain permafrost are unlikely to apply in the relatively near future.

In the special case of permafrost, with its high sensitivity to change as the $0\,°C$ threshold is approached, today's technical solutions must allow for modification as circumstances change. The difficulty of predicting future conditions precisely will require sensitive structures to be closely monitored for thermal change and structural deformation on a continuous or regular basis. Geotechnical risk assessment, based on geothermal modelling and geomorphological process studies, will need to account for climate change and, as already mentioned, must be applied not only to permafrost-based structures but to buildings and infrastructure on lower ground below known permafrost sites.

Regulation Etzelmüller and Frauenfelder (2009) believe that, '[p]ermafrost *in mountains* is a [comparatively] new scientific topic'. Even as recently as 1990 'no systematic design recommendations or regulations concerning environmental protection in mountain permafrost exist[ed]' in Switzerland (Haeberli, 1992, p. 112), and the same undoubtedly applied to many other mountain regions, even though permafrost has been impacting on people (inhabitants and tourists) and structures (e.g. roads, railways, cable-cars, chair-lifts, HEP installations) for many years. In the European Alps many old mountain huts, constructed for convenience, or through lack of alternative sites, on moraines rather than bedrock, were forced to close because of destabilization caused by settlement as ground ice conditions changed. Until recently, the presence of potential permafrost problems has often only become apparent during excavation for foundations or a few years after construction has been completed. However, judging by the amount of academic research and technical investigations that have been published during the last two decades (see for instance references in Harris *et al.*, 2009), and the establishment of PERMOS, Haeberli's (1992) plea for better communication and more collaboration between Swiss scientists and technicians seems to have been heeded. Guidelines for climate change-affected permafrost engineering have been published by the Canadian Panel on Energy Research and Development (Hivon *et al.*, 1998). Instanes (2005; see Chapter 16 of the ACIA) has also made recommendations for construction in Arctic permafrost under the influence of climate change.

7.6 Summary

Permafrost is very extensive and yet, for a number of reasons, its hazardousness is often difficult to determine. It is largely invisible, subject to abrupt change across the $0°C$ threshold, and becomes increasingly sensitive as climate warms. Most permafrost occurs in lowland regions in the North American and Eurasian continents but a significant quantity is also located in mountainous (alpine) areas. This distinction is useful in terms of the different processes responsible for permafrost hazards. A great deal is known about the permafrost environmental system but recent climate change has provided a strong stimulus for research into methods of recording the presence of permafrost and finding ways to mitigate its hazards. Spaceborne remote sensing and geophysical techniques have significantly enhanced knowledge of permafrost distribution. A wide range of technical solutions have been applied to the problem. In the Arctic lowlands, very serious impacts are being experienced in northern Russia and Siberia, largely the consequence of inadequate initial design and subsequent mitigation during a former political regime. Europe appears to be at the forefront of work in mountainous environments, which is not surprising given the high density of the resident and tourist population potentially vulnerable to permafrost hazards. Europe's expertise is urgently required in other, less-developed mountain ranges of the world.

8
Snow Avalanches

8.1 Introduction

At 08.01 on 23 February 1999 the Austrian village of Galtur was struck by a powerful avalanche and 31 people died (BBC NEWS, 1999a and b). These fatalities should not have happened. Although avalanches had occurred in the area, they had never reached the village in living memory. The village was zoned to ensure that the buildings and their inhabitants were beyond significant avalanche impact. So, what went wrong? Why was the population of Galtur taken so fatally by surprise? Essentially, local residents and regional hazard managers did not know as a much as they obviously needed to know about potential avalanches in the Galtur area. This was not a fault, just a fact. No one could remember an avalanche of this scale on this particular slope happening before, so their plans to combat the hazard were based on inadequate knowledge. The preceding weather conditions were exceptional. The snowpack (layers of snow) also seems to have responded unusually to the load put upon it by the weather system, by remaining in place until an exceptional quantity of snow had accumulated. Add all these factors together and it becomes clear that the event was unprecedented. Nothing on this scale had previously been recorded in this location, so there was no reason to expect it or anticipate it. Until records are long enough to include the maximum possible avalanche event at a location, unprecedented avalanche impacts will remain a possibility, if not a probability.

Since this tragic event, laboratory and field experiments have improved our understanding of avalanche dynamics: the structure of some avalanches is more complex than previously understood. Avalanche hazard managers have renewed their efforts to improve their knowledge in order to offer effective advice to vulnerable populations. Locally, in Galtur, various mitigation

Cold Region Hazards and Risks, First Edition.   Colin A. Whiteman.
© 2011 John Wiley & Sons, Ltd. Published 2011 by John Wiley & Sons, Ltd.

measures have been taken; new hard engineering including steel fences and an avalanche dam have been constructed and the hazard zones have been extended (Keiler *et al.*, 2006); but, is Galtur safe? Weather, landforms and relief, snowpack structure and the structure of the avalanches themselves are all complex systems which need to be understood fully before the correct decisions can be taken to reduce the annual toll of avalanche fatalities, between 150 and 250 worldwide. Data on the distribution of avalanche tracks, avalanche return periods, their flow dynamics and their run-out distances are all required for statistical analyses to support forecasting.

However, the avalanche hazard is not just about physical processes and physical impacts. The drama surrounding sudden avalanche events and the scale of fatalities gives them a high media profile. Avalanche hazard zoning and the closure of transport routes, skiing pistes and other land areas create social and economic tensions for political leaders and hazard managers. All these factors require consideration if the avalanche hazard is to be understood fully and its impacts minimized.

8.2 Definition, classification and motion

Avalanches can be generally defined as:

> 'the sudden and rapid mass movement of snow, soil and rock down a slope under the influence of gravity, caused by instability of the snow cover (snowpack), induced by one or more of a range of factors including meteorological conditions, terrain, tectonism, snow structure and human activity' (compiled from Embleton and Thornes, 1979; Cooke and Doornkamp, 1990; Goudie, 1990).

However, a great deal of variety is subsumed within this general definition and avalanches have been further classified as primary (small dry surface runs of snow during snowfall) or secondary (after snowfall and resulting from several possible causes once the 'factor of safety' threshold has been exceeded). Even further classification is possible according to:

- the triggering action (natural, human),
- the type of rupture (slab or loose snow),
- the position of the sliding surface (within or at the base of the snowpack),
- the humidity (wet and dry).

All these different criteria for classifying avalanches have a bearing on their initiation, movement, scale of impact and the ease with which their potential impacts can be mitigated (McClung and Schaerer, 1993). Increasingly, as winter sports enthusiasts ('recreationalists') extend their scope into even more remote off-piste areas, humans are becoming the dominant trigger of avalanches over natural causes. Seismic shocks and the natural weight of

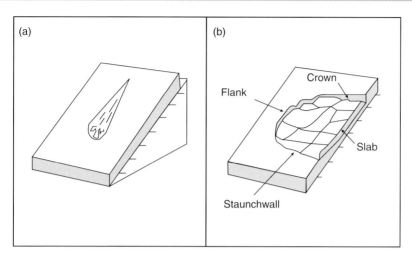

Figure 8.1 Types of avalanche rupture: (a) loose snow, (b) slab avalanche (source: Reynolds, 1992)

snow still initiate avalanches, but data on fatalities (see impacts below) suggest that many avalanches are now initiated in the start zone by recreationalists.

Loose snow avalanches (Figure 8.1) normally begin at a *point* on the slope and, although they gather more snow down slope, the volume is less and the run-out distance shorter than larger slab avalanches which are laterally more extensive. Usually slab avalanches are the most destructive as they involve large layers of snow breaking away from a tension crack across the slope (Figure 8.1). The volume of snow in a slab avalanche depends on the area of the slab and its thickness. The base of the slab is usually a weak layer within the snowpack, or the snow–bedrock boundary. In cold continental areas the slab base is often located at structural weaknesses in old snow. In warmer areas, with oceanic influences, the slab is more likely to consist just of new snow from the most recent big storm, together with wind-blown snow from elsewhere on the mountain, but deeper slabs occur if the snowpack is wetted by thaw or rain (Mears, 1992). The moisture content of the snow will influence its movement: dry, powdery snow is responsible for the classic, billowing avalanches, while less cohesive wet slabs can often deform into a feature more like a debris flow (Mears, 1992).

Sliding dry-snow slabs disintegrate rapidly into fragments and airborne powder and the traditional view was that dry powder avalanches, usually concealed a dense-flow avalanche (DFA) below the powder-snow avalanche (PSA) (Gauer *et al.*, 2008). However, Gauer *et al.* support the view of Schaerer and Salway (1980) that avalanches are often more complex 'mixed' structures with the powder obscuring other layers of debris saltating (bouncing), rolling or sliding closer to the ground. In effect the avalanche is a 'sandwich' with the mobile, fluidized head supplying most of the material for the suspension layer and moving faster than the dense core (Figures 8.2 and 8.3). This is an

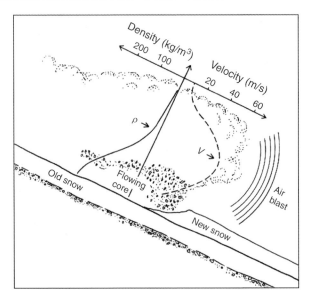

Figure 8.2 Probable avalanche velocity and density structure (Reprinted from Mears, A.I. (1992) Snow-Avalanche Hazard Analysis for Land-Use Planning and Engineering. Colorado Geological Survey Bulletin 49. 55 pp. Fig. 6. Courtesy of the Colarado Geological Survey.)

important finding because it has implications for avalanche dynamics and consequently the design of avalanche defences and avalanche hazard zoning. Actual measurement is difficult but the velocities of some avalanches have been measured by radar and others can be estimated from the amount of

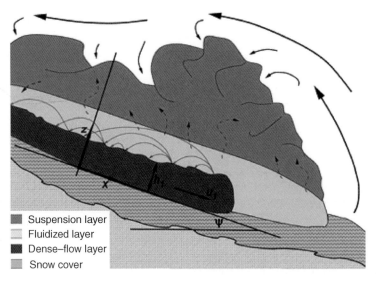

Figure 8.3 Structure of a dry-snow avalanche showing the dense core, the fluidized (saltation) layer and the powder cloud (source: Gauer *et al.*, 2008)

Table 8.1 Typical avalanche maximum velocity estimates (Reprinted from Mears, A.I. (1992) Snow-Avalanche Hazard Analysis for Land-Use Planning and Engineering. Colorado Geological Survey Bulletin 49. 55 pp. Tables 1 and 2 compiled. Courtesy of the Colorado Geological Survey.

Vertical fall (m)	Velocity range (m/s)	
	dry-snow	wet-snow
100–200	20–35	10–20
200–500	35–55	15–30
500–1000	55–70	20–35

damage and the calculation of avalanche dynamics. Velocity is related to the amount of vertical fall (Table 8.1) and is acquired quickly after initiation of the avalanche. This means that even small avalanches can be destructive. One other component of the avalanche event must be considered and that is the air blast which is produced in front of the solid components of the avalanche as the air is put under pressure by the moving mass. Air blast is usually most pronounced where part of the avalanche track is vertical and the falling mass suddenly compresses the air above the ground at the bottom of the fall. The impacts of these different avalanche components will be discussed later. Before that, it is important to understand the key factors that contribute to avalanche formation.

8.3 Factors promoting avalanches

The Avalanche Handbook by McClung and Schaerer (1993) is a very well-illustrated source of information on factors promoting avalanches and has been drawn on extensively for this section. Essentially, avalanches occur when stress on the snowpack exceeds its strength; that is when the *stability of the snowpack* falls below a 'factor of safety' of one. This balance of forces is influenced by the nature of the *terrain*, particularly its steepness and roughness but, of course, nothing will happen until the snow is delivered to the slope and this depends on *meteorological conditions*.

8.3.1 Meteorology

Rapidly fluctuating weather is typical in mountain terrains, especially those in areas of cyclonic, frontal weather systems, and can produce very complex snowpacks with highly variable stability. Meteorological conditions determine how much snow falls, where it falls, and influence where it accumulates. Weather also affects the snow *after* it has fallen by changing its temperature, pressure, moisture content and hence crystal structure and snowpack stability. First of all the atmosphere needs to absorb moisture. The water vapour content of air increases with temperature, so an air mass located over a relatively warm

Table 8.2 Typical properties of the main systems of vertical air motion (source: McClung and Schaerer, 1993, Table 2.2)

Type	Vertical wind speed	Precipitation rate (mm/h)	Duration of precipitation	Horizontal scale (km)
Cyclonic	~1 to 10 cm/s	Up to 2	Tens of hours to several days	1,000
Frontal	~1 to 20 cm/s	~1 to 10	Up to tens of hours	100 to 1000
Orographic	~1 to 200 cm/s	~1 to 5	Up to tens of hours	10 to 100
Convective	~1 to 1000 cm/s	~1 to 30	Minutes to hours	0.1 to 10

ocean, such as the eastern north Atlantic with its warm ocean current, is likely to be a good source of precipitation. Having acquired the moisture, the air needs to be uplifted to a level where its temperature falls sufficiently to induce condensation around ice crystal nuclei which combine into snow flakes.

There are a number of reasons why air is forced to rise: (a) cyclonic convergence in a low pressure system; (b) frontal lifting as a warm air mass rises over a cold air mass; (c) orographic (relief) lifting as air is forced upwards over mountains and (d) convection (thermally induced motion) (McClung and Schaerer, 1993; Table 8.2). Each mechanism tends to generate a different vertical wind speed, precipitation rate and duration of precipitation and distributes the snow over areas of vastly different size. For instance, in cyclonic systems snow falls gradually, perhaps over several days and may cover a large area. In contrast, convective storms are very intense but they are localized and depend on temperature for power. Consequently they tend to contribute little to the avalanche hazard. The dominant methods are frontal lifting (30%) and, orographic lifting (50%), which combine frequency of occurrence with relatively rapid precipitation rates, longish duration and widespread distribution.

In Switzerland, Latenser and Schneebeli (1996) analysed 100 years of interactions between climate and avalanches and recognized a number of distinctive weather patterns ('Grosswetterlagen') related to avalanche frequency. Their data indicates that their weather pattern number 8, 'Nordwestlage, zyklonal' (a northwesterly, cyclonic airflow) is the most influential cause of large avalanches, especially in the eastern and southern areas of the country, although only 15% of the events produce significant avalanches. It was a cyclonic system originating over the Atlantic Ocean and travelling southeastwards towards Austria that brought the snow that crashed onto Galtur, just across the Switzerland–Austria border in western Austria, in 1999. Similar circumstances caused devastating avalanches in Iceland in 1995. Here, a low-pressure area centred to the southeast of Iceland drew in a strong stream of north to northwesterly polar maritime air which delivered large quantities of snow. This triggered a major avalanche cycle in the northwest peninsula of Iceland which caused 34 fatalities in the two villages of Flateyri and Súðavík and stimulated a review of avalanche activity by the Iceland

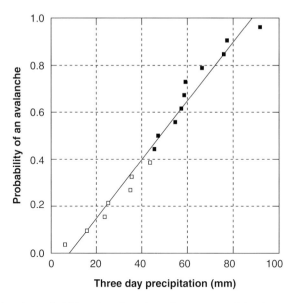

Figure 8.4 Avalanche probability related to three-day snow precipitation at a Norwegian site (source: McClung and Schaerer, 1993, data from S. Bakkehöi)

Meteorological Office (Magnússon, no date). In the northern and western areas of Switzerland, several different weather patterns are responsible for large avalanches. In these areas, winds from both the northwest and the southwest have a 50% probability of causing large avalanches.

The quantity of snow is obviously critical to avalanche occurrence. Norwegian data supplied by S. Bakkehöi to McClung and Schaerer (1993; Figure 8.4) shows a clear, positive correlation between three-day (H_{72}) precipitation totals and avalanche probability. This relationship now forms part of the requirements for avalanche hazard map assessment in Switzerland (Bianchi Janetti *et al.*, 2008; Bocchiola *et al.*, 2008) and is often used elsewhere (e.g. in Italy, Bocchiola, Medagliani and Rosso, 2006). An analysis by Latenser and Schneebeli (1996) indicated that most large avalanche events are the result of heavy, intense snowfalls caused by a strong wind (>50 km/h) along the Alps at temperatures around −5 °C to 0 °C. A surface of smooth old snow with surface hoar or a melt-freeze icy crust, which has seen little skiing across the surface, enhances the likelihood of avalanches. However, the fact that not all large snowfalls result in significant avalanche activity, suggests the influence of additional snow-related factors such as snow settlement rate and internal snow structure, or external factors such as terrain roughness.

In addition to precipitation, other weather-related factors have a bearing on avalanche occurrence. Wind strength and direction influence the location and rate of snow accumulation. Stronger winds bring snow to mountain slopes more quickly. Snow is deposited preferentially on the sheltered lee side of mountain ridges where air decelerates (McClung and Schaerer, 1993, Fig. 2.15). During this process cornices (overhanging masses of snow) often build out dangerously over the steeper slopes. The combination of cornices collapsing onto thick accumulations of snow can be devastating.

The meteorological elements discussed so far, weather systems, precipitation and wind, relate to the delivery and accumulation of snow with the potential to produce avalanches immediately. However, as noted above, this does not necessarily happen if the snow is in a stable condition, depending on snow crystal shape, packing and crystal bonding. The initial shape of snow crystals depends on the temperature of the atmosphere. They may be columns, needles, plates or stellar crystals originally, but arrive on the ground as irregular particles if they have gone through varying growth conditions. Graupel and hail are crystals that have grown by the addition of supercooled water and ice pellets are frozen rain. Each of these types will pack and bond differently with different degrees of stability. Graupel, for example, tends not bond well with adjacent layers.

Whatever the initial state of the snow, post-depositional weather conditions are likely to cause modifications (metamorphism) to snow crystal shape or the addition of new crystals that alter the properties of the snow until an avalanche is triggered. Features described as surface hoar, depth hoar (facetted snow) and radiation recrystallization are potentially the most unstable that can develop in the snowpack (McClung and Schaerer, 1993). Surface hoar is the growth of often large, dendritic (branched) ice crystals, typically 1 mm to over 1 cm in length, on the surface of the snowpack when air carries excess water vapour. They form a weak layer which is very susceptible to shear and avalanche release if subsequently loaded with more snow or a moving object. A cold, clear night with more or less calm conditions just above the snow is ideal. Likely weather conditions might be the passing of a cold front following a damp overcast day.

Crystals conducive to avalanche formation can also form at depth, if there is a temperature gradient of at least 10 °C/m in the snowpack. This will produce a vapour pressure gradient which draws moisture down into the snowpack. Given sufficient pore space, angular grains with stepped and striated facets (flattish surfaces) may form rapidly, eventually producing large cup-shaped crystals referred to as depth hoar. This process is sometimes referred to as constructive metamorphism.

Radiation recrystallization is a term used to describe changes to crystal shape as a result of frequent changes in incoming solar radiation and outgoing long-wave terrestrial radiation, which warms and cools the upper few centimetres of the snowpack, especially on south-facing slopes. If the solar radiation is strong enough, a zone of melting may occur. If this then refreezes an ice crust forms under the weak layer of recrystallized grains, an ideal failure surface when loaded by a later snowfall. Melting may also be caused by the passage of a warm 'föhn' wind down the lee side of a mountain range. Even without these special conditions snow crystals are likely to metamorphose after they have landed as pressure, temperature and moisture conditions in the snowpack will not be the same as those in the atmosphere. The points of classic dendritic crystals, where vapour pressure is high, lose molecules into the surrounding air which are transferred to the central parts of the crystal making the crystals more rounded, a process termed destructive metamorphism. This can reduce internal cohesion and make the snow in general *less* stable and more prone to avalanching. Alternatively, if the points are broken off rather than removed by sublimation, this can make the crystals smaller so that the snow becomes denser and *more* cohesive.

8.3.2 Terrain

In the chapter on mountain permafrost, relief of the landscape was shown to be of great significance in influencing the occurrence of permafrost-related hazards. Terrain is equally involved in the generation of avalanches. Through the long history of avalanche investigations it has become clear that the slope angle of start zones is a critical factor (Table 8.3). Records show that slopes between 15° and 60°, but especially between 25° and 40°, are prone to avalanching. This is because snow cannot accumulate to a great depth on very steep slopes because of their *high* gravitational stress, and is unlikely to lose stability on shallow slopes for the opposite reason, *low* gravitational stress. However, at the very top of a slope, larger masses of wind-blown snow in the form of cornices can be responsible for triggering avalanches lower down the slope when they are dislodged.

Another key element of the terrain is its aspect – that is the direction towards which a slope faces. This can influence the amount of solar radiation reaching a slope. In the northern hemisphere, north-facing slopes are generally cold, while those facing south towards the sun are generally warm. Cold slopes can encourage the development of potentially unstable 'depth hoar' and 'surface hoar'. Snow stability tends to increase in the spring when temperatures rise and 'hoar' is less likely to form. Sunny slopes are warmer and more stable in winter

Table 8.3 Relationship between slope angle and avalanche initiation in the start zone (source: McClung and Schaerer, 1993, Table 5.1)

Slope angle	Avalanche activity
60°–90°	Avalanches rare; snow sloughs frquently in small amounts
30°–60°	Dry loose-snow avalanches
45°–55°	Frequent small-slab avalanches
35°–45°	Slab avalanches of all sizes
25°–35°	Infrequent but often large-slab avalanches; wet, loose-snow avalanches
10°–25°	Infrequent wet-snow avalanches and slush flows

but are likely to become less stable in late winter and spring as moisture due to melting increases.

It is worth mentioning trees in relation to terrain, as they are an important modifier of the land surface. Forests are good stabilizers of the snowpack on slopes providing they are dense and continuous. It has been estimated that 500 conifers per hectare on a gentle slope and 1000 per hectare on a steep slope are sufficient to inhibit normal avalanches (McClung and Schaerer, 1993). Trees also influence the incidence of radiation. They can moderate snow surface temperatures and limit the formation of surface hoar and facetted crystals. However, scattered trees have a limited effect on radiation and are more or less useless as a barrier. Even mature trees can be snapped off or uprooted by very large, rapidly-moving and dense avalanches, leaving a mass of toppled or broken trunks lying parallel to the flow path.

Given the extraordinary complexity of factors involved in avalanche initiation, and the fact that people still live and take their recreation within potential avalanche areas, it is not surprising that avalanches continue to inflict an annual toll of fatalities and damage on the population and its possessions. The next section will discuss these impacts, and the reasons they continue to occur.

8.4 Impacts of avalanches

Avalanches, especially the snow avalanches being discussed in this chapter, are the most common type of alpine, cold-climate hazard. Some small avalanches do no more than transfer snow from one part of a slope to another with no impact on people or their possessions. However, avalanches are at the top of the hazard chart for velocity and shortness of time from initiation to impact. This makes them potentially extremely dangerous because there is little time to react (see Box 8.1 for examples of avalanches that have had significant impacts on society). The largest and most destructive avalanches are capable of

---

Box 8.1   Examples of catastrophic avalanches

On 4 September 1618 the so-called Rodi avalanche buried the town of Plurs, Switzerland, with the loss of over 2427 lives.

The 1720 Galen avalanche claimed 88 lives.

The worst avalanche event in the UK is the Lewes, Sussex, avalanche of 27 December 1836 which killed eight people.

In 1867, 60 to 65 miners were killed at camps at Alta in the Wasatch Range, Utah, USA. In 1937, the first US avalanche observation and research centre was built here.

On the 1 March 1910 an avalanche swept a train from the track in Stevens Pass, Washington, USA, with the loss of 96 lives. Three days later on the

4 March, an avalanche buried a train in Rogers Pass, British Columbia, Canada with the loss of 62 lives. After more than 200 fatalities in Rogers Pass between 1885 and 1911, the authorities rerouted the railway line in a tunnel through the pass.

During World War I many thousands of soldiers are reported to have died in the Alps as a result of avalanches triggered, at least to some extent, by gun fire.

The winter of 1951–1952 has become known as the 'Winter of Terror' when around 649 avalanches were recorded during a three-month period in Austria, France, Switzerland, Italy and Germany in the European Alps which accounted for 265 fatalities. Austria was most severely hit losing thousands of acres of forest, several small villages and more than 100 lives. Switzerland lost 900 buildings and 92 lives.

In 1999, a large avalanche struck Montroc, France, killing 12 people in their chalets under 15 m of snow. The mayor of Chamonix was convicted of second-degree murder for not evacuating the area, but received a three-month suspended sentence. Apparently the avalanche zoning maps were incorrect (The Avalanche Review, 2003).

In 2000 a huge avalanche hit Val d'Isère killing 39 people.

On 8 February 2010 over 170 lives were lost when several avalanches struck a convoy of vehicles crossing Salang Pass, Afghanistan (CBC, 2010).

Source, except where indicated: Thinkquest (no date).

---

burying settlements and their inhabitants. Avalanches also bury and displace motor vehicles and trains and obstruct key transport routes. They destroy large swathes of forestry and smash buildings and other structures to a greater or lesser extent.

Basically, two types of avalanche danger can be distinguished: burial and damage resulting in death and/or loss. The majority of fatalities, 75% in a study by Boyd *et al.* (2009), are due to burial resulting in asphyxia, with trauma (impact) accounting for most of the other 25%. Damage (loss) also results from burial, due to the weight of the snow, especially wet snow, but most destruction is caused by the impact pressure of the flowing avalanche (Table 8.4).

Table 8.4 Impact pressures related to potential damage (source: McClung and Schaerer, 1993)

Impact pressure (kPa)	Potential damage
1	Break windows
5	Push in doors
30	Destroy wood-framed structures
100	Uproot mature spruce trees
1000	Move reinforced concrete structures

Table 8.5 Avalanche flow density (source: McClung and Schaerer, 1993)

Avalanche component	Flow Density (kg/m^3)
Air blast	1
Powder	10
Dry flowing	100–150
Wet flowing	150–300

8.4.1 Damage

In the traditional avalanche, impact pressure (I) is proportional to the product of flow density (ρ (= rho)) (Table 8.5) and the square of the velocity (v^2) (Table 8.1) such that, in the denser flowing avalanches (DFA), $I = \rho v^2$, and in the less dense powder snow avalanches (PSA), $I = 0.5 \rho v^2$. Obviously, the greater damage will be done by the two denser, flowing components of avalanches. The presence of the intermediate layer (Schaerer and Salway, 1980; Gauer et al., 2008) means that greater impacts can be expected in the larger avalanches possessing this flow component, as relatively dense material rises to a higher level.

The cost of damage is related to the amount of property involved, and its value. In the USA property losses have been calculated as US$31 200 peryear (CAIC, 2009) but there is considerable variability from year to year. In 2008, for example, losses were reported as US$1 125 000 (HVRI, 2010). In Canada, McClung and Schaerer (1993) reported property damage as about Can$0.5 m annually, but avalanche control costs were about Can$10 m per year and the cost of static defences some Can$2 m. In Austria, property damage in the two settlements of Galtür and Valzur in 1999 was estimated at €9.3 million (US$12.7million) (Heumader, 2000).

8.4.2 Fatalities

Distribution Avalanches happen wherever steep slopes with accumulations of snow occur. In spite of an increasing range and sophistication of mitigation measures (see below), snow avalanches continue to exact a mean annual *recorded* death toll of at least 150 worldwide (McClung and Schaerer, 1993). However, this obscures considerable annual variation (Figure 8.5), and it is difficult to know how many other fatalities are unrecorded in the more remote mountain areas outside North America and Europe. Although avalanche statistics are easy to obtain in Europe and North America, they are not always compiled accurately in the less-developed parts of the world. In these areas it is often only the larger events, those that attract media attention, that are recorded. For example, in February 2010 over 170

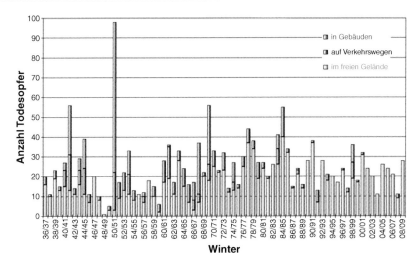

Figure 8.5 Annual avalanche fatalities in Switzerland in buildings (red), on roads etc. (black) and in back-country (blue) (Reproduced courtesy of Colorado Avalanche Information Center.)

people died when several avalanches were released onto a convoy of vehicles crossing the Salang Pass, Afghanistan (CBC, 2010). It is therefore difficult to be certain about the full impact of avalanches worldwide.

In Europe, the most vulnerable area for avalanches is the Alps (France, Austria, Switzerland and Italy) and in North America, the Rocky Mountains (USA and Canada). The Alps, in particular, are densely populated and attract large numbers of tourists. The USA and Canada are both wealthy countries where skiing and other winter sports are popular, and there is a substantial tourist market. In most of these countries the mean annual number of avalanche fatalities has been between about 20 and 30 during the last two decades of the twentieth century. In Canada the average is 12. In Norway, a small, nonalpine, Scandinavian country, it is about five (CAIC, 2009). In recent years, the average in Russia, has been 20 (Seliverstov *et al.*, 2008).

Changes through time The annual number of avalanche fatalities in the USA since the winter of 1950 (Figure 8.6) shows some interesting trends. From 1950 to the late 1960s the annual fatality count was around five. It then increased fairly rapidly to around 15, although there was considerable annual variation. This situation prevailed until the late 1980s when there was a four year dip in the numbers to around eight, but this was followed by another significant rise in annual fatalities to the present level between 25 and 30. The reasons for these changes are complex. It is likely that there was an increase in income during the 1960s following a period of austerity after World War 2. However, data on fatalities in the USA, related to location and activity (Figure 8.7), suggest that other factors also contributed to this late 1960s rise. At about the time that total fatalities increased more fatalities were occurring in back-country locations

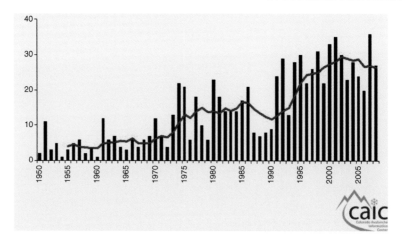

Figure 8.6 Avalanche fatalities in the USA, 1950–2008; red line is five-year running mean (source: CAIC, 2010)

and the hiker/climber fraternity also appeared to be more at risk. The second significant rise, at the beginning of the 1990s, coincided with a substantial rise in fatalities amongst snowmobilers, another group that seem to prefer to travel off-piste.

Swiss data beginning in the mid-1930s shows a similar, though less obvious, pattern of change in terms of total numbers (Figure 8.5). However, while the data suggest that US fatality numbers may have peaked early in the twentyfirst

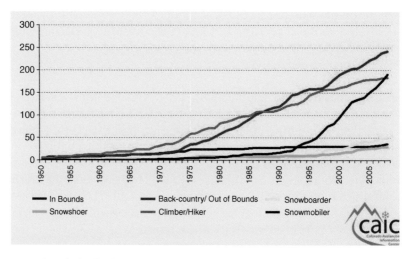

Figure 8.7 Cumulative fatalities by activity in the USA, 1950 to 2008/9 (Reproduced courtesy of Colorado Avalanche Information Center.)

century, the data from Switzerland suggest an earlier downturn in that country, in the 1980s. Interestingly, the Swiss data distinguish the locations where the avalanche accident occurred. Avalanches killed people in buildings during 12 of the years up to 1970. Since 1970 deaths of people in buildings has been recorded in only four years. Since the mid-1980s there has been a significant drop in the number of avalanche fatalities associated with roads and other lines of communication. There is also a hint that back-country travellers are being caught in avalanches less frequently. As there does not appear to be any reduction in the numbers of people visiting winter sports areas, these changes infer that constantly refined and improved mitigation measures, such as hazard forecasting, hazard zoning, hazard warnings, detection and alarm and the education of back-country travellers, have, over the years, succeeded in reducing the level of risk for different groups in some places. Fewer fatalities in buildings suggests that zoning has been more effective. Fewer people caught in vehicles on roads suggests that hazard warnings are more successful and route closures are being respected. Finally, the small drop in back-country fatalities may indicate that education is beginning to have a positive effect on risk awareness and decision-making, especially amongst 20 to 30 year olds who made up 40% of all US avalanche fatalities between 1950 and 2006 (Figure 8.8). Similar patterns are evident in Canadian avalanche statistics (Jamieson and Stethem, 2002; Figure 8.9). The exceptional record in these data for transportation in the period 1900–1909 must relate to the Roger's Pass train disaster of 1910 (see Box 8.1). Few if any transport- or resource industry-

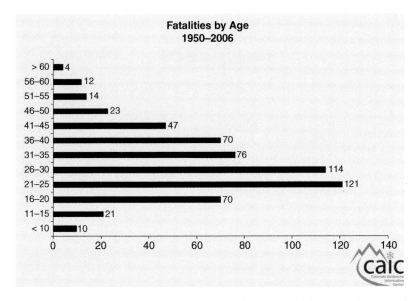

Figure 8.8 US avalanche fatalities by age (Reproduced courtesy of Colorado Avalanche Information Center.)

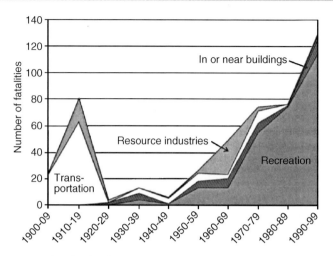

Figure 8.9 Snow avalanche fatalities in Canada by activity 1900–1999 (Reproduced with permission from Jamieson, B. and Stethem, C. (2002) Snow avalanche hazards and management in Canada: Challenges and progress. Natural Hazards, 26, 35–53. Fig. 1 © Springer.)

related (e.g. mining) fatalities have occurred recently which suggests improved mitigation in these sectors. The significant increase, as elsewhere, has been in recreational fatalities.

8.5 Mitigation methods

In view of the destructive and deadly potential of avalanches, a great deal of thought and effort has been put into finding methods for mitigating or even preventing avalanche impacts. The choice of method will depend on the level of risk and its mapped distribution. This requires a knowledge of the terrain, and estimates of avalanche frequency and magnitude, destructive potential (impact pressure) and run-out distance (Jamieson et al., 2002), all of which carry some degree of uncertainty.

Several different approaches to the problem can be recognized (Table 8.6). For example, it is possible to make a distinction between 'hard' and 'soft' techniques, that is between engineered solutions, such as retaining structures and deflection dams, and those which involve observation, analysis and forecasting. Another option is to avoid the problem by warning people or restricting access and movement in potentially vulnerable areas. For this to be effective, avalanche paths must be mapped and the land zoned to control building quantity, type and use, or to forbid all building in the most vulnerable areas. The combination of these methods which best fits a particular hazard situation is often a difficult decision, as it requires the careful analysis of many factors, not only those related to the practicalities of the potential avalanche and the safety of tourists, but also those concerned with the cultural, social and economic interests of the local community.

Table 8.6 Avalanche mitigation strategies and methods (source: McClung and Schaerer, 1993)

Information	Modification			Avoidance
	Terrain engineering	Building modification	Triggers	
Historical database	Support structures	Splitters	Artillery	Zoning
Mapping	Terraces	Reinforcement	Bombs	Detection and alarm
Snowpack stability evaluation	Deflectors		Detonators	Warnings
Run-out modelling	Arresters		Charges	Closure
Forecasting	Retarders		Gas exploders	Evacuation
Numerical avalanche prediction	Splitters		Test skiing	
Hazard rating	Dams		Snow loading	
Safety advice	Snow sheds			

8.5.1 Information

Mapping In the earlier section on factors promoting avalanche formation, terrain was considered to be a key element. Consequently an obvious first step in the process of avalanche mitigation is to indicate on a topographic base map where avalanches occur in an area and where they might be expected to occur. Mears (1992) and Jamieson *et al.* (2002) describe several types of evidence that can be used for avalanche hazard mapping (Table 8.7). Initially, mapping would have been done in the field by noting where avalanches actually run and combining this with observations of other types of evidence for avalanche movement such as vegetation trimlines. Subsequently, these field maps would have been supplemented by aerial photography. With the advent of remote sensing, digital elevation models (DEM) and GIS, it is becoming possible to refine mapping and extend it into more remote areas. This increases the avalanche database and should enable conclusions to be drawn from statistical analyses with more confidence.

The most obvious sign of the passage of an avalanche is a vegetation trimline, the sharp boundary between mature forest and the avalanche track (Figure 8.10). When avalanches of different sizes use the same track vegetation of different ages manages to survive for a time and this will be reflected in the maturity of the vegetation (McClung and Schaerer, 1993; Table 8.8). Frequent small avalanches in the middle of the track may inhibit growth of anything except grass and small, yielding shrubs. Larger, but less frequent events will allow the growth of taller shrubs and young saplings and there may be another stage of semi-mature trees before the mature woodland is reached at the side of the track. Tree ring analysis (dendrochronology) is useful in this situation as it can give an indication of the age of the vegetation in the different areas and

Table 8.7 Evidence and methods for avalanche hazard mapping (source: Jamieson et al., 2002)

Method	Description
Map and air photo interpretation	A good topographic map provides a base for information, evidence of landforms (start zones, tracks and run-out zones) and slope gradients; air photos will also show vegetation patterns and snow accumulation areas
Field study of terrain	Topographic details and start zones, tracks and run-out zones can be confirmed by surveying; subtle terrain features, terrain roughness, surficial deposits and vegetation patterns can be identified
Vegetation and dendrochronology	Vegetation types and patterns provide evidence of extent and timing (freqency) of avalanches; sharp boundaries between damaged and undamaged vegetation is called a trim line and tree rings can help to date large avalanche events (dendrochronology)
Oral and written records	This may include village archives, newspaper reports, interviews with residents and any other similar source of information
Weather and snow records	Recorded snow depths in start zones are related to weather associated with large avalanches
Surficial materials	Avalanches often transport rock, soil and vegetation which remains after snow has melted; vegetation can be carbon-dated, but the extent of coarse debris may be deposited some distance within the absolute maximum of the avalanche powder impact
Topographic-statistical models	Statistical methods can be used to determine extreme run-out distance: the Alpha-Beta model (Lied and Bakkehøi, 1980) and the run-out ratio model (McClung and Mears, 1991); these methods are based on the β point where the track slope angle first decreases to about $10°$
Dynamic models	Dynamic models aim to calculate velocity and stopping position in the run-out zone; the calculation includes terrain measurements (gradient, area, roughness, track shape, etc.) and material properties (boundary friction, viscosity, turbulence, etc.)

hence some idea of the return period of avalanches of different magnitude. Reaction wood, a partial ring formed in response to the bending of the tree by the passage of an avalanche, and scars caused by avalanche impact, are the best means of dating an event, although abrupt changes in growth patterns of trees newly exposed at the edge of an avalanche track may also be helpful. The debris left behind by an avalanche may provide another useful mapping clue, although it is unlikely to indicate clearly the maximum position reached by the largest avalanches, as the heavier, more obvious material will be deposited inside the margin, as soon as the avalanche reaches the less steep terrain of the run-out zone.

Figure 8.10 Avalanche tracks showing sharp boundaries with mature vegetation, Rogers Pass, BC, Canada (source: photographs, C. Whiteman, 2000)

Run-out distance Maximum run-out distance is a critical variable in assessing the probability of avalanche damage and planning land-use boundaries. Ideally, historical observation and records can be used to provide good evidence of maximum run-out distances. This may often be possible in a European setting like Switzerland, with its very long history of avalanche records, but in the USA and Canada with much shorter recording periods, or

Table 8.8 Vegetation as an indicator of avalanche frequency (source: Mears, 1992; McClung and Schaerer, 1993)

Frequency (years)	Vegetation clues
1–2	Alder and willow, shrubs, grass, bare patches; no trees higher than about 1 to 2 m
3–10	No large trees and no dead wood from large trees; some trees higher than 1 to 2 m
10–30	Dense growth of small trees; young trees of climax species (e.g. conifers); core data useful
30–100	Mature trees of pioneer species (e.g. nonconiferous) and young trees of climax species; core data useful
>100	Mature trees of climax species and debris completely decomposed; core data required

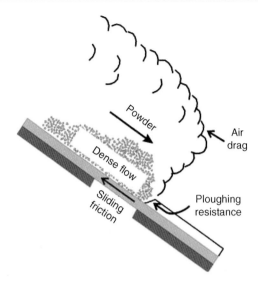

Figure 8.11 Avalanche flow and resistance factors (Reproduced with permission from Jamieson, B., Margreth, S. and Jones, A. (2008) Application and Limitations of Dynamic Models for Snow Avalanche Hazard Mapping. Proceedings International Snow Science Workshop 2008. http://www.ucalgary.ca/asarc/files/asarc/DynModelsAppLim_Issw08_Jamieson.pdf.)

in some Himalayan countries with no records at all from the more remote areas, it often may be necessary to resort to models (Mears, 1992). A dynamics model aims to calculate velocity, run-out distance and even flow depth, deposit depth and lateral extent of an avalanche, based on terrain measurements of gradient, cross section and roughness and avalanche material (snow) properties such as viscosity, turbulence and external boundary friction (Jamieson, Margreth and Jones, 2008; Figure 8.11). In this type of model the run-out position is defined as the point at which zero velocity is indicated by the model.

An alternative technique uses terrain variables (Jamieson *et al.*, 2002; Figure 8.12), in particular the point on the avalanche track where the slope

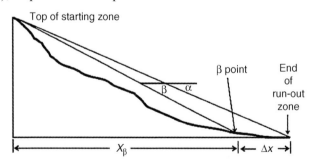

Figure 8.12 Measurements for topographic–statistical run-out models (Reproduced with permission from Jamieson, J.B., Stethem, C.J., Schaerer, P.J. and McClung D.M. (Eds) (2002) Land Managers Guide to Snow Avalanche Hazards in Canada. Canadian Avalanche Association, Revelstoke, BC, Canada. Fig. 4.2 © Springer.)

angle first decreases to 10°. Experience and mathematical analysis shows that large avalanches stop when the gradient falls to around 10° or less, a point that is easily found on the avalanche track. The angle between this point and the top of the start zone is labelled β (beta). The angle from the maximum run-out point to the top of the start zone is labelled α (alpha). It has been widely found that α is proportional to β and therefore α can be calculated and the maximum run-out position defined.

Return period Avalanche risk assessment relies partly on historical precedent. Realistic estimates of risk require knowledge of the frequency and magnitude of *past* avalanche events in terms of their average return period. This can be calculated from historical records (time series), if they exist, such that probability, $P = 1 - (1/T)^L$ where T = the return period and L = length of the observation period. Unfortunately, very long time series, recording avalanche occurrence over centuries rather than decades are scarce. Consequently, the probable 600-year Galtur avalanche event, for example, could not have been foreseen, nor the damaging series of avalanches, resulting in 15 fatalities in northwest Iceland in 1995 (The Iceland Reporter, 1995) because experience of such large avalanches simply did not exist. Both occurrences were a surprise to the authorities: significantly both events have stimulated substantial reviews of the avalanche hazard and its attendant risks in the two countries affected.

In Switzerland, with significant numbers of both avalanches and vulnerable people, some records do extend back for hundreds of years (Tufnell, 1984) and provide excellent time series of data on which to base a statistical analysis. Usually, records cover a much shorter period but most developed countries now maintain a database of avalanche events that can be used for this purpose (e.g. HVRI, 2010). Less developed countries and more remote regions have yet to establish such records, although population stress and the expansion of tourism will probably put pressure on them to do so. Modern remote sensing technology may facilitate progress in these areas (e.g. Bühler *et al.*, 2009; Prokop, 2008).

Snowpack stability The mitigation measures discussed so far relate to consideration of where avalanches *might* occur, how far they *might* extend and how frequently they *might* occur. Of more immediate concern is whether the snow, *now* on the mountainside, is in a stable or unstable condition. The assessment of snowpack stability is therefore a top priority for avalanche hazard managers. Assessment can be determined directly by putting stress on the snowpack to see whether it fails and, if so, how much stress is required. There are several traditional ways of achieving this, including test skiing across a short slope, use of explosives to generate shock stress, the Rutschblock (glide block) test, collapse test, tilt-board test, shear-frame test, the shovel shear test and fracture propagation (see McClung and Schaerer, 1993 for well-illustrated details). New digital snowpack penetrometers, such as the

SnowMicroPen (Schneebeli and Johnson, 1998) have been developed and tested (Pielmeier and Marshall, 2009) to speed up the process of information gathering in the field.

A complementary approach to direct stressing of the snowpack is to record the characteristics of the snowpack and then to make a judgement about the probability of its failure, mainly based on the presence of weak layers and bonds. Essential properties of the snowpack include overall thickness of the snowpack, its layering, the strength of the layers (related to hardness, grain shape and size and density) and the strength of bonds between them. The profile of layers and their properties are best observed and recorded by digging a pit and clearing a vertical face. Gentle brushing will reveal layers, as soft snow is more easily removed. Then, density, hardness, temperature, grain size, grain shape, bond strength and water content can all be recorded. Ideally the plot chosen for this investigation should be characteristic of as large an area as possible. Several countries (e.g. Switzerland, France, Canada and USA) produce stability rating systems for inclusion in public warnings.

Forecasting A knowledge of which areas are currently susceptible to failure is a substantial part of the mitigation process, but whether the slope actually fails will depend largely on future weather. This will determine how the snowpack metamorphoses and whether it is excessively loaded with more snow, either directly or as a result of wind blow. Today massive computers at central meteorological offices have transformed weather forecasting by integrating previous records with the current situation and extending the forecast period to several days to give more time to plan a response. Regional weather forecasts on the synoptic scale cover whole mountain ranges. At a more local scale this regional weather information is supplemented by that from local weather stations to give more detail applicable to individual valleys, ski resorts and transport networks. The avalanche hazard forecast itself will be made by adding snowpack stability information to the weather forecast. Conventional avalanche forecasting relies on experts to collate information on weather and snow-stability measurements and produce a forecast, in which personal experience plays a part. However, given the number of contributory factors which affect avalanches and the enormous amount of data that is being recorded, some on an hourly basis, mathematical analysis is now often employed to make a numerical avalanche prediction. Specific techniques are discriminant analysis and 'nearest neighbour' analysis. Discriminant analysis classifies avalanches into groups and asks: in which group do today's data fall? With 'nearest neighbour' analysis the question is: which sets of past data form a group or cluster around today's data (McClung and Schaerer, 1993). However, although forecasting is constantly being improved, there are many places where avalanche occurrence is (or has been) so common that more permanent ways of protecting the population and their property are required. This usually involves modifying the landscape or controlling the movement of avalanches.

8.5.2 Modification and control

The previous section was essentially concerned with combating the avalanche hazard through the acquisition of data ('soft' mitigation techniques) which could then be used to foresee the occurrence of an avalanche and warn people not to travel in its potential start zone or be in its path. Some of this information, especially maps of regular avalanche tracks and snowpack records, can also be used to locate natural and 'hard' engineered avalanche control and defence systems. These mitigation techniques rely on stopping avalanches happening, or reducing their impact to acceptable levels. Four different approaches will be considered (Table 8.6): terrain engineering, forest control, minimizing risk by design and avalanche triggering.

Terrain engineering Engineered solutions (see, for example, McClung and Schaerer, 1993, Chapter 9) to the avalanche hazard are often costly, unaesthetic and inflexible, but essential where avalanches are frequent, traffic volumes are high or essential structures are located. Structures are designed with several aims in mind: to prevent, deflect, protect, decelerate or stop avalanches. For example, continuous rows of support structures high on a mountainside are designed to withstand snow creep and impact from small avalanches and inhibit large avalanches, so they must be at least as high as the greatest expected snow depth.

Other structures are designed to deal with avalanches after they have started. For instance, long earthen embankments were built on the slopes above Flateyri, northwest Iceland, following a disastrous event in 1995, and these were successful in deflecting subsequent avalanches away from the village (Leah, no date; Figure 8.13). Ideally the initial deflection angle should be less than 20° (McClung and Schaerer, 1993) to reduce the force of the impact on the structure. Another form of deflector, specifically designed to keep snow off roads and railways, is the snow shed, an engineered tunnel over which the avalanche can pass on its way down the mountain. Snow sheds are expensive but essential to protect busy routes in avalanche-prone terrain, such as Roger's Pass, Canada. The surface railway through this dangerous pass was rebuilt in a long tunnel because the relatively slow speed and length of the trains made them more vulnerable to avalanche impact than the more flexible road vehicles. Dams (arresters) are expected to stop avalanches. In Austria, a 300 m long dam was built above the town of Galtur, to prevent avalanches reaching the village following the disaster in 1999. Small ridges of snow or earth are often piled up at roadsides, where avalanche tracks cross the route, to contain small avalanches. Retarders, usually mounds of earth or rock set out in offset rows across an avalanche track, often in the run-out zone, are designed to dissipate the energy of the avalanche and reduce its run-out distance. These structures are most effective against wet snow avalanches and are likely to have little effect against rapidly moving, billowing, dry snow avalanches.

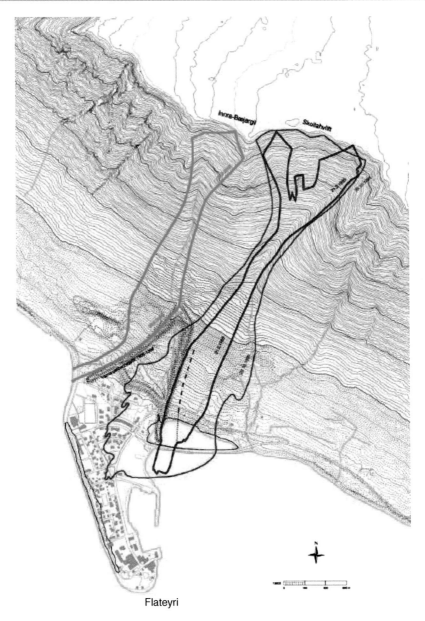

Flateyri

Figure 8.13 Flateyri avalanche dams: 1995 destructive avalanche in red; post dam (1998) avalanches in blue and green, successfully diverted from buildings (Reproduced from http://proceedings.esri.com/library/userconf/proc01/professional/papers/pap439/p439.htm © Tracey Leah.)

Forest control Forests provide a natural form of control over avalanches to some extent, but are obviously not sufficiently robust to maintain growth in the tracks of large avalanches. This is clear from many mountainsides that display easily-mapped avalanche tracks with little or no vegetation. However, there can be no doubt about the stabilizing influence over the snowpack, of a thick forest. The main hotel and information/exhibition centre in Roger's Pass is strategically placed at the foot of a well-forested slope. In some locations, where forest has been removed or reduced by fire, disease, acid precipitation or thoughtless logging, there have been avalanches where none had occurred before (McClung and Schaerer, 1993).

Design   There are cases where structures need to be or have been located in the paths of avalanches. Perhaps situations have changed and what was a safe position is now vulnerable to a new avalanche threat. In these circumstances it is advisable to reinforce or streamline a building so that it can either withstand anticipated avalanche pressures or divert the flow harmlessly around the structure. Sharply angled walls can divert the avalanche around the sides of the buildings. Streamlined concrete reinforcement will help other structures, such as towers and pylons, to withstand avalanche impact pressures.

Triggers and snow compaction   In the USA and Canada artillery is used to generate shock waves which release avalanches in unstable areas before excessively large snow volumes accumulate and while access can be controlled. In Roger's Pass, Canada, there are about 20 sites established where a 105 mm howitzer can be used for avalanche control. However, use of artillery is not accepted everywhere. It is not used in Japan, for instance, and considerable opposition was mounted by environmentalists when the Burlington Northern Santa Fe Railway company wished to use artillery in Glacier National Park (Montana, USA) to protect its 40 daily freight trains as an alternative to expensive new snow sheds (The Washington Post, 2006). According to the report those supporting the National Park argued that 'shelling in the park is aesthetically inappropriate and potentially harmful to wintering mountain goats, elk, deer, wolverines and endangered grizzly bears'. Artillery is also less easy to control in detail.

McClung and Schaerer (1993) discuss a variety of other means of triggering avalanches using explosives, including bombing from helicopters, the placement of detonators and charges to dislodge overhanging cornices and gas exploders. Nonexplosive methods for releasing controlled avalanches include test skiing and snow loading (using bulldozers to push snow from ridges onto unstable slopes). It may be possible, in the start zones of some smaller avalanche tracks, to release small avalanches and, at the same time, compact the snow by skiing or walking, but this is only feasible in small, critical areas.

8.5.3 Avoidance

The third approach to avalanche mitigation (Table 8.6) involves acceptance of the fact that avalanches are, in many situations, uncontrollable, although experience and analysis of data does provide some capacity for anticipating where and when they will occur. It is this ability to anticipate many avalanches, at least the tracks they follow, that provides both residents and tourists with the opportunity to avoid the avalanche hazard altogether. Two types of strategy are available. The long-term option is to make sure that buildings and infrastructure do not lie within known avalanche tracks and run-out areas or, at least, are not located within areas with unacceptable avalanche return periods or pressures. This is generally described as zoning or land-use restriction. The shorter-term options are to react rapidly in response to alarms, warnings and evacuation orders, and to respect the closure of routes and areas which are judged to be at high risk from avalanches.

Long-term option – avalanche hazard zoning Avalanche hazard zoning began in Switzerland, where 65% of the population live in valleys at risk from avalanches, after two catastrophic avalanche periods in January and February 1951, the so-called 'Year of Terror', when a total of 265 people were killed by avalanches in the Alps. The first avalanche hazard maps were produced for the settlements of Gadmen (1954) and Wengen (1960) in the Canton of Bern. Subsequently, guidelines for hazard zoning in Switzerland were set out formally (BFF/SLF, 1984) and were followed by guidelines for the calculation of impact pressures in dense flow avalanches (Salm, Burkard and Gubler, 1990). These two documents provide the basis for making avalanche hazard maps in Switzerland. Other countries have since developed similar schemes (e.g. Mears, 1992, for USA and Jamieson et al., 2002, for Canada). The advent of GIS and digital terrain models (DTMs) has facilitated the development of numerical simulation and has enabled zone mapping to be expanded considerably (Gruber and Haefner, 1995).

Appropriate zoning decisions are made on the basis of key variables relating to the magnitude (destructive potential) and frequency (return period) of the largest avalanches at a particular location and their maximum run-out extent. Magnitude is normally measured as impact pressure (Table 8.9). Run-out extent is generally measured using one of a number of complex avalanche-

Table 8.9 Avalanche hazard zoning criteria for Switzerland and Canada; I = impact pressure and T = return period (sources: BFF/SLF, 1984; McClung et al., 2002)

Zone colour	Swiss criteria	Canadian criteria
Red	$I > 30 \text{ kN/m}^2$ with $T < 300$ years	$I >$ or $= 30$ kPa and/or $T < 30$ years OR product of I and $1/T > 0.1$ for $T = 30–300$ years
Blue	$I < 30 \text{ kN/m}^2$ with T 30-300 years	$I >$ or $= 1$ kPa or product of I and $T < 0.1$ kPa with $T = 30–300$ years
Yellow	$I < 3 \text{ kN/m}^2$ with $T > 30$ years	No direct equivalence
White	No restriction	$I < 1$ kPa with $T > 30$ years OR $T > 300$ years

Note: 1 kPa (kiloPascal) = 1000 N/m² (Newtons per metre squared)

Table 8.10 Design periods related to land uses in Canada (source: Mears, 1992)

Land use	Avalanche design period (years)
Roads and railways*	<1
Ski trails*	<1
Electricity transmission lines	1–10
Telephone lines	1–10
Oil and gas pipelines	10–50
Vehicle parks	10–50
Ski-lift terminals*	10–50
Road and rail structures	50
Residential development, houses	100
Restaurants, schools, hospitals	100–300

* Avalanche control through closure and artificial release is usually used to reduce risk in these areas.

dynamics models such as those mentioned above. An estimate of the return period of avalanches of different magnitudes is made on the basis of known records at the site or at adjacent locations. Swiss authorities are fortunate in being able to consult 300-year old records in some cases. In North America the time period is rarely more than 100 years.

For practical purposes it is useful to consider a 'design avalanche' (Mears, 1992). This is an avalanche of such magnitude that it must be considered in land-use planning and facilities design. 'Design avalanches' are related to different 'design periods' (return periods) (Table 8.10) which are equated to different land uses on the basis of perceived risk. For instance, high-risk schools and hospitals are located in relation to avalanches with long return periods of 100–300 years, whereas some roads and railways, which carry moving objects, may tolerate annual avalanche return periods, although detection, alarm, closure or artificial release mechanisms may be operated in such circumstances to reduce risk. Structures that are unoccupied (e.g. oil or gas pipelines), or are only temporarily occupied, such as parking areas, are assigned an intermediate design period of say 10–50 years.

In Switzerland, communal avalanche hazard maps at scales of 1:10 000 and 1:15 000 are produced based on a compilation of all possible sources of evidence: field observations of terrain and forestry, niveo–meteorological data, historical documents and oral contributions, avalanche dynamics, expert and remote systems, and cadastral (land-use) information according to Swiss guidelines (BFF/SLF, 1984; Gruber, 2001). The Swiss system uses a four-division scale, based on impact pressure and return period, which is represented by different colours: red, blue, yellow and white (Table 8.9, Figure 8.14). Areas in the red zone are very dangerous, where death and destruction are likely. At the other end of the scale, the white zone indicates the area where avalanches do not occur. Similar schemes are in operation in North America but here the zoning

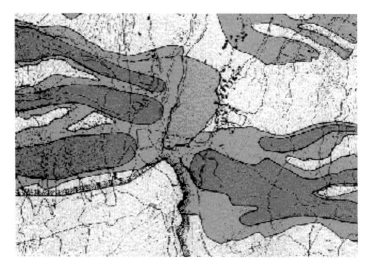

Figure 8.14 Example of a Swiss avalanche hazard map showing the four zones (red, blue, yellow and white) (Reproduced from Gruber, U. (2001) Using GIS for avalanche hazard mapping in Switzerland. http://proceedings.esri.com/library/userconf/proc01/professional/papers/pap964/p964.htm © Urs Gruber.)

has three divisions, red (high risk), blue (moderate risk) and white (low risk) (Table 8.9). The Canadian avalanche hazard zoning scale has been summarized diagrammatically (McClung *et al.*, 2002; Figure 8.15). For instance the blue/white zone boundary is at an impact pressure, I, of 1 kPa for return periods, T, greater than 30 years in contrast to the Swiss scheme which has the blue zone extending to zero impact pressures for return periods from 30 to 300 years.

In Canada new, permanently occupied structures are normally permitted in the white zone. Schools, hospitals and police and fire stations *must* be in the low

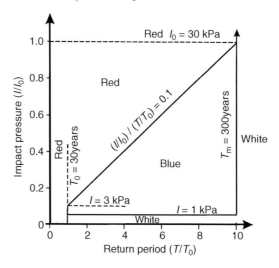

Figure 8.15 Definition of zones for land-use planning (from McClung *et al.*, 2002.)

risk white zone if they are essential. In the blue zone, new buildings such as industrial plant and temporarily occupied buildings may be conditionally permitted; for example if they are reinforced, defended or have adequate evacuation plans in place.

Zoning is not the same as total exclusion. Many pre-zoning buildings remain in inappropriate areas according to subsequent zoning decisions. In some cases there may be no alternative to buildings remaining in hazardous situations, short of moving a whole settlement. In the case of Flateyri in Iceland, for example, the deflector solution mentioned above has been effective so far but some buildings remain within the reach of the largest avalanches, because removal of the whole settlement is not an acceptable option. Even when all static situations have been accounted for, the results of zoning still leave some people (e.g. travellers, maintenance engineers, ski-lift operators) in vulnerable positions if only for limited periods of time. Under these circumstances evacuation, based on forecasts, warnings and alarms, provides a feasible if sometimes unpopular alternative.

Short-term options – detection and alarms, warnings, closure, evacuation In Iceland, after the 1995 disaster, evacuation plans were drawn up for vulnerable settlements based a formal map prepared from new avalanche cycle data, past meteorological conditions and a statistical evaluation of the extreme run-out zones (Magnússon, no date).

Both zoning and evacuation are difficult decisions because to some extent they involve what McClung and Schaerer (1993) refer to as 'intangibles', that is factors that are not easily quantified such as politics, psychology, operation and environment. Social and economic issues are also closely bound up with these decisions. It may be psychologically disturbing to find that you live in the red zone. It may also be economically disadvantageous if house values are lower in that zone. It is not difficult to imagine the pressures felt by the decision-makers responsible for zoning.

Management of the avalanche hazard in relation to roads and railways is generally achieved through forecasting and closure. However, this is likely to result in lengthy delays for trains and motorists as the timing of avalanches is not precise. Control of avalanching using artillery also takes time to set up and check. Jamieson (2001, quoted in Jamieson and Stethem, 2002) reported that the annual direct cost of avalanche-related highway closures in Canada exceeds Can$5m per year and this does not include indirect costs due to business losses. Real-time detection and warning systems should provide a less time-consuming solution to this problem because they respond to the event as it happens.

Automatic avalanche detection is not a new concept: trip wires were used in Utah, USA, in the mid1950s (Rice Jr. *et al.*, 2002). Wires that broke when struck by an avalanche were also set along rail tracks in Europe and Canada and gave a warning to prevent train derailment (Rice Jr. *et al.*, 2002). Trip switches, radar, vibration, sound sensors and photoelectric barriers have all been applied to the avalanche hazard (McClung and Schaerer, 1993). A basic problem with many of these systems is that they require resetting after each event, which is hazardous for maintenance personnel. Some sensitive sensors, such as those measuring detectable sub-audible infrasound signals produced by avalanches, can be affected by wind and other ambient noise. However, these infrasounds propagate over large distances from the avalanche site and can therefore form the basis of a safe avalanche detecting system providing the limitations can be overcome, perhaps by using multiple sensors (Scott et al., 2007).

The Swiss have a remote avalanche warning system based on Doppler radar and geophones linked to warning lights and the public telephone system (Gubler, 1996). At Roger's Pass, Canada, a tensioned cable strung across the avalanche track is tilted by the avalanche and signals a radio or data logger

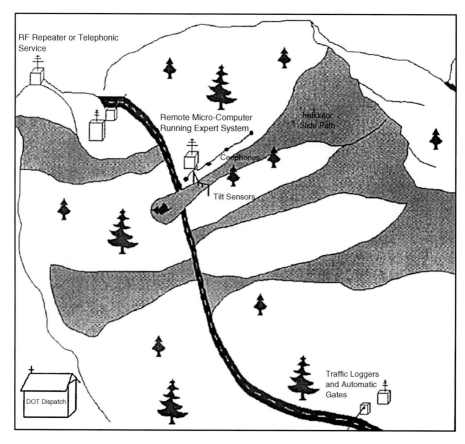

Figure 8.16 Diagram of a corridor avalanche management system (Reproduced with permission from Rice, R. Jr., Decker, R., Jensen, N., Patterson, R., Singer, S., Sullivan, C. and Wells, L. (2002) Avalanche hazard reduction for transportation corridors using real-time detection and alarms. Cold Regions Science and Technology, 34, 31–42. Fig. 1 (C) Elsevier.)

(Statham *et al.*, 1996). Rice Jr *et al.* (2002) report the installation of a 'Corridor Avalanche Management System' (Figure 8.16) installed at Canyon Creek, Idaho, USA. This involves a number of sensing devices, traffic loggers, warning lights and automatic gates to close sections of the road (Figure 8.17). Automatic detection systems are reliant on efficient working of individual elements and they are expensive. However, they may have the potential, eventually, to replace traditional forecasting systems, which also must be funded, take time and entail risk to personnel checking snowpacks or test slopes for stability.

The results of snowpack analyses can be communicated to the public by publishing hazard ratings which classify the degree of danger from low to extreme depending on the particular rating scheme. Before 1993 many European countries had their own scale but in 1993 these were unified and a five-part Avalanche Danger Scale was agreed, later refined at a meeting of the Avalanche Warning Services in 2003 (SLF, 2010; Table 8.11). Symbols representing the degree of danger are used on regular bulletins in Switzerland, for example (Figure 8.18). A similar scale is in operation in the USA (NWAC, no date). Providing the public respect the warnings the best mitigation systems meet their objectives at lowest cost and with acceptable intangibles.

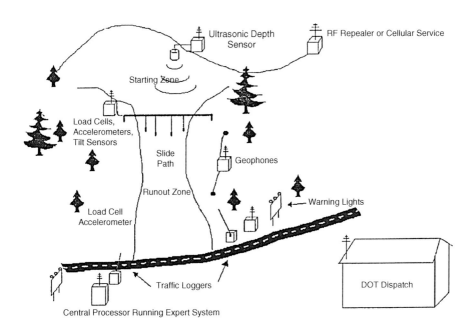

Figure 8.17   Diagram of a time-of-descent avalanche management system (Reproduced with permission from Rice, R. Jr., Decker, R., Jensen, N., Patterson, R., Singer, S., Sullivan, C. and Wells, L. (2002) Avalanche hazard reduction for transportation corridors using real-time detection and alarms. Cold Regions Science and Technology, 34, 31–42. Fig. 2 (C) Elsevier.)

Table 8.11 European avalanche danger scale with recommendations (source: SLF, 2010)

Danger level	Icon	Snowpack stability	Avalanche triggering probability	Consequences for transportation routes and settlements/recommendations	Consequences for persons outside secured zones/recommendations
1 Low		The snowpack is generally well bonded and stable	Triggering is generally possible only with high additional loads ^{2} on very few extreme slopes; only natural sluffs and small avalanches are possible	No danger	Generally safe conditions
2 Moderate		The snowpack is only moderately well bonded on some steep slopes, otherwise it is generally well bonded	Triggering is possible, particularly through high additional loads, mainly on steep slopes indicated in the bulletin; large natural avalanches are not expected	Low danger of natural avalanches	Mostly favourable conditions; careful route selection, especially on steep slopes of indicated aspects and altitude zones

8.5 MITIGATION METHODS 235

3 Considerable	◆	The snowpack is moderately to weakly bonded on many steep slopes	Triggering is possible, even through low additional loads mainly on steep slopes indicated in the bulletin; in certain conditions, some medium and occasionally large natural avalanches are possible	Many exposed sectors are endangered; safety measures recommended in those places	Unfavourable conditions: extensive experience in the assessment of avalanche danger is required; remain in moderately steep terrain/heed avalanche run-out zones
4 High	◆	The snowpack is weakly bonded on most steep slopes	Triggering is probable even through low additional loads on many steep slopes; in certain conditions, many medium and multiple large natural avalanches are expected	Many exposed sectors are endangered; safety measures recommended in those places	Unfavourable conditions: extensive experience in the assessment of avalanche danger is required; remain in moderately steep terrain/heed avalanche run-out zones
5 Very high	◆	The snowpack is generally weakly bonded and largely unstable	Many large natural avalanches are expected, even in moderately steep terrain	Acute danger; comprehensive safety measures	Highly unfavourable conditions; avoid open terrain

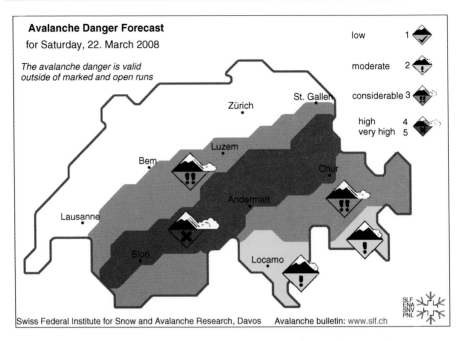

Figure 8.18 Swiss avalanche danger forecast (source: SLF, 2008)

8.6 Summary

Avalanches begin suddenly, move rapidly and can be costly in terms of lives, property and the environment. Avalanches often occur repeatedly along well-defined tracks but the most destructive events may strike in unprecedented locations and/or result from exceptional weather conditions. Although factors responsible for avalanches are generally understood, continuous, overall terrain monitoring and weather forecasting is not yet capable of anticipating all avalanche occurrences and there are many areas, even in developed countries, where funds are not sufficient to monitor all potential avalanches. Incomplete knowledge of physical systems, increasing population pressure, incomplete control of human behaviour and the likely continuing occurrence of unprecedented events, will ensure that major avalanche impacts will continue to occur into the foreseeable future. Avalanche authorities around the world will, from time to time, be obliged to update and improve their avalanche mitigation procedures, as Alpine countries did after the 'Year of Terror' in 1951, Iceland did after 1995 and Austria did after 1999.

9
River Ice – Ice Jams and Ice Roads

9.1 Introduction

Empirical evidence indicates that ice commonly forms on the surface of rivers where mean January temperatures are no higher than $-5\,°C$ (Figures 9.1 and 9.2). In North America this means that most Canadians and Americans as far south as latitude 40°N can expect ice to form on their rivers during the winter. Only the western coastal regions, with a milder, maritime climatic regime, and the eastern seaboard south of Newfoundland, Canada, escape significant river icing. In contrast, the high Rocky Mountains extend the phenomenon south of this general limit. A similar pattern obtains in Eurasia (Figure 9.3). Here, almost the whole of Russia, Outer Mongolia, Northern China and most of Sweden and Finland in the west, are located within the critical $-5\,°C$ January temperature threshold. However, the most densely populated areas of Europe and Asia lie outside this boundary and therefore, today, face very little risk from river-ice hazards.

As the title of this chapter indicates, there are two aspects to the river ice hazard, ice jams and ice roads. Ice jams are essentially masses of broken ice blocks that dam rivers. Their impacts, whether direct or indirect, are almost invariably negative. Although ice jams cause relatively few fatalities their economic and social effects are often severe. In contrast, thick static ice on the surface of rivers and lakes can be prepared to serve as a road during the winter season and can be extremely beneficial to the inhabitants of isolated communities and to the economic success of activities such as remote mining operations. Nevertheless, ice roads are hazardous as their stability cannot always be guaranteed. Ice road truckers (drivers) have died during the course of their activities and, as with ice jams, economic losses caused by ice-road failure and closure can be severe.

Cold Region Hazards and Risks, First Edition.   Colin A. Whiteman.
© 2011 John Wiley & Sons, Ltd. Published 2011 by John Wiley & Sons, Ltd.

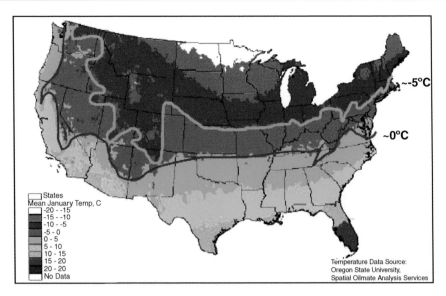

Figure 9.1   Mean January temperature in the contiguous states of the USA (source: CRREL, 2006)

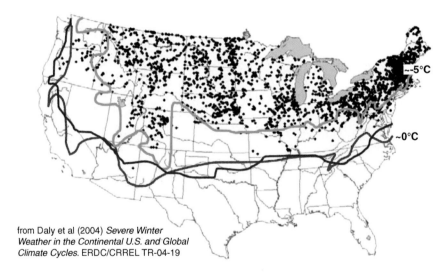

Figure 9.2   Distribution of ice jams and ice covers during the period 1950 to 2001 in the contiguous states of the USA. Blue line represents zero average maximum Air Freezing Degree Days (AFDD) (source: CRREL (2006))

9.2   Ice jams

9.2.1   Introduction

Ice jams result from the accumulation of ice fragments in rivers that build up into a temporary obstruction during the winter or early spring and restrict

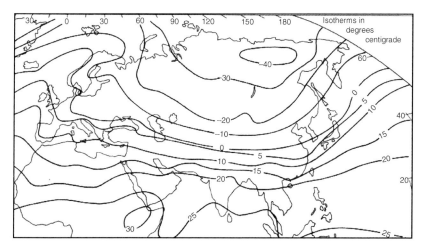

Figure 9.3 Mean January temperature in Eurasia (source: Boucher, K. (1975) *Global Climate*, Figure 1.9. The English Universities Press Ltd, St Paul's House, Warwick Lane, London, EC4P 4AH, ISBN 0 340 15493 4)

the flow of water (Church, 1988). This leads to hazards developing upstream and downstream of the blockage. Ponded water can cause flooding while the eventual sudden release of the water combined with large blocks of ice can cause serious impact damage and further downstream flooding. Unlike avalanches which seem to be well known and are considered in most hazard textbooks, ice jams are little known outside their actual area of occurrence. However, three of the world's largest 27 recorded flows of the last 20 000 years were ice jam floods on the River Lena in Russia in 1948, 1962 and 1967 (O'Connor, Grant and Costa, 2002; see Box 9.1 below for details of the most recent River Lena ice jam events). Other significant historical floods resulting from ice jams include the River Rhine ice jam floods of January 1784 which led to water levels 3 m higher than any known Rhine flood not involving ice jams, and the March 1838 floods on the River Danube which were 2 m higher than non-ice jam floods on this river (Smith and Ward, 1998). Both events date to the period of the 'Little Ice Age' when temperatures were significantly lower than at present (Lamb, 1977). More recently, in 1952, a failing ice jam on the Missouri River in the USA led to discharge rising from about $2100 \, m^3 \, s^{-1}$ to more than $14 000 \, m^3 \, s^{-1}$ in less than 24 hours (O'Connor, Grant and Costa, 2002). The River Lena ice jam flood event around Yakutsk and Lensk in Siberia was possibly the first to be widely reported by the media (e.g. BBC Online, ITAR/TASS News Agency, English Pravda, The Guardian, Interfax News Agency, American Red Cross, The Times). Interest was probably heightened by the dropping of a number of bombs to break the ice!

> **Box 9.1 Yakutsk and the River Lena, Russia**
>
> The River Lena flows northwards across Siberia from central Asia to the Arctic Ocean. Two recent ice jam flood events, in May 1998 and May 2001, were responsible for at least 20 fatalities and millions of dollars of damage. The 2001 event had a return period of at least 100 years, with the river reaching a record high stage of 9.17 m at Yakutsk, the main settlement of the region. Upstream the town of Lensk was also severely affected with almost all its 27 000 inhabitants evacuated. Most of the town's buildings were submerged and 1800 homes destroyed. The 2001 event was caused by exceptional antecedent weather conditions, the combination of severe winter temperatures ($-50\,°C$) resulting in thick river ice, abundant snowfall and an unusually warm spring causing rapid snowmelt in the Sayany mountains to the south, while thick ice remained on the river further north. This resulted in at least seven fatalities, 70 000 displaced and damage or disruption to 396 km of electricity transmission lines, 470 km of communication lines, 163 transformers, 184 km of roads as well as seven health care centres and 26 schools. About 5000 tonnes of oil spilled into the Lena from the Lensk oil base when flood waters penetrated the base and ice blocks damaged pipelines. In response, four bombers and four helicopters dropped bombs on ice jams, a protective dam was constructed in Yakutsk and a causeway was breached north of the city deliberately flooding a large area of low lying land to ease pressure on flood barriers defending Yakutsk's most densely populated areas. In Yakutsk, 10 evacuation centres were set up to house the 3500 people evacuated from inundated areas.

Ice jams and their associated floods are complex phenomena and difficult to predict, since there is such a dynamic interplay of different factors involved in their development such as temperature, precipitation (snowfall and rainfall), ground conditions (including permafrost), river channel morphometry and geometry, natural and artificial channel obstructions and the nature and competence of the ice itself. An analysis (Gerard and Karpuk, 1979; Figure 9.4) of the Peace River, Alberta, Canada, confirms the variability of ice jam flood discharges and suggests that the largest ice jam floods are larger than those of normal summer season floods, at least for return periods greater than about 10 years (Gerard, 1990).

It is only relatively recently that attempts have been made to summarize some of the diverse sources of existing knowledge and data on the subject of ice jams in North America. Studies in the mid-1980s by the Working Group on Ice Jams of the Canadian Committee on River Ice Processes and the Environment culminated in the 1995 publication, River Ice Jams (Beltaos, 1995). About this time CRREL also established its Ice Jam Database (White, 1996), which

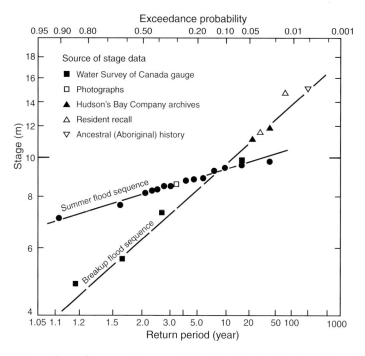

Figure 9.4 Comparison of summer (open water) and ice breakup floods on Peace River, Fort Vermilion, Alberta, Canada (from Church, 1988, after Gerard and Karpuk, 1979)

is accessible through its Ice Jam Clearinghouse website (CRREL, 2008). This section on ice jams draws extensively on both of these sources.

Initiation, development and subsequent loss of the ice and snow cover over rivers are controlled by (a) the flow regime of the river and (b) ambient weather conditions. In rivers flow velocities vary downstream from fast turbulent flow in riffles and rapids to slow laminar flow in larger, deeper pool sections. Transverse to flow, velocity is generally greatest along the thalweg (line of the main channel) and least adjacent to the bank. Above the river, air temperature, humidity and wind velocity influence water temperature, ice formation and the production and distribution of snow cover on the ice which, in turn, influences additional ice growth.

9.2.2 Ice formation and freeze-up

Under laminar flow, water is sufficiently quiescent for overturning to occur as cold dense surface water sinks and is replaced by warmer water from beneath (Beltaos, 1995). This continues until the water temperature throughout is around 4 °C. At this temperature the water attains maximum density and further mixing is inhibited. Subsequently, water temperature cools to freezing point and ice develops, initially at the surface and then at increasing depths as heat is lost from the deeper water. However, when flow is turbulent, as it is in most parts of a river, mixing replaces stratification and overturning of the water column. The degree of mixing is related to the 'densimetric' Froude number (Fr_d):

$$Fr_d = \frac{V}{\sqrt{gd\,\Delta\rho_w/\rho_w}}$$

where V is flow velocity, g is acceleration due to gravity, d is flow depth, ρ_w is bottom water density and $\Delta\rho_w$ is the difference in density between the upper and lower layers. At the large Froude numbers characteristic of rivers, the entire water column cools more or less uniformly, depending on the net surface heat flux, Q_*, across the water–air boundary. Many factors, summarized by the equation below, influence the rate of heat exchange between the river and the atmosphere. Thus

$$Q_* = Q_S + Q_L + Q_H + Q_E + Q_P + Q_F + Q_G + Q_B$$

where Q_S = net flux of short-wave radiation
Q_L = net flux of long-wave radiation
Q_H = sensible heat flux
Q_E = latent heat flux
Q_P = precipitation heat flux
Q_F = flow friction heat flux
Q_G = groundwater heat flux
Q_B = bed heat flux.

The critical factor is radiation, heat from the sun. Gross radiation will naturally decrease during the first part of the winter as solar altitude decreases. Net radiation is influenced by cloud and fog as well as by surface albedo, which is largely dependent on snow cover. Other key influences on the extent and rate of freezing are snow and wind speed. Snow falling directly into rivers has two effects. It introduces sub-zero solid material into the water, and the process of snow melt consumes latent heat and therefore further lowers the overall water temperature. The other factor, wind, is also capable of rapidly reducing water temperature by the transfer of sensible heat from the water surface into the atmosphere. As heat moves from the water into the atmosphere, the atmosphere warms and its cooling effect on the river lessens. However, if this air moves away, new cold air is brought into contact with the water surface and the cooling process can continue. Shallow streams generally cool more quickly than deeper rivers because the ratio of surface area to water volume is larger in the smaller streams and surface heat exchange is more efficient. Once water temperature is reduced to 0 °C further cooling leads to the formation of ice, providing suitable freezing nuclei are available. Pure water can be supercooled

9.2 ICE JAMS

Figure 9.5 Border ice (left) and frazil ice pans ('pancake' ice) (right) (source: ERDC, 2009)

to $-40\,°C$ before ice grows spontaneously from an ice embryo caused by water molecules combining. This would mean that few rivers would freeze significantly. However, most rivers contain impurities that serve as freezing nuclei and so ice forms readily in and on rivers at temperatures significantly above $-40\,°C$. Clearly, complete coverage of a river by ice is a complex process which is made more unpredictable by the natural variation of weather conditions, which may temporarily reverse the freezing process before freezing resumes.

Normally ice forms first along river banks (Figure 9.5), or around islands, boulders or bridge piers, because the denser material of these solid features loses heat more rapidly than river water. However, nonattached sheets of ice, often referred to as skim ice, can form under calm, low-flow conditions. An increase in stream discharge may fragment border ice into individual ice floes which migrate downriver with skim ice if this has formed. If sufficient downstream border ice persists, it may reduce open channel width sufficiently to cause skim ice and the floes to bridge across the surviving water surface and lead to the formation of freeze-up ice jams.

Under normal turbulent flow conditions small ice particles, termed frazil, begin to form at the surface of the river as soon as the temperature drops below $0\,°C$. Ideal hydro-meteorological conditions for frazil ice formation are a cold clear night which facilitates rapid heat loss by long-wave radiation, and a strong wind composed of cold dry air, which produces large convective and evaporative heat losses. The nucleated particles formed under these conditions promote ice crystal growth. The ice crystals may be transferred into the body of the river by turbulence where they grow and agglomerate into frazil 'flocs' (5–100 mm). These combine into larger clusters which float and form frazil slush and eventually ice sheets, or frazil pans, at the water surface (Figure 9.5). Often these pans are swirled around by surface currents, impacting against each other and forming circular masses with upturned edges, appropriately described as 'pancake' ice. These 'pancakes' can form attractive and spectacular components of northern winters in their own right, as Figure 9.5 illustrates.

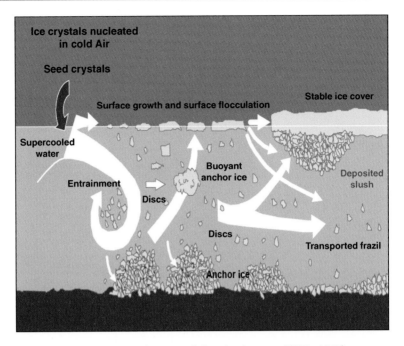

Figure 9.6   Development of river ice (source: CRREL, 2010)

Under very turbulent flow, the supercooled water in which ice crystals form can be transferred to considerable depths where it may help to produce so-called 'anchor' ice, usually adhering to boulders and aquatic vegetation (Figure 9.6).

If appropriate hydro-meteorological conditions persist, ice floes will become so numerous that they bridge or arch across the whole river surface, depending on floe size and velocity, ice transport capacity and the geometry of the channel. As frazil ice continues to flow from upstream, it will accumulate against stationary ice and solidify, as interstitial water between the floes freezes. If flow is sufficiently rapid frazil may submerge beneath an ice cover and re-emerge beyond the covered section. Alternatively, under weaker flow regimes it may attach to the underside of the cover to form a type of jam referred to as a 'hanging dam' (Figure 9.6). If attachment is close to the leading edge of the ice cover, a jam will develop upstream by the process of frontal progression. In summary, freeze-up follows a sequence from initial static border ice through frazil ice formation and its accumulation into pans and flows, to their coalescence into a bridge between the border ice zones which obstructs further downstream ice movement. This sheet of river ice can thicken initially by the freezing-on of floes to its underside or by the gradual growth of the original ice crystals vertically downwards to form clear, columnar black ice.

Over the years the broad temporal and spatial patterns of ice growth and thickness have been recorded. Figures 9.7(a) and (b) illustrate the Canadian situation where, as in the USA, a clear longitudinal difference exists between

9.2 ICE JAMS

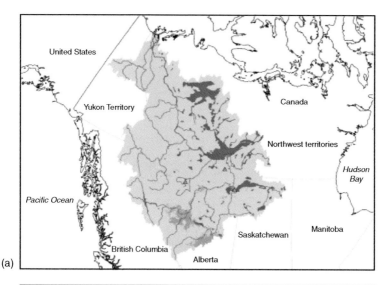

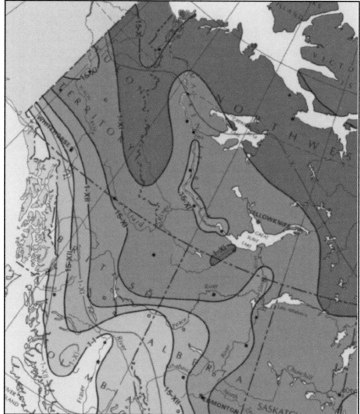

Figure 9.7 Timing of river ice freeze-up (a) and breakup (b) in Canada (source: The Atlas of Canada, 2004, http://atlas.nrcan.gc.ca/site/english/maps/archives/4thedition/environment/water/013_14 [accessed 12 July 2010].)

the west coast, washed by the warm North Pacific Current, and the east coast brushed by the cold Labrador Current. As in the USA, the influence of the Rocky Mountains is clear (Figure 9.7(a)). These general patterns of freeze-up mask the influence of particular conditions within individual river systems and river reaches. Annual bridging locations may be scattered along the length of a river. Freeze-up does not, therefore, necessarily progress systematically upstream. In a southward flowing river freeze-up is likely to occur from the northern tributaries southwards to the main river. Small tributary streams may freeze before large trunk rivers. In the case of Canada's major river, the Mackenzie, freeze-up progresses southwards from the Beaufort Sea at latitude 70°N, to Norman Wells at 65°N but the middle reaches only freeze after the more southerly tributaries have frozen over (Figure 9.7(a)). Thus patterns of freeze-up vary from reach to reach, depending on weather conditions, river geometry and discharge.

9.2.3 Ice breakup

Ice breakup is usually more hazardous than ice freeze-up. During freeze-up ice fragments tend to be small because they are newly formed, and river discharge is falling as precipitation changes from summer rain to winter snow and run-off decreases. Large, resistant ice jams tend to form more readily during ice decay because the contributing ice fragments are often large, having formed from the breakup of existing sheets, and discharge energy is increasing as the snow cover melts. Breakup is a complex process involving melt, fracture, removal and transport of the ice (Beltaos, 2008). It varies both within single rivers and between rivers and from one winter to the next (e.g. 2000–2001, Figure 9.8(a) and (b)). Since 1950, ice jams have occurred at many locations (Figure 9.2) but the record for a single water year such as 2000 (Figure 9.8(a)) generally shows far fewer incidences of both freezing rivers and the impact of jams.

Before the ice begins to melt and decay, any covering snow must be melted. Initially, snow on the surface of river ice insulates the underlying ice from the effects of solar radiation if it is thicker than about 50 cm. Also, fresh snow reflects radiation due to its strong albedo effect which may be as high as 0.95. However, once the snow has been cleared, melting and decay of the ice can begin, especially where black ice with a low-albedo (0.1) formed in calmer reaches of the river. Melting is initially focused around crystal boundaries where impurities concentrated during freeze-up. Eventually individual crystals will detach from the main mass of ice and be carried away by the current in a process called candling. Where white frazil ice dominates, above the steeper, more turbulent sections of the river, decay of the ice cover is likely to take longer, and more competent ice sheets will fracture. Hydrothermal melting, in which the base of the ice sheet melts, may also occur especially if the thermal gradient in the ice is small and the heat transferred from the flowing water to the ice cannot be conducted rapidly away through the ice. Further stress is imparted to the ice by

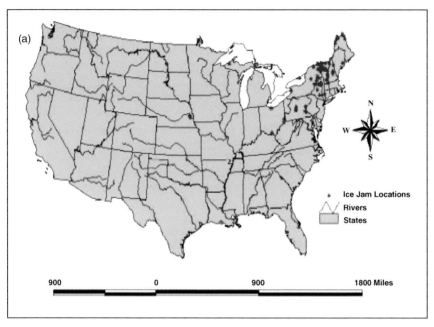

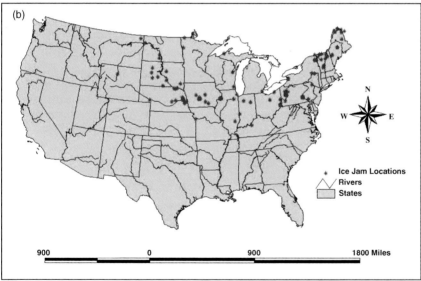

Figure 9.8 Ice jam locations in the USA during the water year (a) 2000, (b) 2001 (source: (a) CRREL, 2000; (b) CRREL, 2001)

the increasing rate of flow following melting. In small streams a crack may form along the centre line of the ice but in rivers there may be several cracks parallel to the bank, especially along the hinge line formed during winter flow. Alternatively the ice may crack in a crazed pattern reflecting its variable strength.

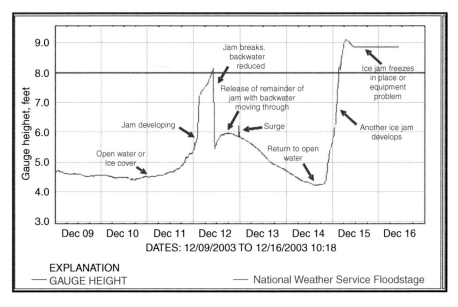

Figure 9.9 Ice jam affected hydrograph (source: NOAA, 2010)

What has been described so far can be termed pre-breakup, because relatively little ice may have moved except in the higher velocity reaches. Full breakup is only achieved when increasing snowmelt and perhaps rainfall, supplemented by floods created by earlier ice jams, raise discharge (Figure 9.9) to a level sufficient to dislodge large areas of intact ice sheet and carry them downstream. Collisions between these large ice masses and against the river banks gradually reduce the size of the fragments to small-diameter brash. This is the material that will, under the right circumstances, converge into a huge pile to form an ice jam, defined by the International Association for Hydraulic Research Working Group on River Ice Hydraulics (IAHR, 1986) as 'a stationary accumulation of fragmented ice or frazil that restricts flow'. It is these concentrations of ice fragments which obstruct the normal flow of the river, causing a variety of hazards such as flooding, scouring of the bed and banks of the channel and damage to or destruction of bridges and riverside property due to ice-block impact.

9.2.4 Ice jam processes and sites

Essentially, ice jams occur during either autumn (fall) freeze-up or spring break-up, when resistance to the flow of water and ice exceeds the ice transport capacity of the river. If the ice floes do not submerge at the obstacle a surface jam will form consisting of a single layer of floes on the water surface eventually forming a solid sheet of ice. If the flow is rapid, floes may submerge. If they are not carried beyond the solid ice sheet they may deposit quickly against the base of the ice as a *thickened* jam, deposit in a low velocity area and accumulate to form a *hanging dam* or, if large enough, become lodged between the obstacle and the bed to form a *grounded* jam (Figure 9.6).

Ice jams may form wherever 'the incoming ice discharge exceeds the local ice transport capacity of the stream' (Beltaos, 1995, p.73). When this happens the ice floes become congested and, if the flow velocity is insufficient to submerge the ice floes, they will form a surface jam. Congestion is especially common during freeze-up in areas of reduced flow velocity. Even strong winds blowing against the flow can induce congestion, especially where large rivers flow into lakes. During breakup, when flow velocities are likely to be higher than during freeze-up due to snow melt, jams are often associated with the arrest of a *breaking front*. This occurs when a mass of ice breaks away from intact river ice and is carried rapidly downstream until it encounters more intact ice. The *breaking front* will then either plough through this intact ice to open water beyond or be arrested within the stationary ice. Arrest of the ploughing mass usually results in a thick jam of ice rubble above and below the level of the ice sheet cover. Given the necessary ice discharge and ice transport capacity conditions, congestion of ice floes and breaking front jams can occur anywhere on a river but, in practice, it has been found that some river configurations are more conducive to ice jam formation than others. Ice jam 'hot spots' tend to be located where appropriate ice discharge and river flow conditions are enhanced. Thus typical river ice jam sites include:

- competent ice cover locations,
- constrictions,
- bends,
- confluences,
- islands,
- bridge piers,
- shallows,
- deep hanging dams,
- anchor ice locations, and
- up-valley wind at river mouths.

Competent ice cover sites provide an obvious obstacle to the downstream passage of floes. Rivers rarely maintain a steady increase in width, as discharge increases downstream, due to the influence of different rock types. Resistant rocks may cause narrow gorges to form but even minor changes in rock hardness may be sufficient to induce variation in channel width and, consequently, ice carrying capacity. Natural islands as well as bridge piers reduce channel width and are common ice jam locations. The degree of constriction will depend on the size of the floes, larger masses of ice being more easily obstructed than smaller floes. Large angular ice sheets are likely to find bends difficult to negotiate. The natural thalweg of the stream will carry the ice towards the outer bank where bank friction will inhibit movement. Constriction may also occur in the vertical plane. On the inside of bends, beyond the thalweg, the river is shallower so grounding of ice floes may occur and jams form. Deep hanging dams and anchor ice both reduce water depth and therefore the ease with which ice floes move. Grounded floes beneath competent ice cover are also likely to induce jams.

Once formed the strength of the jam depends on internal friction and cohesion of the ice blocks and will increase as the jam thickens. The downward growth of black ice beneath a freeze-up jam will add strength to this type of jam. As the jam lengthens flow shear and gravity increase. If the internal strength of the jam is exceeded it may collapse into a thicker, more stable configuration as long as the ice sheet remains intact. Eventually, melt-induced ice deterioration, increased flow stresses or human impacts may cause the jam to break.

The severity of ice jams is influenced by several factors. As already noted, ice jam formation is partially dependent on river discharge. Low flows produce surface jams but greater discharge induces submergence of flows and a build up of ice thickness beneath the surface ice. It follows that wide, steep rivers generate thicker jams and higher water levels. Stronger jams with greater internal friction and cohesion can become thicker providing ice is available. Water temperature and heat transfer are obvious factors influencing the life of a breakup jam. A mature or thermal breakup associated with intense ice deterioration and little run-off is unlikely to create a serious hazard. In contrast, premature or mechanical breakup occurs when the ice is still largely intact but run-off is high and perhaps surging from an earlier upstream jam release. The potential for maximum damage exists when an abrupt warm spell with rain generates substantial run-off over impermeable permafrost and encounters strong ice (Church, 1988). This situation is exacerbated where seasonal melting to the south generates run-off that encounters still-frozen ice to the north. This scenario probably led to the devastating floods on the River Lena in Siberia in 2001 (see Box 9.1), and is one of the factors which contributes to ice jam floods on the Red River in North Dakota, USA, and its lower reaches across the border in Manitoba, Canada. The susceptibility of the Red River to flooding is due to several other factors which enhance the effect of the ice jam itself. The spring thaw and ice breakup to the south provides the discharge to move the ice and cause jams downstream. Downstream the gradient of the river is very low where it crosses a large former lake flat, the ideal floodable area (Schwert, 2009).

9.2.5 Ice jam impacts

Ice jam impacts are widely scattered over space and time for several reasons. Ice jam processes are complex, they form in many different types of location within the river system and they are induced by weather patterns of variable distribution and intensity. In many instances, freeze-up and breakup happen with little or no impact on the landscape, or its inhabitants. Indeed, some evidence suggests that river ice floods are essential for the health of certain ecosystems, such as the Mackenzie River Delta in northwest Canada (Marsh and Hey, 1989). On the other hand some communities, located at prime sites for ice jam formation, face serious ice jam threats almost every year. Montreal, on the St Lawrence River was such a place until, with experience, the authorities developed strategies and emergency plans to cope successfully with the annual ice jam threat. In contrast, communities that experience ice jam impacts infrequently, usually at random intervals, are often less well prepared to cope with an ice jam event when it does occur, especially if the lead time is short. For example, in March 1992, an ice jam developed at 7 a.m. in Montpelier, Vermont, USA. By 8 a.m. the town centre was flooded, eventually to a depth of 1.2–1.5 m, which caused, in less than one day, an estimated US$5 million of damage (Montpelier-Vermont, no date). Given the wide range of potential impacts associated with ice jams (White and Kay, 1996; Table 9.1),

Table 9.1 A classification of ice jam impacts (compiled from White and Kay, 1996)

Type of impact	Details
Fatalities	
Changed discharge and stage	(1) Flooding upstream of jam
	(2) Flooding downstream of jam
	(3) Downstream 'daylighting' (exposure) of water intakes
Structural damage	(1) Undermining or destruction of bridge piers
	(2) Damage to or blockage of water intake systems (e.g. HEP plants)
	(3) Damage to public and private property (buildings)
	(4) Damage to navigation lock gates
	(5) Damage to river training structures
Geomorphological change	(1) River bank erosion or collapse
	(2) Channel erosion and scour
	(3) Channel shifting
Habitat change	(1) Destruction of fish overwintering or spawning grounds
	(2) Modification of bottom habitats by erosion or sedimentation
	(3) Removal of riparian vegetation
	(4) Mobilization of buried toxic sediments
	(5) Sedimentation on agricultural and other land
Indirect impacts	(1) Suspension or delay of commercial navigation
	(2) Disruption of transport nets and delivery schedules
	(3) Economic cost of idle vessels
	(4) Loss of HEP supplies following damage
	(5) Constraints on HEP operations due to requirement to avoid high, jam-inducing discharges
	(6) Social impacts, such as loss of earnings

the high level of economic costs is understandable. Overall, in the USA, ice jams cause more than US$120 million of damage annually (White, Tuthill and Furman, 2007). In contrast, fatalities are now relatively rare, because there is usually some warning of the hazard, but exceptional events, such as two recent ice jam floods on the River Lena in Russia (Box 9.1), are reported to have claimed at least 20 lives (BBC NEWS, 2001). There have been 18 ice jam-related fatalities during 1479 ice events on the Yellowstone River, Montana, USA, since 1894 of which 12 occurred in the 1899 event (NOAA, 2005, 2010).

Recurring ice jams within the Lower Platte River Basin in Nebraska, USA, have caused severe flooding on several occasions (White and Kay, 1996; Figure 9.10). The March 1993 ice jam flood event included jams on several rivers, most notably the Loup River in Columbus and the Platte River below its confluence with the Elkhorn River. This event, which resulted in over US$25 million worth of damage, clearly illustrates the variety of impacts associated with ice jams; in this case, road closures, road failures, flooding of residential, agricultural and industrial areas, and damage to levees, dykes and other river training structures. At Ashland, two major levee breaks resulted in the flooding of 14 000 acres of farmland along the river. More than 74 000 acres were damaged by extensive sediment deposits during the flooding. The 1993 event was the most damaging ice event since February 1978, when ice jam

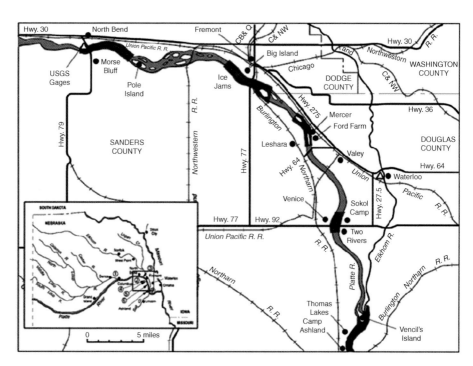

Figure 9.10 Recurring ice jam sites (darker shaded areas) on part of the Platte River, Nebraska, USA (from White and Kay, 1996)

flooding resulted in one death and over US$18 million worth of damage. Flooding was especially severe in the Valley-Fremont area during 1978 due to a levee break that flooded 27 000 acres.

It is clear that flooding due to a rise in the stage of the river is probably *the* major component of the ice jam hazard. The ice jam obstacle raises the water level upstream and its eventual collapse often releases the ponded water violently downstream, potentially creating flooding in both directions. A different river stage problem can arise downstream of a jam where the water level falls below the level of a water intake of a hydroelectric power station or factory, a problem referred to as 'daylighting'. Most of the other problems associated with ice jams are related to ice block impact. This causes structural damage to buildings and infrastructure, geomorphological damage to the bed and banks of rivers and the destruction of habitats. Some of these issues, such as loss of habitat and geomorphological change have received relatively little attention (CRREL, 2006). There can also be significant indirect impacts, especially when communication systems are badly affected.

9.2.6 Ice jam mitigation

It will be obvious from the preceding section on ice jam impacts that their mitigation is unlikely to be a straightforward process. Although the occurrence of ice jams is virtually an annual event for some locations, the fact that ice jams are so dependent on weather and weather is so variable, means that considerable variation in location (Figure 9.2), timing (Figure 9.8) and size is almost inevitable. Consequently, responses to the problem are likely to be equally variable, and over the years a considerable package of mitigation measures, available to national organizations and local communities, has been devised to tackle the impact and cost of the ice jam hazard.

Mitigation methods   Beltaos (1995) and White and Kay (1996) distinguish between structural and nonstructural mitigation methods (Table 9.2). Nonstructural methods can be further subdivided into an information-based sub-group and an ice removal sub-group. The former is concerned largely with procedures for anticipating, monitoring or avoiding potential problems while the latter involves different ways of actually removing the ice itself, once it is in place. Structural methods (Figure 9.11) involve some kind of engineering construction designed to strengthen channels, protect against flooding or control the position of the ice. Each of the structural and ice removal methods can be classified as either a *permanent* measure (e.g. dykes and dams), an *advance* measure put in place usually seasonally prior to the event (e.g. ice booms), or an *emergency* measure taken to relieve an unexpected or rapidly developing situation (e.g. ice breaking, ice dusting, ice blasting). This latter technique was employed in March, 2009, where an ice

Table 9.2 Ice-jam mitigation methods (modified after Beltaos, 1995, and White and Kay, 1996)

Structural methods	Nonstructural methods	
	Information-based	Ice removal
(a) Dykes, levees, floodwalls	(h) Forecasting, warning, database	(l) Thermal control
(b) Dams and weirs	(i) Monitoring and detection by direct observation, aerial photography and remote sensing	(m) Ice breaking, cutting, drilling
(c) Ice booms		(n) Mechanical removal of ice
(d) Retention structures		(o) Dusting
(e) Channel modifications		(p) Blasting
(f) Flood-proofing and sandbagging	(j) Operational procedures	
(g) Ice storage zones	(k) Land management and zoning	

Figure 9.11 Examples of ice jam mitigation measures: (a) temporary barrier; (b) dusting; (c) ice cutting; (d) excavation; (e) permanent weir; (f) blasting (from CRREL including White, 2005, http://www.crrel.usace.army.mil/icejams/tech_files/2006%20Ice%20Jam%20Mitigation.pdf – accessed 17 July 2010)

jam formed when blocks of ice became lodged on a shallow sand bar in the Missouri River, South Dakota, USA (ENS, 2009). Charges were detonated near the State's capital city, Bismark, in order to relieve flooding which had displaced several hundred people. Sandbagging was also used in an attempt to reduce flooding. Salt and sand were dropped by helicopters to help keep a channel open for the flood water to move away.

The choice of method(s) will obviously be influenced by the perceived or actual degree of risk and available funds. There is often a balance to be struck between expensive, but generally more effective structural methods, and less expensive, nonstructural, emergency, ice-removal methods. Those communities experiencing regular occurrences of the hazard will probably have in place some permanent measures against ice jams as well as being ready with seasonal responses to the problem. For example, on the Rideau River, a tributary of the Ottawa River in the city of Ottawa, Canada, there has been an annual ice management operation since the nineteenth century (Reid et al., 1995, quoted in Beltaos et al., 2000). The ice is cut and blasted and removed into the Ottawa River prior to breakup. The annual cost varies between Can$200 000 and Can$400 000 but is justified in comparison with the alternative ice jam-related costs estimated at Can$2 million. Other communities, where ice jams occur sporadically, are likely to take emergency measures when necessary, as the more cost-effective approach to the problem. The methods described so far are essentially 'hard' responses to the ice jam hazard. In contrast, the remaining methods are essentially 'soft' and rely on information.

Information is the basis of judgements about appropriate responses to the hazard. Until recently, information on ice jams in the USA has been scattered widely in many different, often local sources. However, data have become widely available since 1996, when CRREL, within its Ice Engineering Research Program, established a database of ice jam records, now with some 15 000 items (CRREL, 2007), covering all mainland states of the USA. The database can be interrogated for details of ice jam distribution at both national and state level with details related to individual rivers and cities on an annual basis. Individual communities and river authorities probably kept local records appropriate to their situation but that made it difficult to appreciate the true scale of the ice jam problem nationally. The purpose of the database is to provide a national-scale overview of this hazard which can be interrogated, amongst others, by emergency response organizations, and by federal and state personnel with control over policy and budgets. Data for this project have been obtained from several different sources including the United States Geological Survey, the National Weather Service, local authority records, local observers, newspapers and other media outlets. It is important to recognize biases in the data. According to the US Army Engineering and Research Development Centre (ERDC, 2009), 'the most frequently reported ice events are those that occurred between the mid-1930s (when many gauge records were begun) and the early 1960s, and the

> "HAZARDOUS WEATHER OUTLOOK
> NATIONAL WEATHER SERVICE BILLINGS MT
> 500 AM MST MON JAN 5 2009
> With warmer temperatures expected this week, ice on creeks, streams and rivers will begin to move and have the potential to produce ice jams and localized flooding. Persons living in areas that are prone to ice jamming will want to be aware of potential for rising waters if jams form.
> SPOTTER INFORMATION STATEMENT
> Weather spotters are encouraged to report significant weather conditions according to standard operating procedures."

Figure 9.12 Example of a hazardous weather warning concerning ice jams (source: US National Weather Service (NWS), January 2009)

location of ice events included in the database, are primarily the sites of USGS water-stage gauges. Ice events that occur away from USGS gauges are not yet well represented.' Nevertheless, the database does reveal those areas, such as Montana, New York State and Vermont, where ice jams occur most frequently and therefore where mitigation is most necessary, as well as showing how widespread this hazard is in the USA as a whole.

More immediate information is available in the form of warnings, such as those distributed by the US National Weather Service (NWS), or obtained from local ice movement detection systems. A 'hazardous weather outlook', posted on the NWS website in January 2009 (Figure 9.12) provides a flavour of the concern about ice jams and their potential to disrupt the lives of those living near rivers.

Figure 9.13 illustrates a simple ice motion detector system in which a sensor is embedded in an intact ice cover prior to breakup and connected to an

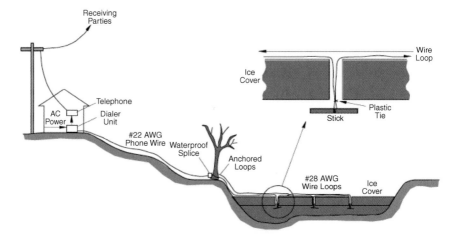

Figure 9.13 Ice motion detector system (from White and Kay, 1996, Figure 22; source: http://www.nlc.state.ne.us/epubs/N1000/B065-1996.pdf)

inexpensive burglar alarm set to automatically dial preselected telephone numbers when the unit detects ice motion (White and Kay, 1996). In some cases, existing US Geological Survey (USGS) river gauges with telemetry transmitters can send ice motion data directly to a local monitoring centre or to State and Federal agencies such as NWS and USGS, either through telephone, radio or satellite (Zufelt, 1993, quoted in White and Kay, 1996).

In many parts of the USA affected by ice jams, networks of volunteer river ice observers or 'spotters' report local ice conditions to the National Weather Service (NWS) via a toll-free telephone number. Following the third worst ice jam-induced flood event in the Susquehanna catchment in 1996, which caused 14 fatalities and US$600 million of damage, NWS set up an ice observers training scheme (SRBC, 2003). This programme was successful and the scheme proved very useful in the Lower Susquehanna catchment in early 2003 when bank to bank ice on the Susquehanna River required constant monitoring. The NWS uses river ice observations in its winter flood forecasts.

Mitigation strategies Based on the available mitigation methods outlined above and listed, (a) to (p) in Table 9.2, White and Kay (1996) proposed a number of different strategies (Table 9.3) for combating the ice jam hazard. These strategies can be viewed as a sequence, from complete avoidance of the problem (1) to a 'last-ditch' emergency measure (6) when serious impacts are imminent. Strategies (1)–(3) and (5) attempt to solve the problem at various stages but *before* it constitutes a hazard. Strategy (4) concedes that jams will happen but seeks to combat or avoid the hazard. Strategy (6) is an attempt to minimize a potentially hazardous situation *after* it has arisen.

Ideally the channel and its river could be modified, and procedures put in place to create a situation with the capacity to convey water and ice through a reach without ever creating an ice jam (1). In practice, a more sophisticated strategy is normally required to deal with the complexities of the ice jam hazard. The next approach (2) is to attempt to reduce the supply of ice and thereby restrict the potential volume of the jam, the rise in stage (water level), and the scale of any subsequent impact. If possible, ice formation can be

Table 9.3 Ice-jam mitigation strategies (modified from White and Kay, 1996)

Ice-jam mitigation strategies
(1) Increase river ice and water conveyance ((e) and (j))
(2) Reduce ice supply ((l), (j), (c), (b), (g) and (d))
(3) Displace ice-jam initiation location ((b), (c), (e), (o) and (m))
(4) Protect surrounding areas from flood damage ((a), (e), (f) and (k))
(5) Control ice-breakup sequence ((i), (c), (d), (o), (m) and (j))
(6) Remove ice from the jam ((l), (m), (n) and (p))

Note: letters in brackets refer to ice-jam mitigation methods listed in Table 9.2.

inhibited or its rate of formation reduced if water temperature can be modified by the controlled release of warm water, usually from power stations or factories. Permanent obstacles, such as weirs and dams, can be constructed in the river, or temporary booms strung across the water surface, to control ice movement and build up. Special areas can be set aside to store ice out of the main channel.

However, such procedures are unlikely to be available to forestall all ice jam events. Assuming therefore that ice jams are inevitable, the next move (3) is to control the location of the jam. This strategy is feasible where long-term records have established exactly where repetitive ice jams occur so that engineered solutions are cost-effective. Again, a variety of obstructions can be placed in the path of the moving ice to induce it to build up away from vulnerable sites. This strategy might require channel modification if the problem is persistent, or more direct methods, such as ice cutting and breaking, to stop the jam forming in an unacceptable location. If modification of the river system itself proves difficult, an alternative strategy (4), involving protecting the surrounding areas from flood damage, may be more appropriate. This is likely to require hard engineering to raise or strengthen levees and dykes, or divert flow away from threatened areas. Alternatively, people could be manoeuvred away from the hazard by managing or zoning floodplain use or, in an emergency, evacuating vulnerable inhabitants.

The strategies considered so far are designed largely to inhibit or control the hazard before and during development. Assuming that the ice is already in place and presenting a potential problem, strategies are then required which provide prompt and effective reaction to an existing situation. A key period in the ice jam scenario is 'breakup', when a continuous ice cover begins to disintegrate into discrete floes which move downstream with the potential to jam at prime sites. The strategy (5) now involves prediction or, at least, observational or electronic detection at an early stage. Ice cutting and breaking or dusting can help to modify the breakup process, and ice booms may be necessary to control flow and restrict ice to safe locations and amounts. Finally, if a large fixed jam has formed with the potential to cause serious flooding and substantial damage, the appropriate strategy (6) includes encouraging melting by thermal control, removing the ice mechanically, or causing its movement by breaking or even blasting using explosives and bombs.

An imaginative strategy was implemented by Montana, the US state which has suffered most ice jam-related fatalities in the USA. This is the 'Montana Ice Jam Awareness Day' (NOAA-NWS, 2010): 21 January 2010 was designated the fourth of these days, designed to raise the profile of this serious Montana hazard in preparation for the main ice jam period in February and March each year.

There is no reason to believe that ice jams will not continue to impact on society, although the long-term situation may change as climate changes. River ice is likely to continue to be most hazardous at the beginning and end of the winter season, during freeze-up and breakup although the precise timing of these events may be different. In the long term those places that already have a marginal ice jam climate may experience the disappearance of this hazard completely. However, there will be other communities that observe this development with concern, because their livelihoods depend on the presence of secure ice. Once solid freeze-up has occurred in the autumn (fall) and the ice has achieved an acceptable thickness, river (and lake) ice actually offers positive benefits as a means of road communication, although not entirely nonhazardous. Ice roads and their hazards are the subject of the rest of this chapter.

9.3 Ice roads
9.3.1 Introduction

Although ice-covered rivers (and lakes) (Figure 9.14) may be unfamiliar to residents of maritime and temperate regions of Eurasia and North America, for those who live in more northerly or continental regions, a frozen river is

Figure 9.14 Oil tanker travelling along a typical ice road across a northern Canadian lake (source: http://www.google.co.uk/imgres?imgurl=http://cache.virtualtourist.com/961796-Ice_Road-Lac_de_Gras.jpg&imgrefurl=http://www.virtualtourist.com/travel/North_America/Canada/Northwest_Territories/Lac_de_Gras-908640/TravelGuide-Lac_de_Gras.html&usg=__quke1703BkMyDhAPDH0Tr73Hzbg=&h=333&w=494&sz=20&hl=en&start=311&um=1&itbs=1&tbnid=k4GRmf0Qe2GFaM:&tbnh=88&tbnw=130&prev=/images%3Fq%3Dice%2Broad%2Bpictures%26start%3D300%26um%3D1%26hl%3Den%26sa%3DN%26rls%3Dcom.microsoft:en-us:IE-SearchBox%26rlz%3D1I7GPEA_en%26ndsp%3D20%26tbs%3Disch:1 – accessed 17 July 2010)

Figure 9.15 Heavy ice road trucking vehicles (sources: Athropolis, 2010)

a familiar sight and for some an essential aspect of their lives. In regions that are remote from the all-weather, surface transport network, this annual freezing of rivers provides a vital means of access for heavy goods vehicles (Figure 9.15) supplying distant settlements and mining communities with essential supplies of fuel and equipment that are too heavy to be airlifted without enormous expense. As soon as river and lake ice is thick enough to bear the weight of large vehicles, ice roads are prepared by bulldozing the surface of the ice clear of loose snow (Figure 9.16) and checking for weaknesses. Once the authorities are satisfied that an ice road is safe it is opened for traffic. Ice on the surface of rivers brings both costs and benefits. Providing drivers follow rules concerning loads and speeds (GNWT, 2010) and keep to the prepared surface, frozen rivers and lakes are a safe means of communication, although they can never be considered entirely risk-free. Ice road seasons are shortening with the impact of climate change (Solomon *et al.*, 2007) and reduced ice thickness is an added danger for heavy loads. Accidents do occur, and climate change is putting increasing stress on remote communities and economic systems by reducing the length of time during which these vital roads can safely be used.

9.3.2 Ice-forming processes and ice types

Ice sheets, even the small ones floating on rivers and lakes, vary widely in terms of their structure, temperature, strength and thickness. CRREL (2008), one of

Figure 9.16 Ice road preparation by scraping (source: The Diesel Gypsy, 2009)

the main research organizations with an interest in cold-climate landscape systems, has issued advice about travelling safely on floating ice sheets and much of this section is based on their work. Ice is structurally very variable because the freezing process often traps air within the frozen crystals. Ice may thus be clear (so-called black ice) or white (due to the presence of air bubbles). For the purposes of assessing load bearing, CRREL recommends that the actual thickness of any white ice is reduced by half in calculating total ice thickness because the presence of air bubbles reduces the overall strength of the ice.

On rivers, ice thickness and quality is often very variable due to frequent bends, riffles (shallows) and tributary junctions. Inflow of warmer water from springs may thin the ice. Near the shores of lakes ice may be thinner due to warmer ground water or snow drifts, or it may be thicker and stronger if it freezes to the bed of the lake. Ice often shows cracks. If these are wet this means that they penetrate right through the ice and are potentially hazardous. If they are dry, they have not reached the base of the ice and are relatively safe. Large recent snowstorms create a hazard by weighing down the ice so that water may reach the ice surface through wet cracks and saturate the base of the snow. This is potentially dangerous until the snow has frozen, when it can be classified as white ice. Rapid, large air-temperature drops may appear to favour an increase in ice strength but, in fact, make the ice brittle and therefore less strong. Temperatures rising above zero will soften ice and reduce its strength. If the temperature is higher than zero for more than six hours more ice is required to support a given load. If above-zero conditions persist for more than 24 hours ice begins to lose strength and normal equations for assessing ice strength against thickness no longer apply.

Figure 9.17   Ice thickness profiling and ice auguring (source: Diavik, 2006)

9.3.3 Construction

From the foregoing it is clear that ice road construction should begin with a careful check that the ice is sufficiently thick and strong, first of all to bear the weight of the bulldozer that will prepare the surface, and then to support the added weight of the heavy freight vehicles that constitute most of the traffic. Assessment of ice thickness and strength is traditionally achieved by drilling or auguring holes through the ice (Figure 9.17), but more sophisticated equipment is now available in the form of ground-penetrating radar (GPR) and global positioning systems (GPS) towed behind a vehicle (Figure 9.17). These electronic devices not only speed up the initial assessment but facilitate rapid, regular checking of the depth and distribution of ice thickness throughout the ice road trucking season. The benefits are obvious, in terms of truckers' safety and remuneration, the well-being of the inhabitants of remote settlements and the enhancement of company profits.

Having determined ice thickness along the route, if some extra 'paving' is required anywhere because the ice is not of sufficient thickness, water trucks can be mobilized to spray water over the freezing surface or water pumps temporarily installed (Figure 9.18). If the river or lake water is under sufficient

Figure 9.18   Pumping and spraying water to increase ice road thickness (source: The Diesel Gypsy, 2009)

pressure, holes may be drilled to allow the water to escape onto the road surface before freezing and adding to the thickness of the road. Once the road bed has been pronounced safe a bulldozer clears loose snow and produces a flat surface. Snow clearance is important for a number of reasons. Loose snow reduces vehicle traction and enhances the risk of skidding especially if water has seeped up into the base of the snow and converted it to slush, but not frozen. Loose snow can also accumulate on vehicles during their journey, for instance between tyres and mudguards, adding weight to the system and slowing progress. Snow 'spray' from leading vehicles reduces visibility for the drivers of those following. However, removal of snow is even more significant for the long-term life of the ice road. Ice covered with snow is not as strong as exposed because it is warmer. Once snow exceeds about 50 cm in thickness it acts as a good insulator, thereby protecting underlying ground or ice from contact with the cold atmosphere which induces freezing. By removing snow from the ice road, the ice is exposed to the cold atmosphere making the roadbed ice both thicker and stronger than the ice around it. Ice roads are normally ploughed wider than they need to be for simple vehicular access, so that storms are less likely to block them with fresh snow and the margin for safety is increased. Speed limits are strictly regulated and generally observed because ice is flexible: trucks can create waves under the ice and if vehicles move too fast (more than 15 mph according to CRREL), waves may burst out at the road margin or around shorelines with risk to both drivers and other people (Athropolis, 2010).

9.3.4 Loading

According to CRREL, the minimum ice thickness required to support a load (that is the combined weight of the vehicle and its load) can be calculated using the simple equation:

$$h = 4\sqrt{P}$$

where h is ice thickness in inches and P is the load, or gross weight, in tons. This is illustrated graphically in Figure 9.19 and tabulated in Table 9.4.

These criteria assume a number of conditions: that the ice is clear and sound, that the load (such as a person on foot, or a wheeled or tracked vehicle) is reasonably distributed over an area of continuous ice, and that the loads are not at or near the edge of a large opening in the ice. A safe distance between two similar loads is about 100 times the ice thickness at the minimum required for that load. In Figure 9.19 the solid diagonal line represents the relationship between ice thickness and load under normal atmospheric and ice conditions. In other words, the ice is free from wet cracks and the air temperature is low enough to maintain minimum ice strength for a particular load. However, if air

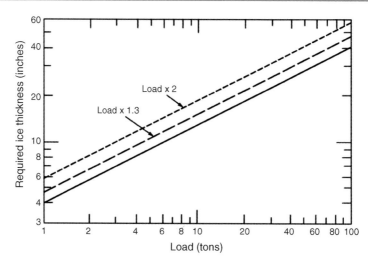

Figure 9.19 Minimum ice thickness required to support a load (Source: http://www.mvp-wc.usace.army.mil/ice/graphics/iceload1.png - Accessed 12 01 09)

temperatures exceed zero for six hours or more some adjustment is required because ice strength will have been reduced necessitating the presence of thicker ice or a smaller load. In this case the line with long dashes is appropriate, equivalent to multiplying the load by 1.3 before using the equation. If wet cracks are encountered, or the load rests on the ice for more than two hours, the line with the short dashes applies, equivalent to multiplying the load by two.

Table 9.4 Ice thickness required for given loads and the distance required between such loads (source: http://www.mvp-wc.usace.army.mil/ice/ice_load.html)

Ice load (tons)	Required ice thickness (inches)	Distance between loads (feet)	Description of safe moving load
x	0.75–1	x	One person on skis
0.1	2	17	One person on foot or skates
x	3	x	One snowmobile
x	3	x	Group of people walking in single file
1	4	34	x
2	6	48	x
3	7	58	Single passenger car
4	8	67	One 2.5 ton truck
5	9	75	One 3.5 ton truck
x	10	x	One 7–8 ton truck
10	13	106	x
20	18	149	x
30	22	183	x
40	26	211	x

Note: x = no data. 1 inch = 2.53 cm; 1 foot = 0.3036 m

9.3.5 Road use

The most extensive ice road networks are, not surprisingly, situated in Canada, Russia and Alaska (USA), where small settlements and economic enterprises are widely scattered across generally inhospitable terrain lacking an overland transport infrastructure. Smaller systems also operate in northern Europe, especially Scandinavia (Finland, Sweden, Norway), but also in Estonia, where the numerous inhabitants of offshore islands sometimes benefit from ice road construction across the Baltic Sea for a limited season.

An indication of the complexities of ice road construction, maintenance and traffic management is provided by Canada's longest and most expensive ice road, running for 567 km from Tibbitt Lake, 70 km east of Yellowknife, to Echo Bay's Lupin Mine, the Ekati diamond mine and the Diavik diamond mine (The Diesel Gypsy, 2009); Figure 9.20). The ice road is managed by the Tibbitt to Contwoyto Winter Road Joint Venture at an annual cost of about $Can10 million (JVTC, 2009). A one-way trip to Lupin mine takes about 20 hours. During construction, ice thickness is measured every day and ice profiling, using ground penetrating radar, continues throughout the operation of the ice road. Routes along lakes follow historical GPS coordinates to avoid, if possible, rocky shoals (shallows) where sub-ice waves generated by truck traffic can erode the base of the ice, reduce its thickness and consequently reduce its load-bearing capacity.

Early in the ice road season, when an ice thickness of 70 cm is achieved over the entire road, very light loads, known as 'hotshots', are dispatched. Later, once the ice has reached 107 cm along the entire road, it is thick enough to support a so-called 'super B' tanker (a tractor (cab section of a lorry) hauling two tanks, as in Figures 9.14 and 9.15), fully loaded with 48 000 to 50 000 l of fuel and weighing approximately 41–42 tonnes. Average ice thickness at the height of the season is usually about 125 cm, sufficient to enable the transport of extremely large pieces of equipment which are sometimes required by the mining industry. Weight configuration on trailers carrying these heavier loads is reviewed by engineers individually to ensure that the loads can meet minimum ice thickness criteria (Figure 9.19) for the particular time when the load is to be transported. As Figure 9.21 shows, it is not cost-effective to risk exceeding limits!

Regular quality control checks of the profiling throughout the length of the ice road are carried out to reduce the risk of such costly accidents. In 2007, 17 members of a security firm, responsible for maintaining safety on the ice road, logged 270 000 km on the group's 10 trucks. Joint Venture representatives, and regulators from the federal Department of Indian Affairs and Northern Development (Environment Canada, Government of the Northwest Territories Highways), as well as the Royal Canadian Mounted Police conduct regular inspections. Radar speed checks are undertaken and unacceptable practices can incur verbal warnings, written warnings, five-day or seven-day

Figure 9.20 Location of the Tibbitt to Contwoyo Ice Road, Northwest Territories-Nunavut, Canada, and the proposed all-weather road between Contwoyo Lake and Bathurst Inlet port (source: http://jvtcwinterroad.ca/ [Accessed 27 09 10].)

Figure 9.21 Not a happy sight for an ice road trucker whichever way you look at it! (source: The Diesel Gypsy, 2009)

suspensions, and a season's ban for drivers who seriously infringe regulations. During the busy 2007 season, the 120 regulation infractions were largely dealt with by verbal warnings but nine five-day suspensions and seven season suspensions were also issued.

In northwestern Europe distances between settlements are generally shorter than in Canada or Russia and it is usually easier to provide all-weather links to most settlements and mines. Also, weather conditions are generally milder and therefore less favourable to ice road construction. However, construction of relatively short ice roads is possible in the more northerly latitudes of Norway, Sweden and Finland. In addition, unprepared frozen river surfaces are commonly used by skidoo drivers as these vehicles are often banned from all-weather highways. Further south, in the Baltic state of Estonia, up to six ice roads, from 3–27 km in length, may be opened between the mainland and larger, more densely populated, offshore islands, such as Saaremaa and Hiiumaa. According to the 2005 Annual Report of the Estonian National Road Administration (ENRA), weather conditions allowed the opening of just one of their six ice roads for 51 days, that connecting Haapsalu with Noarootsi (ENRA, 2005). The following year four roads opened between Haapsalu and Noarootsis and between the main offshore islands (ERA, 2006). Such a variable weather regime means that Estonian ice roads cannot be considered a regular component of their economic system and are, in effect, only a welcome bonus for these island communities.

9.3.6 Hazards

An icy surface floating on water and carrying loads of many tons sounds like a recipe for disaster, but in practice ice roads seem to be remarkably safe. Details of the causes of traffic accident fatalities in the most likely ice road areas of Yukon and Northwest Territories in Canada are not easy to obtain. The precise location of nine traffic collision accidents in Yukon and seven in NWT in 2000 (Transport Canada, 2003), between vehicles or vehicles and pedestrians is not given, so fatalities specifically associated with ice roads cannot be determined; but they are likely to be negligible. For instance, managers of the Tibbitt to Contwoyto Winter Road (described above), recorded only nine accidents, and one minor (shoulder) injury, during the record haul of nearly 11 000 truckloads during the 2007 season. Most of the accidents were collisions and, of these, four occurred on portages (lengths of road crossing land between the river- or lake-based ice road), or in a parking lot; only one accident involved a truck wheel breaking through ice and no one was injured. This safety record is attributed to several factors: the high priority given to safety, high-tech engineering backed by applied expertise, training given to drivers in 'rules of the road' and the strict enforcement of regulations, including the 15 mph limit for full loads.

The science behind ice roads is well understood and traffic, as far as possible, is strictly regulated. Ice road cracks can be healed and, for limited periods and over small areas, roads can be artificially thickened and strengthened by adding water to the surface under appropriate temperature conditions. This enables associated risks and *direct* impacts on humanity to be minimized, and accidents that do occur appear to derive mainly from rule-breaking or unofficial off-road driving. Most problems seem to arise *indirectly*, through late opening or early closure of ice roads. Nothing can be done in the short term to influence the key factors, weather and climate.

Weather is notoriously variable, both within and between years, making it difficult to formulate firm plans for the movement of goods and the holding of stocks of equipment and supplies. For instance, the Tibbitt to Contwoyto Winter Road was open for between 50 and 80 days between 2000 and 2006 (65, 72, 80, 61, 63, 69 and 50 days respectively), and the number of loads transported varied from 3703 in the year 2000 to 7981 in 2001. Such variability and uncertainty puts enormous pressure on businesses that have deadlines to meet and profit margins to secure. CBC (Canadian Broadcasting Corporation) News (see Box 9.2 for details) reported on 27 March 2006 that this ice road, serving the diamond mines, had closed with barely 60% of the normal 7000-odd loads delivered. In view of the state of the road there was nothing that the authorities or industry could do to remedy the situation in the short term. However, longer-term prospects of resolving this particular problem may be better if proponents of two (all-weather) road-and-port projects, designed to serve the mines, are successful in their aims. For example, the business arm of the Kitikmeot Inuit Association has been lobbying for an all-weather road from a port on Bathurst Inlet on the north coast of Canada (Figure 9.20), where access via the Northwest Passage could be a viable medium-term prospect if the extent of Arctic Ocean ice continues to decrease (see Chapter 2). As the president of the Kitikmeot Inuit Association is reported to have said 'it's a good time to say to the (diamond mining) industry and to the (Canadian) government, this is what we need in the Kitikmeot to support the mining industry, having this all-weather road in place' (CBC News, March 27, 2006). A year later, on 30 April, Reuters News Agency (2007) reported growing pressure as warmer weather, larger loads and increasing exploration and mine development in the region put a strain on the ice road route. An alternative long-term plan was required to adequately support a diamond industry which accounts 50% of the gross domestic product of NWT.

Climate is changing in the Arctic (IPCC, 2007) and there is widespread evidence of warming and concomitant shortening of ice road seasons. The Wrigley to Tulita (NWT, Canada) ice road season, for example, has shortened on average from 65 days (17 January – 23 March) to 43 days (2 February – 16 March) (GNWT, 2009) over the last 15 years, while over the same period the average season of the Tulita to Deline ice road has shortened from 65 days (23 January – 27 March) to 52 days (28 January – 20 March). If air temperature is simply not cold enough, for long enough, ice does not form to the required thickness, and trucks and their drivers are forced to remain in the parking lot. As long as this situation persists, those awaiting delivery of vital supplies may be in danger of freezing, starving or even dying for want of fuel or food or medical assistance.

Potential hazards associated with late opening of vital ice roads are often vividly described in reports by the media (see Box 9.2 for detailed examples). For instance, a major crisis faced the First Nation inhabitants of the Wollaston Lake and Fond-du-Lac settlements in northern Saskatchewan, Canada, in January 2006 (CBC, 2006a). Petrol ran out and supplies of propane, used to heat homes and the local school, were very low. Risks were taken on badly cracked roads to obtain extra supplies. Fuel had to be flown to some communities but at considerable expense to the inhabitants. Similarly, on 8 January 2009 it was reported (CBC, 2009a) that some residents of Wrigley, NWT, Canada, had been risking their lives since 12 December in crossing the Mackenzie River on foot in order to reach Fort Simpson, 200 km to the south, to obtain groceries and other necessary supplies, rather than charter an expensive, often unaffordable, aircraft. At the time of this report, the opening of the ice road was at least a month behind schedule because of limited ice thickness and yet, ironically, there had been sufficient ice on the river to force the Campsell Bend ferry to shut down early in late October. Fortunately, on the 9 January, the media were able to report that NWT Department of Transportation officials had announced that the N'Dulee Ice Road on the Mackenzie River near Wrigley was now open to light traffic, to the relief, presumably, of the stranded inhabitants.

Box 9.2   Four Canadian Broadcasting Corporation (CBC) News reports illustrating ice road problems in Canada

1   'Lack of ice roads causing problems in north'

(Last Updated: Tuesday, January 17, 2006 | 9.22 a.m. ET)

Two northern communities that rely on ice roads to get supplies have run out of fuel and are facing a major crisis, community leaders say. Not only have Wollaston Lake and Fond-du-Lac First Nation run out of gasoline, but Wollaston also has a limited supply of propane, which is used to heat many homes and the local school. A third community, Uranium Lake, still has some gas, but without its ice road supplies are expected to run out later in the year. Peter Fern, who runs P & A Gas Bar in Fond-du-Lac, said he's out of gas and has been getting 10 to 20 calls a day from people wondering when he'll be restocked. He's worried he'll go through the ice if he tries to get more. He's already had some close calls. 'I took a risk,' he said. 'The ice was cracked pretty bad.' Each winter, ice roads are cleared to connect the most northern communities to points farther south. The routes cross water, so the ice has to be thick enough to support the trucks. Usually, that's no problem at this time of year but it has been a mild winter and the ice still isn't thick enough. According to Don Deranger, vice-chief of the Prince Albert Grand Council, construction on an ice road can't begin until the ice is at least 20 inches (about 51 cm) thick, but there are areas now where the ice is only 12 inches (30 cm) thick. Fuel is being flown into the communities, but that means people have to pay $2 a litre or more for it. Deranger said the communities just need a couple of weeks of cold weather to get the ice roads built.

Source: CBC, 2006a, http://www.cbc.ca/canada/saskatchewan/story/2006/01/17/ice-roads060117.html, accessed 12 01 09.

2   Lack of ice road creates hardship for Wrigley residents

(Last Updated: Thursday, January 8, 2009 | 4.15 p.m. CT)

Ice conditions on the Mackenzie River near Wrigley, N.W.T., have made it difficult, even risky, for residents to get groceries and other necessary supplies over the past two months. The NWT community of Wrigley is about 200 km downstream from Fort Simpson.. Ice on the river forced the ferry at Campsell Bend, near the community of 122, to shut down early at the end of October. However, the opening of the ice road across the river is at least one month behind schedule. 'There's been a problem with the ice freezing to a thick enough depth where traffic can cross it safely', Paul Nadjiwan of the Pehdzeh Ki First Nation in Wrigley told CBC News. The NWT Department of Transportation says temperatures have dropped recently and it expects the ice road to open next week. But having no ice road in the meantime has prompted anxious residents to find other ways of getting the food and supplies they need in Fort Simpson, about 200 km south of Wrigley along the river. 'They have to either charter a plane, which a lot of people often can't afford, or what they have been doing for almost a month now is taking chances by walking over the Mackenzie River at Campsell Bend,' Nadjiwan said. Among those taking their chances is Chief Darcy M. Moses, who had just returned to Wrigley from Fort Simpson on Wednesday. Moses had walked on the river in temperatures around $-40\,°C$. 'Ever since December 12, we've been walking across ... hauling food across, sometimes we have to go get our groceries to make the costs reasonable for the community,' Moses said, adding that some residents have also shuttled supplies across the river with snowmobiles. Usually at this time of year, a nurse would be stationed in Wrigley until the ice road is open. However, no nurse was present this year, leaving the community's only health worker to try to resuscitate an elder who had stopped breathing. 'The thought was that there may have been a possibility had there been a nurse stationed here that that elder's life may have been saved,' Nadjiwan said.

Source: CBC, 2009a, http://www.cbc.ca/canada/north/story/2009/01/08/wrigley.html?ref=rss, accessed 12 01 09.

3 NWT opens ice road to Wrigley

(Last Updated: Friday, January 9, 2009 | 4.24 p.m. CT)

The winter ice road linking Wrigley, NWT, to the rest of the territory finally opened Friday, giving relief to residents who were isolated for two months because the ice had been too slow to form. Officials with the territorial Department of Transportation announced Friday that the N'Dulee Ice Road on the Mackenzie River near Wrigley is now open to light traffic. People in the community of 122 had been cut off from the rest of the NWT since October, when ice on the river forced the ferry at Campsell Bend to shut down at the end of that month. But that ice formed slowly, meaning an ice road was not ready for traffic until Friday, about a month later than usual. As a result, some residents – including the community's chief – walked across the somewhat-frozen river to get groceries and supplies in Fort Simpson, about 200 km south of Wrigley. The lack of an ice road for so long also raised health care issues in Wrigley after an elder died last month without a nurse in the community at the time. 'Just before Christmas there, one of the elders had died in his home. The community health worker was not able to revive the senior,' said Paul Nadjiwan, band manager with the Pehdzeh Ki First Nation in Wrigley.

Source: CBC, 2009b, http://www.cbc.ca/canada/north/story/2009/01/09/wrigley-road.html, accessed 12 01 09.

4 Warm weather may shorten life of NWT ice road

(Last Updated: Wednesday, March 8, 2006 | 9.35 a.m. CT)

Truck drivers hauling supplies to three diamond mines in the Northwest Territories along a 500-km ice road say they're frustrated by new restrictions and reduced speed limits. Warm weather and deteriorating conditions have led to more restrictions on the number of trucks allowed on the road and the speed they are allowed to travel. The drivers say they are facing delays of up to 24 hours as they wait to get clearance to travel the road. It is built each winter over frozen lakes and land portages. The committee in charge of the road denies rumours it's going to close early but does say the life of the road will be limited if drivers don't stick to the speed limits. Independent trucker Rob McAllister says it doesn't look good. 'The way it's

> going right now we can get two weeks out of it, I think, but if it turns cold, maybe the end of the month,' says McAllister. 'It all depends on the drivers following the rules.' This season 9000 truckloads of construction equipment, fuel and supplies are slated to be hauled on the road.
> Source: CBC, 2006b, http://www.cbc.ca/canada/north/story/2006/03/08/truckers-iceroad08032006.html, accessed 12 01 09)

9.4 Summary

River ice has its 'pros' and 'cons'. Although fatalities are not normally a major outcome of ice jams, the annual cost of damage inflicted by this hazard is considerable and has stimulated a wide range of methods to combat its effects. The problem at regular ice jam sites can usually be solved by structural means. Sometimes, more severe problems can occur where ice jams are not anticipated or are only sporadic in their occurrence. The basic difficulty is the very varied timing and location of ice jams, but a range of strategies are now available to tackle this hazard.

In northern cold regions of the world, especially, ice roads are a fundamental component of economic life in remote areas beyond the reach of conventional transport systems. Ice road networks are dependent on climate and weather for their temporary, annual existence. Sub-zero temperatures for several weeks or months are necessary to enable ice to develop to sufficient thickness to support vehicles and their heavy loads. Careful preparation of the surface is then required and traffic regulations, strictly enforced, are necessary to minimize risks. As climate warms the Arctic, the problems faced by the inhabitants of remote settlements and mines are less concerned with the technical aspects of ice road safety, though these remain, and more to do with the ever-decreasing length of many ice road seasons. Ultimately, this latter problem will only be resolved by the construction of all-weather roads or the abandonment of inaccessible settlements.

10
Winter Storms – Ice Storms and Blizzards

10.1 Introduction

Steel pylons supporting electricity transmission lines are not supposed to bend so that the top touches the ground (Figure 10.1)! However, this is what happened to many of these tall structures, in eastern Canada and north-eastern USA, during the early days of January 1998. Over the course of about one week, this area experienced unprecedented weather conditions, during which the surfaces of trees, structures, stationary vehicles, buildings and monuments (Figure 10.2) were rapidly coated with up to 10 cm of ice. Excessive loadings caused the collapse of structures, millions of trees were damaged or destroyed and driving and walking became a hazardous exercise. Such severe accumulations of ice are referred to as ice storms. These phenomena are not uncommon in areas where temperatures remain below freezing point for weeks or months during the winter and yet are still accessible to warm air masses delivering the moisture which is a critical part of the problem.

Another, more familiar, winter hazard is the winter storm which often incorporates a blizzard, a combination of strong winds and falling or blowing snow capable of delivering large quantities of snow and bringing otherwise vibrant cities to a standstill. Unlike the hazards discussed in previous chapters, ice storms and blizzards are essentially products of atmospheric cyclonic weather systems and largely independent of landscape.

These weather systems bring together moist sub-tropical air and cold Arctic air and this convergence produces precipitation in a variety of forms, of which snow and ice will be discussed here. Under 'normal' circumstances snow and ice can be dealt with efficiently by responsible authorities and the public. Problems arise when *excessive* amounts of snow and/or ice accumulate very

---

Cold Region Hazards and Risks, First Edition.   Colin A. Whiteman.
© 2011 John Wiley & Sons, Ltd. Published 2011 by John Wiley & Sons, Ltd.

Figure 10.1 A line of Hydro-Quebec high voltage transmission towers, Quebec, Canada which collapsed during Ice Storm '98' and their replacements alongside. (source: (IEEE, 2004) http://www.ieee.org/portal/cms_docs_pes/pes/subpages/meetings-folder/PSCE/2004Presentations/281/Jean-Marie-HQ-PSCE-oct13rd2004v4.pdf [Accessed 27 09 2010].)

rapidly and hazard mitigation systems are unable to cope. Fatalities involving both people and animals are not uncommon, but the greatest impact is usually economic in nature and concerns damage or disruption to utility, transport and agricultural systems and domestic properties. Hail, which is not necessarily associated with cold temperatures at ground level, and extreme cold temperatures will not be considered.

10.2 Definitions

Definitions of winter storms, blizzards and ice storms are summarized in Table 10.1 (NWS-NOAA, 2010a; Environment Canada, 2007a). In North America winter weather is defined slightly differently in the USA and Canada. In Canada there are also slight differences regionally to take account of the significant latitudinal range of that country (Environment Canada, 2007a), but overall there is broad similarity and it is clear what constitutes a blizzard and an ice storm wherever they occur.

Figure 10.2 Impacts of 'Ice Storm '98' in Canada: (a) source: Windupradio (1998); (b) source: ENCS Concordia (1998)

Table 10.1 US National Weather Service and Environment Canada winter weather warnings (Reproduced from US National Weather Service (NWS 2010) http://news.bbc.co.uk/1/hi/world/americas/7137682.stm and Environment Canada (EC, 2007) http://www.pnr-rpn.ec.gc.ca/air/severewthr/warning.en.html. - accessed 23 02 10)

Weather	USA Definition	Canada Definition
Winter storm warning	Storm producing, or is forecast to produce, heavy snow or significant ice accumulations	Storm is likely to produce a combination of weather elements; one winter warning criterion is met or expected (snow, wind, windchill, freezing rain) and another is likely to be met
Ice storm warning	Freezing rain produces significant and possibly damaging accumulation of ice	Freezing rain is expected to last for 1 hour or more, or slippery or hazardous walking or driving conditions due to freezing precipitation is expected
Blizzard warning	Sustained wind or frequent gusts to 35 mph (56 km/h) or greater with considerable falling and/or blowing snow and visibility <0.25 miles (<0.4 km) for more than 3 hours	Snow or blowing snow with winds of 40 km/h or more, visibility <1 km in snow and windchill of $-25\,°C$ or colder for at least 4 hours

10.3 Weather systems and processes

Winter storms (including blizzards) and ice storms are more closely associated with atmospheric conditions than any of the other hazards being considered in this book. They are caused by some form of precipitation, generally snow or freezing rain, although other forms, such as ice pellets and freezing drizzle may make a contribution. Snow falls from the atmosphere directly. Ice accumulates on cold surfaces when they are hit by freezing rain. These different types of precipitation are generated in migratory, cyclonic (counter-clockwise), low-pressure weather systems in which air rises and cools until condensation occurs around nuclei. This may be in the form of ice crystals, which combine into snowflakes, or raindrops both of which fall as precipitation. NSIDC (2010a) provides a simple explanation of how the system works in relation to North America.

10.3.1 Air pressure patterns

Surface weather is very closely related to upper atmosphere winds at an altitude of about 5 500 m where pressure is around 500 millibars (mb) (it is usually near 1000 mb at the ground surface). At this higher level it is possible to recognize elongated areas of low pressure, called troughs (NSIDC, 2010b; Figure 10.3), where air moves southwards from the north before swinging northwards again further east. These upper air winds have the effect of strengthening the circulation of the surface low-pressure systems beneath. The surface systems appear as round or oval patterns of isobars (NSIDC, 2010c; Figure 10.4) in contrast to the elongated trough-like pattern in the upper atmosphere. The most intense surface storms usually occur in two places: just east of the upper-trough axis, where the air is being lifted most effectively and the surface 'low' becomes strongest, and to the north-west of the surface storm track, where northerly winds bring in cold, Arctic air. With the aid of both 500 mb and surface pressure maps, it should be possible to forecast where strong storms and blizzards are likely to form and subsequently move.

The position of the upper trough is obviously important in directing storms to different parts of the continent. For instance, the upper troughs associated with a devastating blizzard on the east coast of the USA on 6–8 January 1996 and a winter storm on 11–13 January 1996 were located over the eastern states, *east* of the Mississippi river. In contrast, the blizzard which struck the northern and central Midwest plains a few days later, on 17–18 January, was linked to an upper air trough located to the *west* of the Mississippi river. The associated storm track ran up the centre of the continent rather than the eastern seaboard (Atlantic coast).

Although winter storms and ice storms are both associated with cyclonic 'lows', and there is considerable overlap in their distribution, the heartlands of these two hazards are different as two recent analyses, covering all or most of

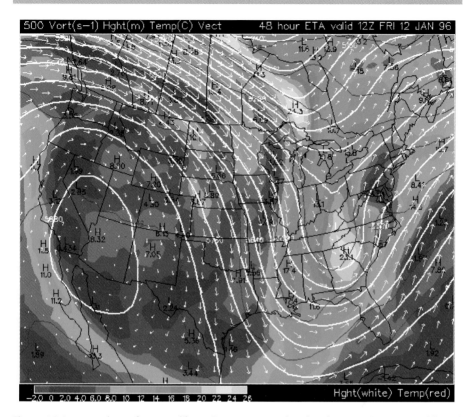

Figure 10.3   500 mb map forecasted for 12 January 1996, related to the 1996 US east-coast blizzard (Reproduced courtesy of The National Snow and Ice Data Center, University of Colorado, Boulder.)

the second half of the twentieth century have shown. Although these two analyses are based on different sources of evidence, this probably does not invalidate the comparison. Schwartz and Schmidlin (2002) used information from *Storm Data*, a publication of the National Weather Service, to produce a map of the annual probability of blizzards in the USA based on blizzard frequency (Figure 10.5). This highlights North and South Dakota and western Minnesota as the dominant area with a 50%–76% probability of an annual blizzard. Changnon and Changnon (2002) used records of loss compiled by the property-casualty insurance industry to produce maps showing the distribution of ice storms and associated losses (Figure 10.6(a)–(d)). These maps emphasize the east and north-east states as the main focus of the ice storm hazard. It is likely that this difference is related to differences in the source of cyclogenesis, and hence the structure and temperature of 'lows'.

Whittaker and Horn (1981) recognized three main areas of cyclogenesis. First, the Gulf of Mexico coast in winter and the US east coast produce the so-called 'Nor'easter' storms that can intensify over the Atlantic Ocean and bring high winds and much snow or ice to the east coast of North America.

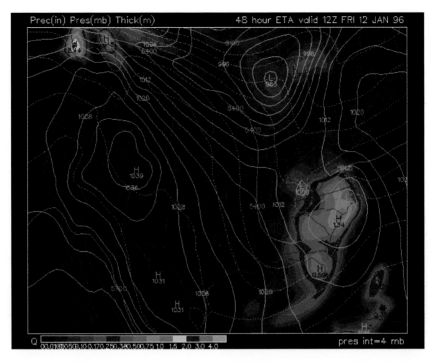

Figure 10.4 Surface pressure map (blue lines – mb) with distribution of precipitation in inches (Reproduced courtesy of The National Snow and Ice Data Center, University of Colorado, Boulder.)

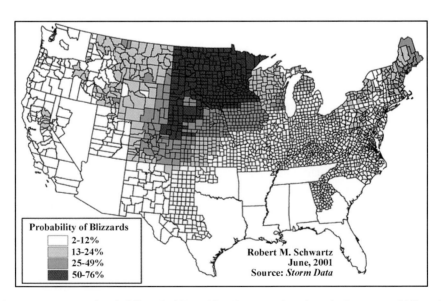

Figure 10.5 Annual probability of a blizzard by US county, based on the frequency of blizzards during the winter periods 1959/60 to 1999/2000 (from Schwartz and Schmidlin, 2002, Figure 4)

Figure 10.6 Ice storm occurrence in the 48 contiguous states of the USA, 1949–2000: (a) Number of major ice storms by climatic region – values in parentheses indicate number of local storms; (b) number of times each state incurred losses from ice storms; (c) contoured plot showing average annual number of days with freezing rain; (d) losses (US$million in 2000 values) from ice storms in each climate region – values in parentheses give average losses per event (Reproduced with permission from Changnon, S.A. and Changnon, J.M. (2002) Major ice storms in the United States, 1949-2000. Environmental Hazards, 4, 105-111. Figs 1–4.)

Sea surface temperatures in the western Atlantic Ocean appear to control the relative amount of snow and ice. A warmer ocean converts more snow into freezing rain while a cooler ocean reduces the air temperature and produces mostly snowfall (Da Silva *et al.*, 2006). Second, Colorado and the Great Basin produce storms that affect eastern USA and Canada. Third, Alberta and the Northwest Territories in Canada, produce the so-called 'Alberta-clipper' storms over the Canadian Rocky Mountains which eventually affect the central and eastern provinces of Canada, the Great Lakes region and north-east USA. Of these three locations, the Colorado region appears to be the greatest source of winter storms in North America.

While the overall source and trajectory of the winter storms is controlled largely by air pressure in the upper and lower levels of the atmosphere, the actual precipitation type that is directly responsible for the blizzard and ice storm hazards is related more directly to air temperature.

10.3.2 Air temperature

In cyclonic, low-pressure systems, masses of air are drawn in towards the core before rising in a counter-clockwise direction. In North America cold arctic air is drawn in from the north and east and warm, subtropical air from a southerly direction. Being less dense the warm air circulates over the cold air, gradually cooling as it rises until ice crystals and snow form. The leading edge of this wedge of warm air rises over the cold air and the sloping boundary between the two types of air is the warm front. The trailing boundary at the other side of the warm wedge is the cold front. The cross section through the warm front (Figure 10.7) illustrates different vertical distributions of warm and cold air and the different types of precipitation that can be expected at any location.

To the left of the figure, virtually the whole of the air column is occupied by the warm sector. Here, the snow at the top of the column melts as it falls through the warm air. The cold wedge is too thin to freeze the precipitation and so it hits the ground as rain. Further up the warm front the snow again melts but this time the rain enters a thicker layer of cold air where it may become supercooled (exist as a liquid at sub-zero temperatures) if there are insufficient nuclei for the water to crystallize around. These supercooled droplets form freezing rain, which immediately freezes on contact with surfaces that have been cooled by the cold air, to form a clear, dense glaze. Depending on the amount of precipitation several centimetres of ice can accumulate on surfaces, sometimes with disastrous results (see Box 10.1 for examples). (Incidentally, an interesting variation on the freezing rain theme is 'freezing spray', which is produced when strong onshore winds lift and carry droplets from the ocean inland. For instance, the NOAA Hazard Assessment for 4 February 2010 carried a 'heavy freezing spray warning' for the Alaskan Peninsula (NOAA, 2008)). Higher on the warm front, the cold air layer is too thick and the rain droplets, formed from the melted snow, refreeze to a varying extent

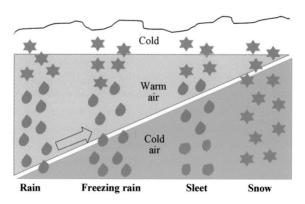

Figure 10.7 Types of precipitation associated with a warm front in a cyclonic low pressure circulation (note: names relate to the lowest symbol in each column)

Box 10.1 Examples of snow and ice storm impacts

1888: 112 deaths and many cattle killed in the Northern Plains of the USA and southern Canada (Gray and Male, 1981).

1941: 39 deaths in eastern North Dakota and 76–90 in the USA overall (Gray and Male, 1981).

30 January to 8 February 1947: Worst blizzard in Canadian railway history. A 10-day blizzard buried towns and trains from Winnipeg to Calgary, causing some Saskatchewan roads and rail lines to remain plugged with snow until spring. Children stepped over power lines to get to school and built tunnels to get to the outhouse. A Moose Jaw farmer had to cut a hole in the roof of his barn to get in to feed his cows (Environment Canada, 2010).

27 to 31 January 1966: 257 cm of snow fell at Oswego, New York State, in one storm. This is downstream of the Great Lakes, an area which regularly have 'snow bursts' due to moisture being evaporated from the lakes. Also during 1966, 130–160 km/h blizzard winds buried trains and collapsed buildings; 36 cm of snow fell in Winnipeg and required a street cleaning bill of Can$1million (Gray and Male, 1981).

4 March 1971: Montreal's Snowstorm of the Century – Montreal's worst snow storm killed 17 people and deposited 47 cm of snow on the city with winds of 110 km/h producing second-storey drifts. Winds snapped power poles and felled cables, cutting electricity for up to 10 days in some areas. In total, the city hauled away 500 000 truckloads of snow (Environment Canada, 2010).

February 1978: 60 deaths and US$1billion damage from a snow storm in New England. Massachusetts manufacturing, retail and service industries lost US$441million (Gray and Male, 1981).

8 February 1979: A blizzard isolated Iqaluit, Baffin Island (now capital of Nunavut). Weather with $-40\,°C$ temperatures, 100 km/h winds and zero visibility in snow kept residents of Iqaluit indoors for 10 days (Environment Canada, 2010).

13 April 1984: Newfoundland glaze storm cut power to 200 000 residents of the Avalon Peninsula who were without electricity for days when cylinders of ice as large as 15 cm in diameter formed on overhead wires. The severe, two-day ice storm covered all of south-eastern Newfoundland with 25 mm of glaze (Environment Canada, 2010).

Ice storm of 9–13 February 1994 in south-east USA: Unusual for its large extent (it began in Mississippi, Tennessee and Alabama where the impact was worst and spread north-east into the Carolinas, Virginia and Kentucky) and intensity – up to 6 inches (15.2 cm) of ice was unprecedented in southern areas. Two million people were without electricity for some time. The ice storm resulted in over US$3 billion in damage (Lott and Ross, 1994).

and hit the ground as sleet or ice pellets. This precipitation is usually less hazardous. Finally, at the edge of the depression, the wedge of warm air is very thin or absent and the high-level snow falls all the way to the ground without melting.

Of course, natural systems like these low-pressure areas tend not to behave perfectly according to a diagram in a book. They vary as they move across the landscape and different types of precipitation are deposited in a particular area over time. This is illustrated by the records kept during the severe ice storm in south-eastern Canada and north-eastern USA in early January, 1998 (Environment Canada, 1998; Figure 10.8(a)–(c)). Several other types of precipitation occurred at the two locations during the course of the storm, but it was freezing rain that dominated and which was the major cause of severe impacts.

10.4 Impacts

It is difficult to grasp the magnitude and intensity of a single blizzard that is reckoned to have caused 4000 fatalities, but this is reported to have happened in Iran between 3 and 9 February 1972, when whole villages and their occupants were buried by huge snow drifts (NOAA News, 2010). This was an exceptional event, with perhaps 10 times the number of fatalities recorded in the worst North American event (400 in 1888; Gray and Male, 1981). Winter storms are not unusual and are widely distributed around the northern hemisphere. For instance, Mongolia became a focus of attention in 2000 when blizzards during the winter of 1999–2000 killed about 2 250 000 livestock, a resource on which many of Mongolia's population depend. Many other countries suffer similarly from snow storms and ice storms, but the rest of this discussion of winter storm impacts is focused on North America (USA and Canada) where data and case studies are readily available.

Two of the most damaging blizzards in North America occurred in the mid1990s. 1993, between 12 and 15 March, recorded what has been termed the 'Storm of the Century'. An unusually deep upper trough developed with warm, moist air drawn in from the Gulf of Mexico and Arctic air circulating around a Midwest high-pressure area. The result was a huge storm with hurricane force winds over 26 eastern US states and much of eastern Canada, in which 270–300 people lost their lives (Lott, 1993; Table 10.2). Only three years later, between 6 and 8 January 1996, a major blizzard caused 154 deaths across 17 US states from Arkansas to New York, of which 80 occurred in Pennsylvania alone. The storm deposited 121.4 cm (48 inches) of snow at Snowshoe, West Virginia. An additional 33 deaths were attributed to subsequent floods as the snow melted, and the total damage was reported to be US$3billion. More recently, the City of Buffalo in New York State, notorious for its heavy snowfalls, received a record 610 mm of snow in just 16 hours. The combination of very cold Arctic air meeting exceptionally warm (15.5 °C), moist air over Lake Erie was the cause

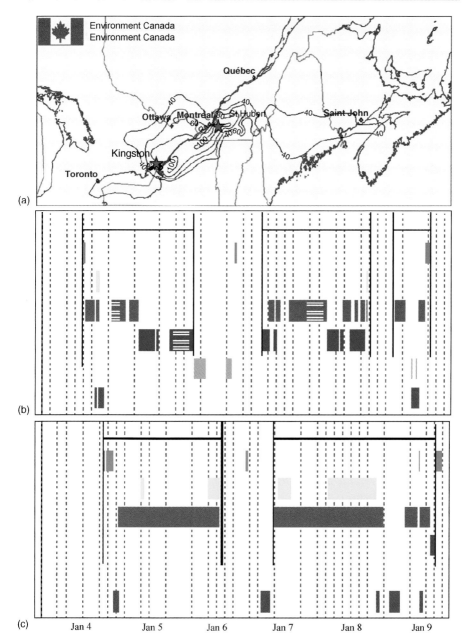

Figure 10.8 (a) Preliminary map of freezing rain accumulations in mm between 4 and 10 January 1998 in Eastern Canada (updated on 4 March 1998); (b) timing of different types of precipitation at Kingston, Ontario, Canada; (c) timing of different types of precipitation at St Hubert, Quebec, Canada; green = snow; yellow = ice pellets; red = freezing rain; blue = drizzle; grey = freezing drizzle (source: modified from Environment Canada, 1998)

Table 10.2 Summary of Blizzard 1993 impacts (source: Lott, 1993)

Category	Impact
Duration	12–15 March 1993
Extent	26 US states and much of eastern Canada
Snow thickness	1.1 m of lying snow; snow drifts to about 10.7 m
Wind speed	Hurricane force wind gusting to 144 mph on Mt Washington, NH
Temperature	Record low March temperature (-12 degrees F) at Burlington, Vermont
Population affected	100 million
Electricity	Loss of power to 3–10 million people
Airports	Every airport between Atlanta, Georgia, and Halifax, Nova Scotia, closed for some time; for 2 days 25% US flights cancelled
Losses	Property damage >US$3billion Overall costs US$6–10 billion
Fatalities	270–300 (cf. 1888 storm = 400)

and resulted in at least 10 fatalities, widespread damage to trees and power lines, 80% of roads impassable and the closure of schools and the airport (NWS-NOAA, 2006; Figure 10.9(a) and (b)).

Ice storms can be just as devastating as snow storms. Between 4 and 10 January 1998, an ice storm, on a scale unprecedented in Canada, crossed Ontario, Quebec, and the Maritime Provinces of New Brunswick and Prince Edward Island, and the north-east states of the USA (McCreedy, 2004; RMS, 2008; NWS-NOAA, 2008; Figure 10.10; Table 10.3). There were 35 fatalities, and the economic cost, like the 'Storm of the Century', was very high, amounting to Can$5.1billion in Canada alone. Even huge electricity transmission line pylons folded under the weight of accumulated ice and 4.7 million people lost their power supply, some for several weeks.

In the USA, between 1949 and 2000, freezing rainstorms resulted in losses totalling US$16.3billion (Changnon, 2003). Further examples of the impacts of snow and ice storms can be found in Box 10.1, and in Table 10.4, which gives an impression of winter storm impacts in one country, Canada, across a whole year.

On average, 105 snow producing storm systems hit the 48 contiguous states of the USA each year (NOAA Economics, 2010). The impacts of snow storms on the scale of whole cyclonic depressions can be substantial (NWS-NOAA, 2000; Table 10.5) and often complex, with many 'knock-on' effects that may be difficult to quantify. Ice storms have generally had less impact than blizzards and are less frequent (Table 10.6).

From the descriptions of winter storms impacts can be conveniently classified under five headings:

(a) *fatalities and injuries to humans*;
(b) *physical damage*: structures, buildings, vehicles, trees, crops, livestock and flood damage during melt;

10.4 IMPACTS

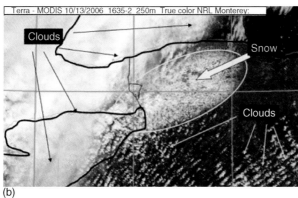

Figure 10.9 (a) Distribution of snow (inches) around the eastern end of Lake Erie, New York State, USA; (b) satellite image showing extent of snow on 13 October 2006 (source: http://www.erh.noaa.gov/buf/storm101206.html – accessed 27 02 10; Photo courtesy of the Naval Research Laboratory web site's National Polar Orbiting Environmental Satellite System (NPOESS))

(c) *economic losses:* costs of removal of ice and snow, lost retail trade, lost wages and lost tax revenue;
(d) *transport and traffic disruption*: roads, railways and ports;
(e) *power loss* (outage) with widespread 'knock-on' effects across both domestic and industrial sectors.

10.4.1 Fatalities and injuries

According to NWS-NOAA (2010b), most winter storm-related deaths are not due directly to the impacts of the storm but to indirect effects such as

Figure 10.10 Pattern of high-level winds controlling flow of warm moist air into north-eastern USA and Canada during Ice Storm '98 – G o M = Gulf of Mexico (source: NWS-NOAA, 2008)

traffic accidents on icy roads, heart attacks while shovelling snow and hypothermia from prolonged exposure to cold. Of those related to exposure to cold, 50% are people over 60 years old, more than 75% are males and only 20% occur in the home. Of those related to ice and snow, about 70% occur in cars and about 25% are people caught out in the storm. Again the majority are males over 40 years old. Falling objects, especially trees, are another cause of fatalities.

10.4.2 Physical damages

It is obvious from the descriptions of storms that the amount of damage is related to storm intensity, length and extent on the one hand, and population density on the other. It is the very large, often unprecedented storms, such as the 1993 'storm of the century' (Table 10.2) or the 1998 ice storm (Table 10.3), that impact most heavily. Unprecedented storms are obviously difficult if not impossible to forecast accurately because they are beyond peoples' experience. However, generally, as forecasting has become more sophisticated and systems of mitigation have improved with experience, losses from damage have been reduced in some areas. For example, it is unusual to suffer very large losses of animals these days as most farmers are adequately equipped with indoor facilities.

10.4.3 Economic losses

Except for the direct cost of clearing snow and ice, economic losses are often less obvious than physical damage because they result from indirect effects.

Table 10.3 Summary of Ice Storm '98 impacts (sources: McCreedy, 2004, and RMS, 2008)

Category	Impact
Duration	4–10 January 1998; freezing rain or drizzle for 80 hours (average annual 45–65 hours in periods of a few hours only) arriving in three consecutive waves
Extent	S. Ontario to Nova Scotia in Canada; New England states and New York State in USA
Ice thickness	Up to 100 mm south of Montreal
Population affected	18% of Canadian population
Electricity	1000 pylons wrecked, including 130 major structures at Can$100 000 each; 30 000 poles down; 1850 km of network down; total rebuild of electricity network Can$1billion; 4.7 million people lost electricity, some for several weeks before system was restored
Employment	2.6 million (19% of local population) failed to get to work or were late
Agriculture	36% of Quebec and 22% of Ontario in 40 + mm zone; 13.5 million litres of milk dumped worth Can$7.8 million; 1.5 million litres processed in USA and returned to Canada; unmilked cows prone to disease
Trees	1 m trees including valuable sugar maple trees destroyed (Quebec supplies 70% of world maple syrup); industry lost Can$25 million; takes 30–40 years to restock
Losses	Short-term loss of economic output estimated Can$1.6 + billion; total storm-related costs in Canada Can$5.1billion
Insurance	600 000 domestic and industrial insurance claims worth over Can$1 billion
Shelter	100 000 people required shelter, mainly due to loss of lighting and heating
Assistance	16 000 army personnel assisted; largest peacetime deployment
Fatalities	35 (Canada 28, USA 7) mainly trauma and hypothermia (and 945 injuries)

Economic losses are usually incurred by businesses and employees, but the state also loses through receiving reduced tax revenues.

10.4.4 Transport and traffic disruption

Losses and inconvenience experienced through the interruption of road and rail services are mostly incurred by those in the affected area, but in the case of airports 'knock-on' effects can be experienced far beyond the impacted area. For example, during the blizzards of February 2010 major Washington, DC airports were closed (BBC News, 2010c) and flights to Washington, DC from London airports, on the other side of the Atlantic Ocean in the UK, were also cancelled.

Table 10.4 Aggregate impact of snow on one country, Canada, during one year, 1966 (source: Mergen, 1997)

Quebec	January	115 km/h winds; traffic disruption and some deaths in Montreal
British Columbia	January	Electricity transmission towers collapsed; snow slides, heavy snow and freezing rain
British Columbia	January	Rogers Pass, Rockies, BC, main transport route with seasonal snow of 1481 cm closed for 583 hours (24+ days) due to snow and avalanches compared to average of 194 hours; Can$200 000 spent on artillery release of avalanches
Newfoundland	January	Blizzard in Newfoundland stopped a ferry
Alberta	February	100 km/h winds deposited 10 cm of snow
Saskatchewan	February	100 km/h winds deposited 10 cm of snow
Ontario	February	13 cm in 80 km/h winds which reduced visibility to zero and caused drifting and impassable roads
Quebec	February	13 cm in 80 km/h winds which reduced visibility to zero and caused drifting and impassable roads
Manitoba	March	80 km/h winds deposited up to 35.6 cm of snow and paralysed S. Manitoba for 3 days; street clean-up bill in Winnipeg was Can$1 million
New Brunswick, Prince Edward Island	March	St Lawrence Valley without electricity for a week following a 72 hour snowstorm and freezing rain; 300–600 cm of snow fell
Alberta	April	Storm affected Edmonton in Alberta

Table 10.5 Impacts of winter storms in the USA, 2000–2008 (Compiled from NOAA NWS, Summary of Natural Hazard Statistics for 2000 in the United States, 2000., (http://www.economics.noaa.gov/?file=bibliography#noaa.2000c) and individual years to 2008.)

Year	Fatalities	Injuries	Property damage (US$ million)	Crop damage (US$ million)	Total damage US$ million)
2008	21	121	931.9	19.7	951.6
2007	9	159	101.0	0.2	101.2
2006	17	109	571.0	0.0	571.0
2005	34	72	293.8	0.1	293.9
2004	28	190	183.5	0.2	183.7
2003	28	112	499.4	8.6	508.0
2002	17	105	752.0	0.0	752.0
2001	18	173	103.6	0.1	103.7
2000	41	182	1035.3	0.0	1035.3
1999	39	316	61.9	0.0	61.9
1998	64	285	525.9	1.6	527.5
Total	316	1824	5040.6	30.5	5071.1
Average	28.7	165.8	458.2	2.8	461.0

(http://www.economics.noaa.gov/?file=bibliography#noaa.2000c) – accessed 02 02 10 and individual years to 2008. NOAA-NCDC (1998, 1999): http://www.economics.noaa.gov/?goal=climate&file=events/snow&view=costs – accessed 17 02 09.

Table 10.6 Impact of ice storms in the USA, 2000–2008

Year	Fatalities	Injuries	Property damage ($ million)	Crop damage ($ million)	Total damage ($ million)
2008	0	0	104.1	0	104.1
2007	7	11	1379.8	0	1379.8
2006	0	0	0	0	0
2005	0	0	0	0	0
2004	0	0	0	0	0
2003	0	0	0	0	0
2002	0	0	0.1	0	0.1
2001	0	0	0.4	0	0.4
2000	0	0	0.3	0	0.3
Total	7	11	1484.7	0	1484.7
Average	0.8	1.2	164.9	0	164.9

Compiled from NOAA NWS, Summary of Natural Hazard Statistics for 2000 in the United States, 2000. (http://www.economics.noaa.gov/?file=bibliography#noaa.2000c) and individual years to 2008.

10.4.5 Power loss

This is often a significant impact because of the juxtaposition of power lines and trees (Figure 10.2(b)). Disentangling the mess is time-consuming and it often takes several days or even weeks before power is restored to everyone (Table 10.3). This is a 'bone of contention' between home-owners and utility companies (see mitigation section below). In the case of all the power transmission pylons that collapsed in 1998, the lack of precedence for such a large storm is obviously relevant. However, when ice-sensitive structures fail in an ice storm it should not be automatically assumed that ice loading is the cause, as collapse may have been initiated by a single component failing because of *previous* damage or deterioration (Jones and Mulherin, 1998).

10.5 Mitigation

Some attempts have been made to manipulate the atmosphere, such as inducing rainfall to overcome drought. North America even has an Interstate Weather Modification Council (NAIWMC, 2010), and a number of active projects in the western half of the USA, but although some success has been claimed, there is still a good deal of scepticism (American Meteorological Society, 1998; Nature, 2008b). Perhaps this will be addressed in the USA, if the new Weather Mitigation Research and Development Policy Authorization Act of 2009 (Bill, S. 601) is successful in establishing a Weather Mitigation Research Program in the National Science Foundation to develop a national research strategy and grant programme on weather mitigation.

However, atmospheric manipulation is certainly not a realistic approach to the mitigation of blizzards and ice storms at the present time, in view of the large scale of snow- and ice-bearing weather systems. Ice storms and blizzards will continue to happen for the foreseeable future and their mitigation is more likely to be achieved by:

(a) effective forecasting,
(b) detailed forward planning on the part of the authorities and the public,
(c) efficient and effective handling of the problem to minimize its effects after it has arisen,
(d) a sensible response by the public to the associated dangers,

rather than by attempts to interfere with complex, fast-moving natural systems. In other words successful mitigation relies on effective *responses* to the hazard before, during and after the event, rather than on attempts to manipulate the problem before it has turned into a hazardous event. Call (2005) recognizes a range of factors that contribute to the effective mitigation of winter storms: meteorological factors, the response of government, the response of the public, the relationship between meteorologists and the media and a number of additional factors (Table 10.7).

10.5.1 Forecasting

Forecasting is based on predictive models of atmospheric behaviour developed from meteorological science and from records of past events. According to NOAA (2006) the 'Storm of the Century' in 1993 was the first severe weather

Table 10.7 Factors affecting snow events (modified from Call, 2004)

Meteorological variation	Governmental response	Actions of the general public	Meteorologists and the media	Additional factors
Amount of snow	Equipment	Parked cars	Accuracy of forecasts	City layout and terrain
Intensity of snow	Preparation	Behaviour of motorists	Lead time	Credibility of government, meteorologists and media
Timing	Use of contractors	Choice to stay home or not	Tone of coverage	Experience with past events
Temperature	Politics	Adherance to driving bans	Amount of coverage	
Snow density	Outside aid	Criminal behaviour		
Wind strength	Budget	Clearing footpaths		
Wind duration				

event that was successfully forecast, five days in advance, using numerical weather prediction models that had benefitted from advances in global analyses, numerical modelling and computing power to increase their resolution and accuracy. Storm and blizzard warnings were given two days in advance, to the 100 000 000 people in the eastern third of the United States and, more particularly, to the responsible authorities. The Governors of the New York and the New England states, for instance, decided to declare states of emergency *prior* to the event and to take actions to mitigate a potential disaster. It is impossible to know the extent to which damage and deaths were reduced in 1993 and since by this forecasting capability, but it must make some difference to the ability of authorities and the public to plan their campaign of mitigation.

Today, the NOAA's National Weather Service (NWS) continuously updates a 'Hazards Assessment' for the USA that can be viewed online as both a map and text (NOAA, 2008). Its stated aim is

> 'to provide emergency managers, planners, forecasters and the public advance notice of potential hazards related to climate, weather and hydrological events' [by] 'integrat[ing] existing NWS official medium (3–5 day), extended (6–10 day) and long-range (monthly and seasonal) forecasts and outlooks, and hydrological analyses and forecasts...use[ing] state-of-the-art science and technology in their formulation.'

10.5.2 Forward planning

Forward planning is based on expectation and anticipation. Given the long history of previous storms, it is not unreasonable to expect more in the future. It is better to anticipate what might happen, and prepare for it, than to react after the event. In many places in north-east USA and south-east Canada where ice storms are frequent, there have been enough storms in the past to allow the probability of future events to be calculated. For instance, after the 1998 ice storm, CRREL personnel (Jones and Mulherin, 1998) carried out an evaluation of the severity of the storm in the US sector. From the 1998, and earlier, data they estimated return periods for large ice loads (Table 10.8). For upstate New York and north-west Vermont, for instance, an ice storm capable of depositing 1 inch (25 mm) of ice, which itself would be capable of collapsing a tower, could be expected *on average* every 65 years (Table 10.9); but two events of this magnitude could occur in consecutive years if the region was very unfortunate.

Presumably the high probability of a severe event justifies larger expenditure on preparation and equipment, given the greater scale of the adverse consequences that can be expected. Obviously forward planning is constrained by available funds and the ability of the authorities to foresee problems in their areas and develop appropriate procedures to tackle them when they arise.

Table 10.8 Return periods (years) for large ice loads in north-east USA (from Jones and Mulherin, 1998, Table 8)

Uniform ice thickness inches (mm)	Upstate New York and NW Vermont	Coastal and central Maine and New Hampshire
0.50 (12.7)	15	5
0.75 (19.0)	35	20
1.00 (25.3)	65	40
1.25 (31.6)	100	85
1.50 (38.0)	145	165

In Montreal, where freezing rain typically falls between eight and 18 times a year (CRIACC, 2008), in short bursts depositing only a few millimetres of ice, trained road crews are ready to use de-icing material to rapidly reduce the risk of minor traffic collisions and pedestrian accidents. Electricity transmission pylons, lines and other equipment are built according to demanding standards since large ice-accumulation events have happened many times. After the 1961 ice storm over Montreal deposited 30–40 mm of ice, Quebec Province authorities raised the building standards for this equipment. In spite of this improvement, as noted earlier, the 1998 event devastated the power transmission network in the same area. The problem is that Ice Storm '98 was unprecedented in Canada. This raises the question, how high can the standards be set above previous maximum impact levels when every increase in standards is likely to increase costs which are usually politically sensitive. Based on their analysis of the 1998 ice storm in the US sector, Jones and Mulherin (1998) proposed the raising of the 50-year return period ice load from 0.6 inches (15.2 mm) to 0.9 inches (22.8 mm). In other words larger impacts should now be expected as frequently as the smaller ones previously. On the basis of this

Table 10.9 Probability of exceeding the load for a given return period at least once, $P_n = 1 - (1 - P_1)^n$ (from Jones and Mulherin, 1998)

Lifetime (years)	Return period for load		
	50 years	100 years	300 years
1	0.02	0.01	0.003
2	0.04	0.02	0.01
5	0.10	0.05	0.02
10	0.18	0.10	0.03
20	0.33	0.18	0.06
40	0.55	0.33	0.13
50	0.64	0.39	0.15
100	0.97	0.63	0.28
300	1.00	0.95	0.63

analysis, they revised the ice-load map for a manual of minimum design loads for buildings and other structures. Nevertheless, impacts can be reduced by introducing and maintaining effective engineering standards. The applicable code for overhead lines is the National Electrical Safety Code (NESC) and the standard for communication towers is the Structural Standards for Steel Antenna Towers and Antenna Supporting Structures (Wireless Estimator, 2006).

However, no amount of forward plans will be of any use if the event turns out to be beyond the capacity of the public and authorities to deal with it, or to execute appropriate plans effectively.

10.5.3 Effective procedures

Authorities in most winter storm-prone localities have plans for dealing with excessive snow and ice. Good leadership is essential (Call, 2005). When this breaks down or political differences affect snow mitigation operations, Call argues that minor snow *storms* can quickly become major snow *events*. His summary of an article in the *Buffalo* [New York State] *News*, 18 March 1936, emphasizes this point (Call, 2005, p. 1787):

> 'Among the cities studied, Buffalo [average snow depth per year, 229 cm] has experienced the most politically enhanced snow events. A budget dispute between Mayor George Zimmerman and Buffalo Common Council was already simmering when the St Patrick's Day 1936 storm hit the city. Because the snow budget was exhausted, Zimmerman did not order crews to begin [ploughing] until the evening, well after the snow was already settling into a dense, slushy mess. Council blamed the lack of funds on fiscal mismanagement by the mayor, while the mayor-appointed public works director accused the council of failing to provide sufficient funds for equipment.'

By 2005, Call believed that snow mitigation in Buffalo is efficient, the public are experienced and forecasters skilled, but that does not mean that particularly intense, poorly timed (e.g. rush hour) snow accumulations cannot reduce a city to chaos as it did on so-called 'Gridlock Monday', 20 November 2000, when 63 mm of snow fell.

The efficiency problem is not confined to Buffalo, or to the 1930s. This is part of another newspaper article posted in the '*Sentinel and Enterprise*' in Fitchburg, Massachusetts, on 1 February 2010 (Devlin, 2010):

> 'David Keese says fielding angry phone calls from Fitchburg residents who want city sidewalks cleared after snowstorms is part of his job as dispatcher for the Department of Public Works (DPW).... DPW Commissioner Lenny Laakso detailed the department's staffing and equipment limitations last week. There are 15 DPW workers who take to the streets for snow and ice removal when a

storm hits, plus a few workers are borrowed from the Parks Department to shovel around city buildings, for a total of about 18. That's a far cry from the 140 workers who would be taking on the same work before Proposition 2 1/2 passed in the early 1980s, which limits the amount cities can hike [raise] property taxes unless they vote to override, according to Laakso. And the DPW right now relies on just one sidewalk [plough] for clean up, plus a few snow blowers, Laakso said.'

Across the USA, in Oklahoma City in 2006, there was a more positive reaction to snow. Here, after an emergency meeting of the Oklahoma County Board of County Commissioners on 19 January 2006, it was decided to designate north-to-south and east-to-west 'super snow routes' to facilitate travel within Oklahoma County (Oklahoma County, 2006).

Doubtless, many more examples of the human input to ice and snow hazards could be found to support Call's (2005) thesis that the impact of snow and ice storms cannot be adequately analysed unless the human element is properly addressed. This is also an essential component of the 'Northeast Snowfall Impact Scale', recently devised by Kocin and Uccellini (2004) and then adopted by the National Climatic Data Centre (Squires and Lawrimore, 2006)(Table 10.10 and Section 10.5.5 below) to categorise the magnitude of snowfall impacts.

In the worst cases, like the 1998 ice storm in the US sector, mitigation was enhanced by a contribution from US Central Government, which declared a 'disaster area' and called in the National Guard in the USA. In Canada, the Government mobilized 16 000 troops as a contribution to the hazard mitigation effort.

10.5.4 Public response

It is clear from the Massachusetts newspaper report, above, that most citizens expect to see snow and associated obstructions, such as fallen trees and structures, removed from roads and pathways (sidewalks) as soon as possible so that they can carry out their normal business. However, the public also have a part to play in the effective mitigation of these hazards (Call, 2005). Often, in the early stages of an event, the safest place to be is at home, as long as this is

Table 10.10 NESIS (Northeast Snowfall Impact Scale) categories, corresponding values and descriptive adjectives (Reproduced from http://lwf.ncdc.noaa.gov/snow-and-ice/nesis.php#overview.)

Category	Value	Description
1	1–2.49	Notable
2	2.5–3.99	Significant
3	4–5.99	Major
4	6–9.99	Crippling
5	>10	Extreme

weatherproof and the roof does not collapse under excessive loads of snow or ice. Outside, slippery surfaces are the cause of accidents to pedestrians and vehicles. Vehicles, either parked or moving erratically on highways, can obstruct snowploughs and delay clearance operations.

In their report on the 1998 ice storm in north-east USA, Jones and Mulherin (1998) emphasized the fact that, while there was little damage to *high-voltage* transmission lines and communication towers, *lower-voltage* transmission lines and distribution lines were often vulnerable to damage from broken branches and fallen trees, a point made clearly in a Department of Energy report (Commonwealth Associates, 1979) two decades earlier. The recommendation at that time for reducing the number of power outages, was 'more frequent and thorough tree trimming and better planning in the planting of new trees near power lines', but the matter of aesthetics is a consideration for people who believe that trees enhance a neighbourhood. An alternative possibility is that electrical cables could be buried to prevent ice build up (Public Safety Canada, 2009). Either way, expense is involved and an early satisfactory solution to this particular, widespread problem seems unlikely.

10.5.5 North-east Snowfall Impact Scale (NESIS)

Perhaps the response of authorities, the public, insurance companies and other interested parties to the blizzard hazard (and, potentially, ice storms) would be enhanced if they had a measure of the scale of the problem, similar to the Fujita (tornadoes) and Saffir–Simpson (hurricanes) Scales that classify other important types of hazard. This is one of the aims of the North-east Snowfall Impact Scale (NESIS) developed by Kocin and Uccellini (2004) of the National Weather Service. Its development also reflects the importance of the north-east region of the USA in terms of knock-on transportation and economic effects on the rest of the country.

In essence, the scale is a measure of snow volume combined with the population affected. The scale ranks (1-5; Table 10.10) high-impact north-east snowstorms with large areas having at least 25 cm (10 inches) of snowfall. Unlike other storm indices (e.g. Zielinski, 2002, which relies exclusively on meteorological data), NESIS adds population information to meteorological measurements in order to indicate impacts on society. NESIS values are therefore calculated from three variables (snowstorm area, snow thickness and the size of the population within the area of the storm) within a geographical information system (GIS) (Figure 10.11). Since being proposed (Kocin and Uccellini, 2004), it has been refined by the National Climatic Data Centre in order to formalize quality control for a public service (see Squires and Lawrimore, 2006, for details). NESIS values vary from around one for 'notable' storms, to over 10 for 'extreme' storms, the end members of five

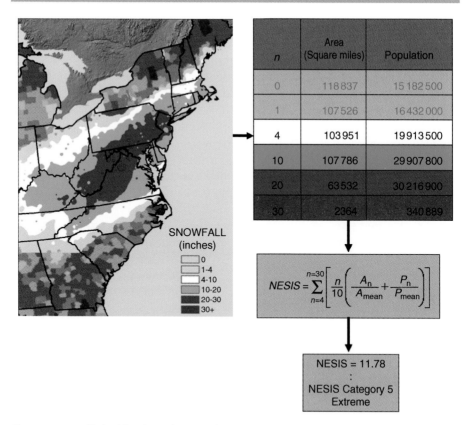

Figure 10.11 Method for determination of an example of the NESIS Category *NESIS Algorithm*: n = snowfall category (4 for >4 inches, 10 for >10 inches, 20 for >20 inches, 30 for >30 inches) A_n = area of snowfall greater than or equal to category n (square miles) P_n = population affected by snowfall greater than category n (2000 census) A_{mean} = mean area of >10" snowfall within the 13-state Northeast region (91 000 square miles) P_{mean} = mean population affected by snowfall >10" within the 13-state north-east region (35. 4 million). The mean area and population constants are for 30 historical storms from 1956 to 2000 for the 13 north-eastern states; these constants calibrate this index to the north-eastern storms (Reproduced from NOAA/NCDC, 2010 (http://lwf.ncdc.noaa.gov/snow-and-ice/nesis.php#overview – accessed 01 02 10) (after Kocin and Uccellini, 2004 and Squires and Lawrimore, 2006, http://lwf.ncdc.noaa.gov/snow-and-ice/docs/squires.pdf.)

categories (Table 10.10). Exemplar maps of the five NESIS categories are shown in Figure 10.12(a)–(e). Extreme category 5 storms are those which produce thick snowfalls over large areas that encompass major metropolitan centres (e.g. Figure 10.12(a).

10.5.6 Impact of climate warming

Little convincing evidence has been presented which suggests that climate warming has been responsible for an increase in the frequency of winter

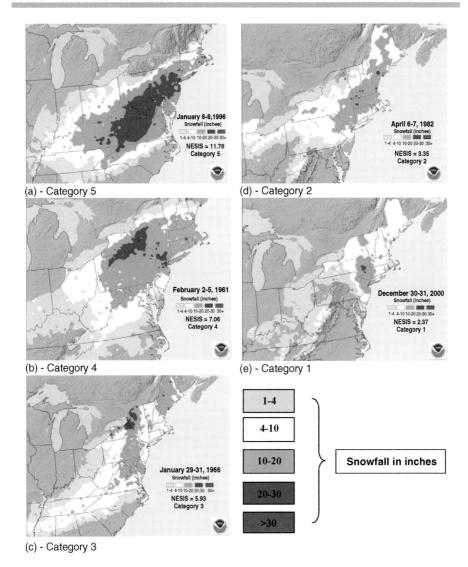

Figure 10.12 Maps of snowfall thickness distribution illustrating examples of NESIS Categories 1–5 (Reproduced from NOAA/ NCDC, 2010 (http://lwf.ncdc.noaa.gov/snow-and-ice/nesis.php#overview.)

storms, although the intensity may have been enhanced. In the 52 year period from 1949 to 2000 the average number of severe ice storms in the USA changed from 1.1 per year in the period 1949 to 1976 to 1.6 from 1977 to 2000 (Changnon and Changnon, 2002). In a similar study on US blizzards from 1959 to 1999, a linear regression showed a positive trend with a modelled increase from 6.6 blizzards per year at the beginning of the period to 15.2 per year at the end (Schwartz and Schmidlin, 2002). However, it is pointed out that whether this change is a true climatological increase or due to more efficient reporting is unknown. Interestingly, a Canadian study (Lawson, 2003) has shown a decreasing trend (90% significance or above) in the number of blizzards in the western Prairie Ecotone in the prairie provinces of Alberta, Saskatchewan and Manitoba. This is in line with several other studies (e.g. Agee, 1991; McCabe, Clark and Serreze, 2001; Serreze *et al*, 1997) which found a decrease in Northern Hemisphere (especially 30–60°N) cyclones during the second half of the twentieth century. The major driving force behind prairie blizzards is a powerful outbreak of cold Arctic air behind migratory cyclones. However, the significant increase in temperature in the Canadian Arctic (IPCC, 2007) is likely to have reduced the barometric gradient between southern and northern Canada and thereby the strength of cyclogenesis.

10.6 Summary

Blizzards and ice storms are common and widespread across temperate and polar latitudes. Unlike the other hazards discussed in this book, they have the power, especially blizzards, to bring a region, even in the most developed of countries, to a virtual standstill. They still occur wherever low temperatures coincide with high precipitation. The nature of these hazards, and their scale, make it almost impossible to control many of their most damaging impacts. These are very wide-ranging, striking at people, their property, their transport networks and their power supplies. In response to these dynamic atmospheric systems, plans for mitigating their impacts can be put in place, but the mitigation of these hazards is more a matter of response than of prevention or avoidance. Only the total inhibition of freezing temperatures will remove winter storm hazards completely.

11
Conclusions – The Future

The preceding chapters have demonstrated the wide scope and complexity of existing cold region hazards, in which ice plays the key role. The impacts of some of these hazards – avalanches, icebergs, ice storms, blizzards and GLOFs, for example – are direct and immediate. Other hazards impact humanity and its environment in less direct yet still significant ways and could, in the long run, turn out to be more problematic. Fortunately, awareness of the nature of cold region hazards, especially the big, longer-term issues such as ice sheets and permafrost, has increased substantially during the last few decades. This is in line with the advance of scientific methods, not least remote sensing, and the use of increasingly sophisticated field techniques. For example, it is now possible to detect sea ice thickness and consequently calculate ice volume in order to predict potential impacts more accurately. The development of a number of spatial and temporal databases has enabled hazard managers to focus attention on key areas and problems and, more easily, inform fund holders and decision makers. At least some small Himalayan countries are now more aware of the location and full extent of their key GLOF problems following data collection exercises, even though they may not yet possess the funds and facilities to mitigate the problem without external assistance.

However, while these advances have increased our understanding of hazards, *as they exist today*, present knowledge and mitigation strategies may require substantial, and on-going revision, if current trends in climate persist. Inevitably climate warming, if it continues, will impose significant change on the cryosphere because this is so dependent on a narrow range of climate and weather conditions. After all, fluctuations of temperature around the 0 °C threshold are fundamental to the presence, absence and behaviour of ice, and hence its associated hazards. There is abundant evidence in the geological record to indicate that global climate change is a normal state of affairs and that the most recent period, the Quaternary, has experienced more frequent and

---

Cold Region Hazards and Risks, First Edition. Colin A. Whiteman.
© 2011 John Wiley & Sons, Ltd. Published 2011 by John Wiley & Sons, Ltd.

rapid changes than most other periods of geological time (see, for example, any issue of the *Journal of Quaternary Science*). The scientific community is still learning about these changes, and especially about the change that is currently taking place (e.g. Solomon et al., 2007). Although a human contribution to climate warming is indisputable, and unlikely to be significantly reduced quickly, judging by the present rate of progress, the future remains uncertain because of the difficulty of predicting the scale and timing of the contribution of *natural* forces to climate change. For the last 900 000 years (Shackleton and Opdyke, 1973; Shackleton, Berger and Peltier, 1990) a pattern of climate change has been established in which 10 000 year long warm 'interglacials' alternate with cooler 'glacials' lasting up to 100 000 years. Assuming that this pattern is continuing, the Earth should now be cooling from the present Holocene interglacial into another glacial stage. Although natural systems tend to fluctuate more frequently than this broad pattern, as seen in the Arctic sea ice data (Chapter 1, Figure 1.3), the current natural trend in temperature should be downwards in opposition to the current upward trend due to human forcing. Consequently, attempting to predict how geocryogenic hazards will alter in the future is fraught with difficulty.

Nevertheless dramatic predictions have already been made and some interesting questions are now being asked. The IPCC 2007 Report (Parry et al., 2007, AR4, WGII) predicts that 'glaciers will experience a substantial retreat during the twentyfirst century... [with] small glaciers... disappear [ing and] larger glaciers... suffer[ing] a volume reduction between 30% and 70% by 2050.' Zemp et al. (2006) posed the question 'Alpine glaciers to disappear within decades?' and concluded that 'the probability of Alpine glaciers disappearing within the coming decades is *far from slight*' (this author's italics). This would make life in Switzerland *less* dangerous – no more crevasses, seracs or ice avalanches – but possibly also less productive in terms of tourist numbers! A similar fate has been predicted for Himalayan glaciers: not as soon as 2035, as mistakenly reported (Parry et al., 2007; The Guardian, 2010), but possibly in a few more centuries, depending on what actually happens to the climate. With the Himalaya described as the 'water tower' of the plains, the complete disappearance of glaciers from these mountains would likely put a massive strain on water supplies in parts of the Indian subcontinent.

'Does Arctic sea ice have a tipping point?' is another question, posed by Winton (2006). As Chapter 2 made clear, current trends in the Arctic suggest that this ocean could be open to nonreinforced shipping during the summer, within just a few decades, but Winton's (2006) modelling suggests that a more substantial change in temperature is required to tip the balance in favour of a completely ice-free Arctic Ocean. With the Arctic acting as a substantial heat sink within the context of global climate, this is not an insignificant question to ask. It will be interesting to see how this issue proceeds in the light of Lenton et al.'s (2008) predictions on Arctic and Antarctic tipping points.

If current warming climate trends continue, many of the geocryogenic hazards discussed here will undoubtedly show significant changes in terms of the scale and location of their impact, but whether these can be accurately predicted remains to be seen. The risks associated with some hazards, such as small glaciers, may decrease quickly and disappear altogether. Other risks (GLOFs, for example) may increase and then decrease as glaciers shrink and cease to exist. Probably the most difficult task is to predict timescales. Hazards associated with the great ice sheets are likely to persist for many hundreds of years, or thousands in the case of the EAIS. The global population should have plenty of time to plan their response to a potential global sea level rise of 70 m!

For the present, the main aim must be to enhance our understanding of cold region hazards and the geocryogenic systems in which they occur. In the context of rapidly changing climate humanity will need to adapt as different situations develop. This may require drastic measures such as migration, or will technology eventually come to our aid? Ultimately, of course, all cold region hazards *will* cease to exist if no ice survives, but has the Earth yet reached the tipping point to this situation? If it has not, then perhaps the expected natural cooling climate cycle will reassert its dominance over the current trend of climate change, the Earth will move inexorably towards another 'glaciation' and the largest hazards will be expanding ice sheets and abandoned coastlines, but these are not problems requiring our *immediate* attention.

References

Abdalati, W. and Steffen, K. (2001) Greenland ice sheet melt extent, 1979–1999. *Journal of Geophysical Research*, **106**, 33983–33988.

ACECRC (2009) http://www.sealevelrise.info/access/repository/resource/e50f775a-acdd-102c-bf59-005056996a56/PA04_Ice%20Sheets_FIN_090610.pdf [accessed 25 March 2010].

ACGR (Associate Committee on Geotechnical Research) (1988) Glossary of permafrost and related ground ice terms. Permafrost Subcommittee, National Research Council of Canada, Ottawa, Technical Memorandum 142.

ACIA (2004) *Impacts of a Warming Arctic*, Arctic Climate Impact Assessment, Cambridge University Press, Cambridge, UK.

ACIA (2005) *Arctic Climate Impact Assessment, Scientific Report*. Cambridge University Press, Cambridge, UK.

Agee, E.M. (1991) Trends in cyclone and anticyclone frequency and comparison with periods of warming and cooling over the Northern Hemisphere. *Journal of Climate*, **4**, 263–267.

Ainley, D.G., Clarke, E.D., Arrigo, K., *et al.* (2005) Decadal-scale changes in the climate and biota of the Pacific sector of the Southern Ocean, 1950s to the 1990s. *Antarctic Science*, **17**, 171–182.

Alean, J. (1985) Ice avalanches – some empirical information about their formation and reach. *Journal of Glaciology*, **31**, 324–333.

Allan, S. (2002) *Media, Risk and Science*. Open University Press, Buckingham, UK.

Allard, M., Wang, B. and Pilon, J.A. (1995) Recent cooling along the southern shore of Hudson Strait, Québec, Canada, documented from permafrost temperature measurements. *Arctic and Alpine Research*, **27**, 157–166.

Allen, S. K., Schneider, D. and Owens, I. F. (2009) First approaches towards modelling glacial hazards in the Mount Cook region of New Zealand's Southern Alps. *Natural Hazards and Earth System Sciences*, **9**, 481–499.

AMAP (1998) *AMAP Assessment Report: Arctic pollution issues*. Arctic Monitoring and Assessment Programme, Oslo, 859 pp.

American Meteorological Society (1998) http://www.ametsoc.org/policy/wxmod98.html [accessed 26 February 2010].

Anderson, A. (2009) *After the Ice*. Virgin Books, London, 298 pp.

Anisimov, O.A. and Lavrov, C.A. (2004) Global warming and permafrost degradation: risk assessment for the infrastructure of the oil and gas industry. *Technologies of Oil and Gas Industry*, **3**, 78–83.

ANP (2008) *The European Union and the Arctic: Policies and Actions*. Nordic Council of Ministers, Copenhagen. http://www.norden.org/da/publikationer/publikationer/2008-729/at_download/publicationfile [accessed 7 March 2010].

Arctic Foundations (2009) http://www.arcticfoundations.com/index.php?option=com_content&task=view&id=28&Itemid=56 [accessed 7 December 2009].

Associated Press (2006) http://www.worldproutassembly.org/archives/2006/01/arctic_ocean_ic.html [accessed 27 March 2010].

Athropolis (2010) http://www.athropolis.com/arctic-facts/fact-ice-road.htm [accessed 23 February 2010].

Aveco (1986) *Tuktoyaktuk Shore Erosion Protection*. Aveco Infrastructure Consultants.

Baker, T.H.W. and Goodrich, L.E. (1990) Heat-pump chilled foundations for buildings on permafrost. *Geotechnical News*, **8**, 26–28.

Ballantyne, C.K. (2002) Paraglacial geomorphology. *Quaternary Science Reviews*, **21**, 1935–2017.

Bamber, J.L., Vaughan, D.G. and Joughin, I. (2000) Widespread complex flow in the interior of the Antarctic Ice Sheet. *Science*, **287**, 1248–1250.

Barrow webcam (2010) http://seaice.alaska.edu/gi/observatories/barrow_webcam [accessed 23 November 2010].

BBC NEWS (1999a) http://news.bbc.co.uk/1/hi/uk/285895.stm [accessed 25 February 2010].

BBC NEWS (1999b) http://www.google.co.uk/search?q=BBC+News+horizon+galtur+avalanche&rls=com.microsoft:en-us:IE-SearchBox&ie=UTF-8&oe=UTF-8&sourceid=ie7&rlz=1I7GPEA_en&redir_esc=&ei=iBfsTIubM42IhQeywpTNDA [accessed 23 November 2010].

BBC NEWS (2001) http://news.bbc.co.uk/1/hi/world/europe/1344039.stm [accessed 22 April 2010].

BBC NEWS (2002) http://news.bbc.co.uk/1/hi/sci/tech/1880566.stm [accessed 10 April 2010].

BBC NEWS (2005) http://www.bbc.co.uk/1/hi/world/europe/4533945.stm [accessed 10 April 2010].

BBC NEWS (2007) *Arctic summers ice-free 'by 2013'*. http://news.bbc.co.uk/1/hi/7139797.stm [accessed 16 April 2010].

BBC NEWS (2010a) http://news.bbc.co.uk/today/hi/today/newsid_8540000/8540344.stm [accessed 26 March 2010].

BBC NEWS (2010b) http://news.bbc.co.uk/1/hi/sci/tech/8609428.stm [accessed 10 April 2010].

BBC NEWS (2010c) http://news.bbc.co.uk/1/hi/world/americas/8502700.stm [accessed 26 February 2010].

Bell, W.W., Donich, T., Groves, K.L. and Sytsma, D. (1999) Tsho Rolpa GLOF Warning System Project http://www.iahr.org/membersonly/grazproceedings99/doc/000/000/175.htm [accessed 14 March 2010].

Beltaos, S. (ed.) (1995) *River Ice Jams*. Water Resources Publications. Highlands Ranch, Colorado, USA.

Beltaos, S. (ed.) (2008) *River Ice Breakup*. Water Resources Publications. Highlands Ranch, Colorado, USA.

Benn, D.I. and Evans, D.J.A. (1998) *Glaciers and Glaciation*. Edward Arnold, London.

Bentley, M.J. (2010) The Antarctic palaeo-record and its role in improving predictions of future Antarctic Ice Sheet change. *Journal of Quaternary Science*, **25**, 5–18.

BFF/SLF (1984) Richtlinien zur Berücksichtigung der Lawinengefahr bei raumwirksamen Tätigkeiten. *Bundesamt für Forstwesen (BFF) und Eidg. Inst. für Schnee- und Lawinenforschung (SLF)*. Eidg. Drucksachen- und Materialzentrale (EDMZ), Bern. Quoted in Gruber, U. http://www.slf.ch/lawineninfo/lawinenbulletin/nationale_lawinenbulletins/index_EN [accessed 25 February 2010].

Bianchi Janetti, E., Gorni, E., Sovilla, B. and Bocchiola, D. (2008) Regional snow-depth estimates for avalanche calculations using a two-dimensional model with snow entrainment. *Annals of Glaciology*, **49**, 63–70.

Bieri, D. (1996) *Abschätzung von Gletschergefahren im Raum Grindelwald – Lauterbrunnental – Lötschental: eine regionale Studie mittels empirischer Faustregeln*. Diplomarbeit (MSc thesis) Department of Geography, University of Zurich, Switzerland.

Blair, R.W. (1994) Moraine and valley wall collapse due to rapid deglaciation in Mount Cook National Park, New Zealand. *Mountain Research and Development*, **14**, 347–358.

Bocchiola, D., Janetti, E.B., Gorni, E. *et al.* (2008) Regional evaluation of three day snow depth for avalanche hazard mapping in Switzerland. *Natural Hazards and Earth System Sciences*, **8**, 685–705.

Bocchiola, D., Medagliani, M. and Rosso, R. (2006) Regional snow depth frequency curves for avalanche hazard mapping in central Italian Alps. *Cold Regions Science and Technology*, **46**, 204–221.

Bolin, I. (2001) When Apus are losing their white ponchos: environmental dilemmas and restoration efforts in Peru. *Development and Cooperation*, **6**, 25–26.

Box, J.E. (2002) Survey of Greenland instrumental temperature records: 1873–2001. *International Journal of Climatology*, **22**, 1829–1847.

Boyd, J., Haegeli, P., Abu-Laban, R.B. *et al.* (2009) Patterns of death among avalanche fatalities: a 21-year review. *Canadian Medical Association Journal*, **180**, 507–512.

Braun, M. and Fiener, P. (1995) *Report on the GLOF Hazard Mapping Project in the Imja/Dudh Kosi Valley, Nepal*. Snow and Glacier Hydrology Project, Nepal Department of Hydrology and Meteorology, Kathmandu, Nepal.

Briggs, D. and Smithson, P. (1992) *Fundamentals of Physical Geography*. Routledge. UK.

Broecker, W.S. (1994) An unstable superconveyor. *Nature*, **367**, 414–415.

Brower, C.D. (1960) *Fifty Years Below Zero: A Lifetime of Adventure in the far north*. Dodd Mead, New York, USA.

Brown, J., Ferrians, O.J., Heginbottom, J.A. and Melnikov, E.S. (1997) *Circum-Arctic Map of Permafrost and Ground Ice Conditions*. United States Geological Survey, Circum-Pacific Map Series, CP-45, scale 1:10,000,000.

Budd, W.F. (1975) A first simple model of periodically self-surging glaciers. *Journal of Glaciology*, **14**, 3–21.

Bühler, Y., Hüni, A., Christen, M. *et al.* (2009) Automated detection and mapping of avalanche deposits using airborne optical remote sensing data. *Cold Regions Science and Technology*, **57**, 99–106.

CAIC (2009) http://avalanche.state.co.us/acc/accidents_stats.php [accessed 21 February 2010].

CAIC (2010) http://avalanche.state.co.us/acc/acc_images/Slide9.JPG [accessed 21 February 2010].

Calgary Herald (2008) http://www.canada.com/calgaryherald/news/calgarybusiness/story.html?id=663b91ce-7090-4de1-9647-6948a5f10ef0 [accessed 7 March 2010].

Call, D. A. (2004) *Urban snow events in Upstate New York: An integrated human and physical geographical analysis*. Unpublished M.A. Thesis, Department of Geography, Syracuse University.

Call, D. A. (2005) Rethinking snowstorms as snow events: a regional case study from upstate New York. *Bulletin of the American Meteorological Society*, **86**, 1783–1793.

Canatec Consultants (1999) *Compilation of Iceberg Shape and Geometry Data for the Grand Banks Region*. PERD/CHC Report 20–43, Calgary, Alberta, Canada. Report prepared for the National Research Council of Canada, Ottawa. ftp://ftp2.chc.nrc.ca/CRTreports/PERD/Visual_00.pdf [accessed 9 July 2010].

Carey, M. (2005) Living and dying with glaciers: people's historical vulnerability to avalanches and outburst floods in Peru. *Global and Planetary Change*, **47**, 122–134.

Carey, M. (2008) The politics of place: inhabiting and defending glacier hazardzones in Peru's Cordillera Blanca, in *Darkening Peaks* (eds B. Orlove, E. Wiegandt and B.H. Luckman), pp. 229–240. University of California Press, Berkley, USA.

Cas.cn (2009) *Eighth International Symposium on Permafrost Engineering*. http://English.careeri.cas.cn/ns/icn/201001/t20100120_50218.html [accessed 20 April 2010].

Catania, G.A., Neumann, T.A. and Price, S.F. (2008) Characterizing englacial drainage in the ablation zone of the Greenland ice sheet. *Journal of Glaciology*, **54**, 567–578.

Cazenave, A. and Llovel, W. (2010) Contemporary Sea Level Rise. *Annual Review of Marine Science*, **2**, 145–173.

CBC (2006a) http://www.cbc.ca/canada/saskatchewan/story/2006/01/17/ice-roads060117.html [accessed 12 January 2009].

CBC (2006b) http://www.cbc.ca/canada/north/story/2006/03/08/truckers-iceroad08032006.html [accessed 12 January 2009].

CBC (2009a) http://www.cbc.ca/canada/north/story/2009/01/08/wrigley.html?ref=rss [accessed 12 January 2009].

CBC (2009b) http://www.cbc.ca/canada/north/story/2009/01/09/wrigley-road.html [accessed 12 January 2009].

CBC (2010) http://www.cbc.ca/world/story/2010/02/10/afghan-avalanche.html [accessed 13 February 2010].

CBC News (2006) http://www.cbc.ca/canada/north/story/2006/03/27/nor-ice-road-closes-27032006.html [accessed 22 April 2010].

Changnon, S.A. (2003) Characteristics of ice storms in the US. *Journal of Applied Meteorology*, **42**, 630–639.

Changnon, S.A. and Changnon, J.M. (2002) Major ice storms in the United States, 1949–2000. *Environmental Hazards*, **4**, 105–111.

Chernetsov, V.A., Malyutin, A.A. and Karlinsky, S.L. (2008) Floating production platform for polar seas designed to resist iceberg impact. *Proceedings of the Eighteenth International Offshore and Polar Engineering Conference*, **1**, 679–685.

Christensen, T.R., Friborg, T. and Johansson, M. (2008) Trace gas budgets of High Arctic permafrost regions (Plenary paper) in *Proceedings of the Ninth International Conference on Permafrost, Vol. 1* (eds D.L. Kane and K.M. Hinkel), Institute of Northern Engineering, University of Alaska Fairbanks, USA, pp. 251–256.

Church, M. (1988) Floods in cold climates, in *Flood Geomorphology* (eds V.R. Baker, R.C Kochel and P.C. Patton), pp. 205–229. John Wiley and Sons, Inc., New York, USA.

Church, M. and Ryder, J. M. (1972) Paraglacial sedimentation, a consideration of fluvial processes conditioned by glaciation. *Geological Society of America Bulletin*, **83**, 3059–3072.

Chwedorzewska, K. J. (2009) Terrestrial Antarctic ecosystems in the changing world: An overview. *Polish Polar Research*, **30**, 263–276.

CIS (2003a) http://www.ec.gc.ca/glaces-ice/default.asp?lang=En&n=84F6AA59-1&wsdoc=FE5C2688-21A8-4165-8FFB-5D28B2A1D943 [accessed 5 July 2010].

CIS (2003b) http://ice-glaces.ec.gc.ca/App/WsvPageDsp.cfm?Lang=eng&lnid=27&ScndLvl=yes&ID=239 [accessed 7 March 2010].

CIS (2003c) http://ice-glaces.ec.gc.ca/content_contenu/images/driftchart.gif [accessed 7 March 2010].

CIS (2009a) http://ice-glaces.ec.gc.ca/App/WsvPageDsp.cfm?ID=1&Lang=eng&Clear=true [accessed 1 March 2010].

CIS (2009b) http://www.ice-glaces.ec.gc.ca/App/WsvPageDsp.cfm?Lang=eng&lnid=3&ScndLvl=yes&ID=167 [accessed 2 April 2010].

CIS (2009c) http://www.ice-glaces.ec.gc.ca/App/WsvPageDsp.cfm?ID=1&Lang=eng&Clear=true [accessed 2 April 2010].

Clague, J.J. (1979) An assessment of some possible flood hazards in Shakwak Valley, Yukon Territory, in *Current Research, Part A, Geological Survey of Canada, Paper 79-1B*.

Clague, J.J. and Rampton, V.N. (1982) Neoglacial Lake Alsek. *Canadian Journal of Earth Sciences*, **19**, 94–114.

Commonwealth Associates (1979) *Investigation into Outages of Electric Power Supply as the Result of Ice Storms*, DOE/RG/6674 T1.

Cook, A. J., Fox, A. J., Vaughan, D. G. and Ferrigno, J. G. (2005) Retreating glacier fronts on the Antarctic Peninsula over the past half century. *Science*, **308**, 541–544, doi:10.1126/science.1104235 [accessed 17 March 2010].

Cook, A. J. and Vaughan, D. G. (2010) Overview of areal changes of the ice shelves on the Antarctic Peninsula over the past 50 years, *The Cryosphere*, **4**, 77–98. www.the-cryosphere.net/4/77/2010/ © Author(s) 2010. This work is distributed under the Creative Commons Attribution 3.0 License [accessed 17 March 2010].

Cooke, R.U. and Doornkamp, J. C. (1990) *Geomorphology in Environmental Management: A New Introduction*, 2nd edn. Clarendon, Oxford, UK.

Copland. L. and Mueller, D.R. (eds) (2009) *Arctic Ice Shelves and Ice Islands*. Springer, New York, USA.

CRIACC (2008) http://www.criacc.qc.ca/climat/suivi/verglas98_e.html Centre de Resources en Imacts et Adaptation au Climat et à ses Changements [accessed 26 February 2010].

CRREL (2000) http://www.crrel.usace.army.mil/ierd/tectran/IERD29.pdf [accessed 23 February 2010].

CRREL (2001) http://www.crrel.usace.army.mil/ierd/tectran/IERD32.pdf [accessed 23 February 2010].

CRREL (2006) http://www.crrel.usace.army.mil/icejams/tech_files/2006%20Ice%20Jams%20Intro.pdf [accessed 20 April 2010].

CRREL (2007) http://www.crrel.usace.army.mil/icejams/index.htm [accessed 20 April 2010].

CRREL (2009) http://www.crrel.usace.army.mil/icejams/icesafety/ [accessed 20 April 2010].

CRREL (2010) http://www.crrel.usace.army.mil/icejams/index.htm [accessed 12 January 2010].

Cruikshank, J. (2005) *Do Glaciers Listen? Local Knowledge, Colonial Encounters and Social Imagination*. UBC Press, Vancouver, Canada.

Cuffey, K.M. and Marshall, S.J. (2000) Substantial contribution to sea-level rise during the last interglacial from the Greenland ice sheet. *Nature*, **404**, 591–594.

Czudek, T. and Demek, J. (1970) Thermokarst in Siberia and its influence on the development of lowland relief. *Quaternary Research*, **1**, 103–120.

Damen, M. (1992) *Study on the potential outburst flooding of Tsho Rolpa Glacier Lake, Rolwaling Valley, east Nepal*. International Institute for Aerospace Survey and Earth Sciences, ITC, Enschede, The Netherlands.

Da Silva, R., Bohrer, G., Werth, D. *et al.* (2006) Sensitivity of ice storms in the southeastern United States to Atlantic SST: insights from a case study of the December 2002 storm. *Monthly Weather Review*, **134**, 1554–1564.

Davies, M.C.R., Hamza, O. and Harris, C. (2001) The effect of rise in mean annual temperature on the stability of rock slopes containing ice-filled discontinuities. *Permafrost and Periglacial Processes*, **12**, 137–144.

Davies, M.C.R., Hamza, O. and Harris, C. (2003) Physical modelling of permafrost warming in rock slopes, in *Proceedings of the Eighth International Conference on Permafrost* (eds M. Phillips, S.M. Springman and L.U. Arenson), Balkema, Lisse, The Netherlands, pp. 169–174.

Davis, C.H., Li, Y.H., McConnell, J.R. *et al.* (2005) Snowfall-driven growth in East Antarctic ice sheet mitigates recent sea-level rise. *Science*, **308**, 1898–1901.

DEC Alaska (2003) http://www.dec.state.ak.us/SPAR/images/gallery/overseasohio/pages/Ohio01.htm [accessed 4 March 2010].

Deem, J.M. (2008) *Bodies from the Ice: Melting Glaciers and the Recovery of the Past*. Houghton Mifflin, USA.

Demek, J. (1994) Global warming and permafrost in Eurasia: a catastrophic scenario. *Geomorphology*, **10**, 317–329.

DesJarlais, C. (2004) *S'adapter aux Changements Climatiques*. Ouranos, Montreal, Canada.

Devlin, E. (2010) http://www.sentinelandenterprise.com/ci_14309415 [accessed 4 February 10].

DHS&EM (2006) http://fc.ak-prepared.com/dailysitrep/FOV5-0000D6AC/FOV5-00017578/I00CF7C0E [accessed 27 March 2010].

Doake, C.S.M. and Vaughan, D.G. (1991) Rapid disintegration of the Wordie Ice Shelf in response to atmospheric warming. *Nature*, **350**, 328–330.

DPNET Nepal (no date) http://www.dpnet.org.np/focus/golf.php [accessed 26 August 2009].

Durner, G.M., Douglas, D.C., Nielson, R.M. *et al.* (2009) Predicting 21st-century polar bear habitat distribution from global climate models. *Ecological Monographs*, **79**, 25–58.

EBA (2002) *Hamlet of Tuktoyaktuk Shore Erosion and Community Impact Study*. EBA Engineering Consultants Ltd./IEG.

EBA Engineering (2010) http://www.eba.ca/news.asp?a=322 [accessed 23 November 2010].

Echelmeyer, K., Clarke, T.S. and Harrison, W.D. (1991) Surficial Glaciology of Jakobshavns Isbrae, West Greenland. 1. Surface-Morphology. *Journal of Glaciology*, **37**, 368–382.

Eckardt, A. (2005) http://www.msnbc.msn.com/id/8432120/ [accessed 30 January 2010].

Edlund, J., Gordon, D. and Robinson, W.P. (1998) A model mine shows its cracks: an independent report on environmental problems at the Kubaka Gold Mine in the Russian Far East. http://www.pacificenvironment.org/downloads/model_mine.pdf [accessed 7 December 2009].

Eik, K. (2008) Review of experiences within ice and iceberg management. *Journal of Navigation*, **61**, 557–572.

Eik, K. (2009) Iceberg drift modelling and validation of applied metocean hindcast data. *Cold Regions Science and Technology*, **57**, 67–90.

Embleton, C. and Thornes, J. (1979) *Process in Geomorphology*. Edward Arnold, London.

ENCS Concordia (1998) http://users.encs.concordia.ca/~grogono/icestorm.html [accessed 27 November 2010].

ENRA, (2005) http://www.mnt.ee/atp/failid/mnt_2005aastakogumik_eng_.pdf [accessed 2010].

ENS, (2009) http://www.ens-newswire.com/ens/mar2009/2009-03-26-091.asp [accessed 22 April 2010].

Environment Canada (1998) http://www1.tor.ec.gc.ca/nwsd/OtherPub/ice_storm/hiostory2_e.cfm [accessed 17 November 1999].

Environment Canada (2003) http://ice-glaces.ec.gc.ca/WsvPageDsp.cfm?Lang=eng&lnid=4&ScndLvl=yes&ID=10164 [accessed 9 March 2010].

Environment Canada (2007) http://www.pnr-rpn.ec.gc.ca/air/severewthr/warning.en.html [accessed 23 February 2010].

Environment Canada (2010a) http://www.ec.gc.ca/glaces-ice/default.asp?lang=En&xml=217719AA-F5C9-4D3E-B1DA-8F8227D0B9DA [accessed 9 July 2010].

Environment Canada (2010b) http://www.ec.gc.ca/meteo-weather/default.asp?lang= En&n=6A4A3AC5-1 [accessed 27 November 2010].

ERA (2006) http://www.mnt.ee/atp/?id=3636 [accessed 22 April 2010].

ERDC (2009) http://www.crrel.usace.army.mil/icejams/tech_files/2006%20Ice%20Jams%20Intro.pdf [accessed 22April 2010].

Esch, D.C. and Osterkamp, T.E. (1990) Cold regions engineering: climate warming concerns for Alaska. *Journal of Cold Regions Engineering*, **4**, 6–14.

ETH Life (2008) http://www.ethlife.ethz.ch/archive_articles/0810XX_PermaSense/Index.EN [accessed 27 November 2010].

Etzelmüller, B. and Frauenfelder, R. (2009) Factors Controlling the Distribution of Mountain Permafrost in the Northern Hemisphere and Their Influence on Sediment Transfer. *Arctic Antarctic and Alpine Research*, **41**, 48–58.

Etzelmüller, B., Ødegård, R.S., Berthing, I. and Sollid, J.L. (2001) Terrain parameters and remote sensing data in the analysis of permafrost distribution and periglacial processes: principles and examples from Southern Norway. *Permafrost and Periglacial Processes*, **12**, 79–92.

EU (2008) *Climate Change and International Security*. Council of the European Union, Commission and the Secretary-General/High Representative, ANNEX 7249/08, pp. 1–8.

Evans, S.G. and Clague, J.J. (1988) Catastrophic rock avalanches in glacial environments. *Landslides. Proceedings of the Fifth International Symposium on Landslides, Vol. 2*, (ed. C. Bonnard), pp. 1153–1158.

Evans, S.G. and Clague, J.J. (1994) Recent climatic-change and catastrophic geomorphic processes in mountain environments. *Geomorphology*, **10**, 107–128.

Exxonmobil (no date) http://www.exxonmobil.com/corporate/energy_project_arctic_grandbanks.aspx [accessed 7 March 2010].

Ferrians, O.J., Kachadoorian, R. and Green, G.W. (1969) Permafrost and Related Engineering problems in Alaska. *United States Geological Survey, professional papers* **678**, 37 pp.

Fleming, F. (1998) *Barrow's Boys*. Granta Books, London, UK.

Ford, J.D. (2009) Dangerous climate change and the importance of adaptation for the Arctic's Inuit population *Environmental Research Letters* **4**, 1–9 doi:10.1088/1748-9326/4/2/024006 [accessed 29 March 2010].

Ford, J.D., Smit, B. and Wandel, J. (2006) Vulnerability to climate change in the Arctic: a case study from Arctic Bay, Canada. *Global Environmental Change – Human and Policy Dimensions*, **16**, 145–160.

Fowler, A.C. (1987) A theory of glacier surges. *Journal of Geophysical Research*, **92**, 9111–9120.

Freitas, C., Kovacs, K.M., Ims, R.A. and Lydersen, C. (2008) Predicting habitat use by ringed seals (*Phoca hispida*) in a warming Arctic. *Ecological Modelling*, **217**, 19–32.

French, H.M. (1996) *The Periglacial Environment*, 2nd edn. Addison Wesley/Longman, Harlow, UK.

French, H.M. (2007) *The Periglacial Environment* 3rd edn. John Wiley and Sons, Ltd., Chichester, UK.

French, H.M. and Slaymaker, O. (1993) *Canada's Cold Environments*. McGillQueen's University Press, Montreal, Canada.

Frenot, Y., Chown, S.L., Whinam, J. *et al.* (2005) Biological invasions in the Antarctic: extent, impacts and implications. *Biological Review*, **80**, 45–72.

Gauer, P., Issler, D., Lied, K. *et al.* (2008) On snow avalanche flow regimes: inferences from observations and measurements, *Proceedings International Snow Science Workshop, Whistler, Canada*, 717–723.

Geology UK (2003) http://www.geologyuk.com/mountain_hazards_group/pdf/Chapter_7.pdf [accessed 14 March 2010].

George, J.C., Huntington, H.P., Brewster, K. *et al.* (2004) Observations on shorefast ice dynamics in arctic Alaska and the responses of the Inupiat hunting community. *Arctic*, **57**, 363–374.

Gerard, R. (1990) Hydrology of floating ice. Northern Hydrology: Canadian Perspectives. *NHRI Science Report No. 1*, National Hydrology Research Institute, Environment Canada, Saskatoon, Canada, pp. 103–134.

Gerard, R. and Karpuk, E. (1979) Probability analysis of historical flood data. *ASCE Journal of Hydraulics Division*, **105(HY9)**, 1153–1165.

Giani, G.P., Silvano, S. and Zanon, G. (2000) Avalanche of 18 January 1997 on Brenva Glacier, Mont Blanc Group, Western Italian Alps: an unusual process of formation. *Annals of Glaciology*, **32**, 333–338.

Gillman, P. (ed.) (1993) *Everest*. Little Brown, London, UK.

Glaciorisk (2003) http://www.nimbus.it/glaciorisk/GlacierList.asp?vista=paese&paese=Norway [accessed 27 October 09].

GNWT (2009) http://www.dot.gov.nt.ca/_live/pages/wpPages/Open_Close_Dates_Ice_Bridges.aspx [accessed 22 April 2010].

GNWT (2010) http://www.thetruckersplace.com/IceRoadTrucking.aspx [accessed 22 April 2010].

Goering, D.J. (2004) Thompson Drive to contain permafrost protection measures: modern technology keeps the ground frozen year-round. *Alaska Business Monthly*. http://www.arcticfoundations.com/index.php?option=com_content&task=view&id=72&Itemid=39 [accessed 21 November 2009].

Goodrich, L.E. and Plunkett, J.C. (1990) Performance of heat pump chilled foundations, in *Proceedings of the Fourth Canadian Permafrost Conference*, (ed. H. M. French), pp. 409–418. Calgary, National Research Council Canada, Ottawa, Canada.

Goudie, A. (ed.) (1990a) *The Encyclopaedic Dictionary of Physical Geography*, Basil Blackwell Ltd., Oxford, pp. 528.

Goudie, A. (1990b) *The Human Impact on the Natural Environment*, 3rd edn. Blackwell, Oxford, UK.

Government of Canada (1976) *Shore Erosion and Protection Study – Stage, Tuktoyaktuk*. Department of Public Works, Western Region.

Gray, D.M. and Male, D.H. (eds) (1981) *Handbook of Snow: Principles, Processes Management and Use*. Pergamon, Toronto, Canada.

Grebenets, V.I. (2003) Greocryological-geoecological problems occurring in urbanised territories in Northern Russian and methods for improvement and restoration of foundations, in *Proceedings of the Eighth International Conference on Permafrost 1* (eds M. Phillips, S.M. Springman and L.U. Arenson), Balkema, Lisse, The Netherlands, pp. 303–308.

Grebenets, V.I., Fedoseev, D.B. and Lolaev, A.B. (1994) Geotechnical aspects of environmental violations in cryolitic zone, in *Proceedings of theFirst International Congress of Environmental Geotechnics, Edmonton, 14–17 July 1994* (ed. W.D. Carrier), pp. 118–134.

Grebenets, V.I., Kerimov, A.G.-O. and Savtchenko, V.A. (1997) Stability of foundation in cryolitic zone under the condition of technogenic inundation and salting. *Proceedings of the 14th International Conference on Soil Mechanics and Foundation Engineering, Hamburg, 6–12 September 1997* (ed. Publications Committee of the 14th ICSMFE), pp. 113–114.

Grida (no date) http://maps.grida.no/go/graphic/ice-avalanches-of-the-nevados-huascar-n-in-peru [accessed 25 January 2010].

Grinsted, A., Moore, J.C. and Jevrejeva, S. (2010) Reconstructing sea level from paleo- and projected temperatures 200 to 2100 AD. *Climate Dynamics*, **34**, 461–472.

Grove, J. M. (1988) *The Little Ice Age*. Methuen, London, UK.

Gruber, S. (2005) Mountain permafrost: transient spatial modelling, model verification and the use of remote sensing. PhD Thesis, Universität, Zürich, Switzerland, 121 pp.

Gruber, S. and Haeberli, W. (2007) Permafrost in steep bedrock slopes and its temperature-related destabilisation following climate change. *Journal of Geophysical Research*, **112**, F02S18. Doi:1029/2006JF000547.

Gruber, S. and Hoelzle, M. (2001) Statistical modeling of mountain permafrost distribution: local calibration and incorporation of remotely sensed data. *Permafrost and Periglacial Processes*, **12**, 69–77.

Gruber, S., Hoelzle, M. and Haeberli, W. (2004) Rock-wall temperatures in the Alps: modelling their topographic distribution and regional differences. *Permafrost and Periglacial Processes*, **15**, 299–307.

Gruber, U. (2001) *Using GIS for avalanche hazard mapping in Switzerland.* http://proceedings.esri.com/library/userconf/proc01/professional/papers/pap964/p964.htm [accessed 24 February 2010].

Gruber, U. and Haefner, H. (1995) Avalanche hazard mapping with satellite data and a digital elevation model. *Applied Geography*, **15**, 99–113.

Gubler, H. (1996) Remote avalanche warning, alarm- and control systems: fundamentals, applications and experience. pp. 165–172. *Proceedings of the International Snow Science Workshop, Banff, Canada*.

Haeberli, W. (1983) Frequency and characteristics of glacier floods in the Swiss Alps. *Annals of Glaciology*, **4**, 85–90.

Haeberli, W. (1992) Construction, environmental problems and natural hazards in periglacial mountain belts. *Permafrost and Periglacial Processes*, **3**, 111–124.

Haeberli, W. (2008) *Changing views of changing glaciers* in *Darkening Peaks* (eds B. Orlove, E. Wiegandt and B.H. Luckman), pp. 23–32. University of California Press, Berkley, USA.

Haerberli, W. and Beniston, M. (1998) Climate change and its impacts on glaciers and permafrost in the Alps. *Ambio*, **27**, 258–265.

Haeberli, W., Huder, J., Keusen, H.-R., *et al.* (1988) Core drilling through rock glacier permafrost. *Proceedings of the Fifth International Conference on Permafrost*, vol. 2, Tapir, Trondheim, Norway, pp. 937–942.

Haeberli, W., Clague, J.J., Huggel, C. and Kääb, A. (2009) Hazards from glaciers, permafrost and lakes in high mountain regions: processes and interactions. http://www.baunat.boku.ac.at/fileadmin/_/H87/H872/4Veranstaltugnen/Haeberli.pdf [accessed 5 July 2010].

Haeberli, W. and Hohmann, R. (2008) Climate, glaciers and permafrost in the Swiss Alps 2050: scenarios, consequences and recommendations, in *Proceedings of the Ninth International Conference on Permafrost, Vol. 1* (eds D.L. Kane and K.M. Hinkel), Institute of Northern Engineering, University of Alaska Fairbanks, USA. pp. 607–612.

Haeberli, W. and Holzhauser, H. (2003) Alpine glacier mass changes during the past two millennia. *PAGES News*, **1**, 13–15.

Haeberli, W. and Gruber, S. (2008) Research challenges for permafrost in steep and cold terrain:an alpine perspective, in *Proceedings of the Ninth International Conference on Permafrost Vol. 1* (eds D.L. Kane and K.M. Hinkel), Institute of Northern Engineering, University of Alaska Fairbanks, USA. pp. 597–605.

Haeberli, W., Huggel, C., Kaab, A. *et al.* (2003) Permafrost conditions in the starting zone of the Kolka-Karmadon rock/ice slide of 20 September 2002 in North Osetia (Russian Caucasus) in *Extended Abstracts, Proceedings of the Eighth International Conference on Permafrost* (eds W. Haeberli and D. Brandova), pp. 49–50.

Haefeli, R. (1965) Note sur la classification, le Mécanisme, et le contrôle des avalanches de glaces et des crues glaciaires extraordinaires. Symposium International sur les Aspects Scientifique des Avalanches de Neige, Davos. *In Extrait de la publication de l'A. I. H. S.*, **69**, 316–325.

Hansom, J.D. and Gordon, J.E. (1998) *Antarctic Environments and Resources*. Addison Wesley/Longman Ltd, Harlow, UK.

Harper, J.R. (1990) Morphology of the Canadian Beaufort Sea Coast. *Marine Geology*, **91**, 75–91.

Harris, C., Arenson, L.U. *et al.* (2009) Permafrost and climate in Europe: Monitoring and modelling thermal, geomorphological and geotechnical responses. *Earth-Science Reviews*, **92**, 117–171.

Harris, C., Haeberli, W., Vonder Mühll, D. and King, L. (2001) Permafrost monitoring in the high mountains of Europe: the PACE project in its global context. *Permafrost and Periglacial Processes*, **12**, 3–11.

Harris, C., Luetschg, M.A., Davies, M.C.R. and Smith, F.W. (2007) Field instrumentation for real-time monitoring of periglacial solifluction. *Permafrost and Periglacial Processes*, **18**, 105–114.

Harwood, D.M., Winter, D. M. and Srivastav, A. (1994) Climatic implication for the absence of early Pliocene Antarctic sea-ice: a discussion. *EOS, Transactions American Geophysical Union*, **75**(16) Suppl. (abstr. 041A-12).

Hauck, C. and Vonder Mühll (2003) Evaluation of geophysical techniques for application in mountain permafrost studies. *Zeitschrift für Geomorphologie*, **132**, 159–188 NF., Suppl.

Hauck, C. and Kneisel, C. (eds) (2008) *Applied Geophysics in Periglacial Environments*. Cambridge University Press, Cambridge, UK.

Hauck, C., Guglielmin, M., Isaksen, K. and Vonder Mühll, D. (2001) Applicability of frequency-domain and time-domain electromagnetic methods for mountain permafrost studies. *Permafrost and Periglacial Processes*, **12**, 39–52.

Hay, J.E. and Elliott, T.L. (2008) New Zealand's glaciers: key national and global assets for science and society in *Darkening Peaks* (eds B. Orlove, E. Wiegandt and B.H. Luckman), pp. 185–195. University of California Press, Berkley, USA.

Heinrich, H. (1988) Origin and consequences of cyclic ice-rafting in the Northeast Atlantic Ocean during the past 130,000 years. *Quaternary Research.* **29**, 142–152.

Heinrichs, T. A., Mayo, L. R., Echelmeyer, K.E., and Harrison, W.D. (1996) Quiescent-phase evolution of a surge-type glacier: Black Rapids Glacier, Alaska, U.S.A. *Journal of Glaciology*, **42**, 110–122.

Heumader, J. (2000) Die Katastrophenlawinen von Galtür und Valzur am 23 und 24.2.1999 im Paznauntal/Tirol. *Proc. Internationales Sympos. Interpraevent*, Villach, Austria, 26–30 June 2000, **2**, 397–410.

Hilberg, S. and Angel. J. (2006) The Cold Hard Facts about Winter Storms. Illinois State Water Survey. http://www.sws.uiuc.edu.

Hill, B. (2004) *Database of ship collisions with icebergs.* http://www.docstoc.com/docs/16842176/DATABASE-OF-SHIP-COLLISIONS-WITH-ICEBERGS [accessed 7 March 2010].

Hill, B. (2008) http://www.icedata.ca/ [accessed 18 April 2010].

Hivon, E., Ladanyi, B., Lavender, B. *et al.* (1998) *Climate Change Impacts on Permafrost Engineering Design.* http://cgc.rncan.gc.ca/permafrost/communities_e.php [accessed 20 April 2010].

Hoelzle, M., Mittaz, C., Etzelmüller, B. and Haeberli, W. (2001) Surface energy fluxes and distribution models of permafrost in European mountain areas: an overview of current developments. *Permafrost and Periglacial Processes*, **12**, 53–68.

Holland, D.M., Thomas, R.H., De Young, B. *et al.* (2008) Acceleration of Jakobshavn Isbrae triggered by warm subsurface ocean waters. *Nature Geoscience*, **1**, 659–664.

Holland, M. M. and Bitz. C. M. (2003) Polar amplification of climate change in coupled models. *Climate Dynamics*, **21**, 221–232.

Holland, M.M., Serreze, M.C. and Stroeve, J. (2010) The sea ice mass budget of the Arctic and its future change as simulated by coupled climate models, *Climate Dynamics*, **34**, 185–200.

Holubec, I., Jardine, J. and Watt, B. (2008) Flat loop evaporator thermosyphon foundations: Design, construction and performance in the Canadian permafrost regions, in *Proceedings of the Ninth International Conference on Permafrost Vol. 1* (eds D.L. Kane and K.M. Hinkel), Institute of Northern Engineering, University of Alaska Fairbanks, USA. pp. 735–740.

Huebert, R. and Yeager, B.B. (2008) *A New Sea: The Need for a Regional Agreement on management and Conservation of the Arctic Marine Environment.* WWF International Arctic Programme, Oslo, Norway.

Huggel, C., Haeberli, W., Kääb, A. *et al.* (2004) An assessment procedure for glacial hazards in the Swiss Alps. *Canadian Geotechnical Journal*, **41**, 1068–1083.

Huggel, C., Zgraggen-Oswald, S., Haeberli, W. *et al.* (2005) The 2002 rock/ice avalanche at Kolka/Karmadon, Russian Caucasus: assessment of extraordinary avalanche formation and mobility, and application of QuickBird satellite imagery. *Natural Hazards and Earth System Sciences*, **5**, 173–187.

Hughes, K.A. and Convey, P. (2010) The protection of Antarctic terrestrial ecosystems from inter- and intra-continental transfer of non-indigenous species by human activities: a review of current systems and practices. *Global Environmental Change- Human and Policy Dimensions*, **20**, 96–112.

Hughes, T.J., Denton, G.H. and Fastook, J.L. (1985) Is the Antarctic Ice Sheet an analogue for Northern Hemisphere palaeo-ice sheets?, in *Models in Geomorphology* (ed. J.J. Wodenburg), pp. 25–72. Allen & Unwin, Boston, USA.

Hulbe, C. L., Scambos, T. A., Youngberg, T. and Lamb, A. K. (2008) Patterns of glacier response to disintegration of the Larsen B ice shelf, Antarctic Peninsula. *Global Planetary Change*, **63**, 1–8, doi:10.1016/j.gloplacha.2008.04.001 [accessed 17 March 2010].

Hume, J.D. and Schalk, M. (1964) The effects of ice push on Arctic beaches. *American Journal of Science*, **262**, 267–273.

Hurrell, J. W., and Deser, C. (2009) North Atlantic climate variability: The role of the North Atlantic Oscillation. *Journal of Marine Systems*, **78**, 28–41.

HVRI (2010) http://webra.cas.sc.edu/hvriapps/sheldus_setup/sheldus_login.aspx [accessed 23 February 2010].

IAATO (2006) http://image.zenn.net/REPLACE/CLIENT/1000037/1000116/application/pdf/TourismSummarybyExpedition.pdf [accessed 18 April 2010].

IAATO (2007) http://www,iaato.org/docs/Boot_Washing07.pdf [accessed 18 April 2010].

IAHR (1986) River ice jams: a state of the art report. *Proceedings, International Association of Hydraulics Research International Ice Symposium*, Iowa City, USA, Vol. **3**, 561–594.

IARC (2006) http://www.iarc.uaf.edu/news/news_shorts/ivu.php [accessed 27 March 2010].

Iceland Meteorological Office (2010) http://andvari.vedur.is/snjoflod/haettumat/log_49_1997_e.pdf [accessed 27 November 2010].

IDEAM (2004) Boletín Julio 12 al 16 de 2004. Colombia.

IEEE (2004) http://www.ieee.org/portal/cms_docs_pes/pes/subpages/meetings-folder/PSCE/2004Presentations/281/Jean-Marie-HQ-PSCE-oct13rd2004v4.pdf [Accessed 27 September 2010].

IISD (2010) http://www.iisd.org/climate/vulnerability/inuit.asp [accessed 11 March 2010].

Industry Canada (2009) http://www.ic.gc.ca/eic/site/ic1.nsf/eng/home [accessed 8 March 2010].

Instanes, A. (2005) Infrastructure: buildings, support systems and industrial facilities, in *Arctic Climate Impact Assessment Scientific Report*, pp. 908–944. http://www.acia.uaf.edu/PDFs/ACIA_Science_Chapters_Final/ACIA_Ch16_Final.pdf [accessed 20 April 2010].

Inuvik, (no date) http://www.inuvik.ca/townhall/utilidors.html [accessed 13 December 2009].

IPCC (2001) *Climate Change 2001: The Scientific Basis. Contribution of Working Group I to the Third Assessment Report of the Intergovernmental Panel on Climate Change* (eds J.T. Houghton, Y. Ding, D.J. Griggs *et al.*), Cambridge University Press, Cambridge, UK. http://www.grida.no/climate/ipcc_tar/wg1/pdf/WG1_TAR-FRONT.pdf [accessed 5 July 2010].

IPCC (2007) *Climate Change 2007: Synthesis Report. Contribution of Working Groups I, II and III to the Fourth Assessment Report of the Intergovernmental Panel on*

Climate Change (eds Core Writing Team, R.K. Pachauri and A. Reisinger) IPCC, Geneva, Switzerland.

Ives, J.D. (1986) *Glacial Lake Outburst Floods and Risk Engineering in the Himalaya*. ICIMOD Occasional Paper No. 5, International Centre for Integrated Mountain Development, Katmandu, Nepal.

Jamieson, B. (2001) Snow avalanches, in *A Synthesis of Geological Hazards in Canada* (ed. G.R. Brooks), Geological Survey of Canada Bulletin 548, pp. 75–94.

Jamieson, B., Margreth, S. and Jones, A. (2008) Application and limitations of dynamic models for snow avalanche hazard mapping. *Proceedings International Snow Science Workshop 2008*. http://www.ucalgary.ca/asarc/files/asarc/DynModels AppLim_Issw08_Jamieson.pdf [accessed 19 February 2010].

Jamieson, B. and Stethem, C. (2002) Snow avalanche hazards and management in Canada: challenges and progress. *Natural Hazards*, **26**, 35–53.

Jamieson, J.B., Stethem, C.J., Schaerer, P.J. and McClung D.M. (eds) (2002) *Land Managers Guide to Snow Avalanche Hazards in Canada*. Canadian Avalanche Association, Revelstoke, BC, Canada.

Jay, C.V., and Fischbach. A.S. (2008) Pacific walrus response to Arctic sea ice losses. *U. S. Geological Survey Fact Sheet 2008-3041*. http://pubs.usgs.gov/fs/2008/3041/ [accessed 10 March 2010].

Johnson, K., Solomon, S., Berry, D. and Graham, P. (2003) Erosion progression and adaptation strategy in a northern coastal community, in *Permafrost [1]* (eds M. Phillips, S.M. Springman and L.U. Arenson), pp. 489–494. 8th International Conference on Permafrost, Zurich, Switzerland.

Jones, B.M., Arp, C.D., Jorgenson, M.T. *et al.* (2009) Increase in the rate and uniformity of coastline erosion in Arctic Alaska. *Geophysical Research Letters*, **36**, L03503, doi:10.1029/2008GL036205 [accessed 29 March 2010].

Jones, B.M., Hinkel, K.M., Arp C.D. and Eisner. W. R. (2008) Modern erosion rates and loss of coastal features and sites, Beaufort Sea coast, Alaska, *Arctic*, **61**, 361–372.

Jones, K.F. and Mulherin, N.D. (1998) *An Evaluation of the Severity of the January 1998 Ice Storm in Northern New England: Report for FEMA Region 1*. U.S. Army Cold Regions Research and Engineering Laboratory.

Joughin, I., Howat, I.M. Fahnestock, M. *et al.* (2008) Continued evolution of Jakobshavn Isbrae following its rapid speedup. *Journal of Geophysical Research-Earth Surface*, **113**, F04006.

Joughin, I., Tulaczyk, S., Bamber, J.L. *et al.* (2009) Basal conditions for Pine Island and Thwaites Glaciers, West Antarctica, determined using satellite and airborne data. *Journal of Glaciology*, **55**, 245–257.

JVTC (2009) http://jvtcwinterroad.ca/ [accessed 22 April 2010].

Kääb, A. (2008) Glacier volume changes using ASTER satellite stereo and ICESat GLAS laser altimetry. A test study on Edgeøya, Eastern Svalbard. *IEEE Transactions on Geoscience and Remote Sensing*. **46**, 2823–2830.

Kääb, A. Wessels, R., Haeberli, W. *et al.* (2003) Rapid ASTER imaging facilitates timely assessment of glacier hazards and disasters. *EOS, Transactions of the American Geophysical Union*, **84**, 117 and 121.

Kääb, A., Huggel, C., Fischer, L. *et al.* (2005) Remote sensing of glacier- and permafrost-related hazards in high mountains: an overview. *Natural Hazards and Earth System Sciences*, **5**, 527–554.

Kamensky, R.M. (ed.) (1998) *Geocryological Problems of Construction in Eastern Russia and Northern China. Proceedings, International Symposium*, 23–25 September, Chita, Russia. SB RAS, Yakutsk, vol. 1, 255 pp., vol. 2, 197 pp.

Kattelmann, R. (2003) Glacial lake outburst floods in the Nepal Himalaya: a manageable hazard? *Natural Hazards*, **28**, 145–154.

Kawerak (2009) http://www.kawerak.org/ledps/Unalakleet.pdf [accessed 9 March 2010].

Kawerak (no date) http://www.kawerak.org/tribalHomePages/shishmaref/ [accessed 30 March 2010].

Keiler, M., Sailer, R., Joerg, P. *et al.* (2006) Avalanche risk assessment – a multi-temporal approach, results from Galtur, Austria. *Natural Hazards and Earth System Sciences*, **6**, 637–651.

Khrustalev, L. (2000) Allowance for climate change in designing foundation on permafrost grounds. In: *Proceedings, International Workshop on Permafrost Engineering*, Longyearbyen, Svalbard, Norway, 18–21 June (ed. K. Senneset). Norwegian University of Science and Technology (NTNU)/University Courses on Svalbard (UNIS), pp. 25–36.

Klimes, J., Vilimek, V and Omelka, M. (2009) Implications of geomorphological research for recent and prehistoric avalanches and related hazards at Huascaran, Peru. *Natural Hazards*, **50**, 193–209.

Kneisel, C., Hauck, C., Fortier, R. and Moorman, B. (2008) Advances in geophysical methods of permafrost investigations. *Permafrost and Periglacial Processes*, **19**, 157–178.

Knight, P.G. (1999) *Glaciers*. Stanley Thornes Ltd., Cheltenham. UK.

Kocin, P. J. and Uccellini, L. W. (2004) A snowfall impact scale derived from northeast storm snowfall distributions. *Bulletin of the American Meteorological Society*, **85**, 177–194. www.ncdc.noaa.gov/oa/climate/research/snow-nesis/kocin-uccellini.pdf [accessed 17 April 2010].

Kooyman, G.L., Ainley, D.G., Ballard, G. and Ponganis, P.J. (2007) Effects of giant icebergs on two emperor penguin colonies in the Ross Sea, Antarctica. *Antarctic Science*, **19**, 31–38.

Kovacs, A. and Sodhi, D.S. (1980) Shore ice pile-up and ride-up: field observations, models, theoretical analyses. *Cold Regions Science and Technology*, **2**, 210–288.

Kwok, R., Cunningham, G. F., Wensnahan, M. *et al.* (2009) Thinning and volume loss of the Arctic Ocean sea ice cover: 2003–2008, *Journal of Geophysical Research*, **114**, C07005, doi:10.1029/2009JC005312. [accessed 10 March 2010].

Kwok, R., and Rothrock D. A. (2009) Decline in Arctic sea ice thickness from submarine and ICESat records: 1958–2008. *Geophysical Research Letters*, **36**, L15501, doi:10.1029/2009GL039035. [accessed 10 March 2010].

Lachenbruch, A.H. (1968) Permafrost, in *Encyclopaedia of Geomorpholog* (ed. R.W. Fairbridge), Reinhold, New York, USA, pp. 833–838.

Ladurie, E. Le R. (1972) *Times of Feast, Times of Famine: A History of Climate since the Year 1000*, George Allen and Unwin, UK.

Laidre, K.L. and Heide-Jørgensen, M.P. (2005) Arctic sea ice trends and narwhal vulnerability. *Biological Conservation*, **121**, 509–517.

Lamb, H.H. (1977) *Climate: Present, Past and Future*. Methuen, London, UK.

Lamb, H.H. (1982) *Climate, History and the Modern World*. Methuen, London, UK.

Lamb, H.H. (1995) *Climate, History and the Modern World*, 2nd edn. Routledge, London, UK.

LANT (no date) http://www.assembly.gov.nt.ca/_live/pages/wpPages/maptuktoyaktuk.aspx [accessed 30 March 2010].

Latenser, M. and Schneebeli, M. (2002) Temporal trend and spatial distribution of avalanche activity during the last 50 years in Switzerland. *Natural Hazards*, **27**, 201–230.

Lawson, B.D. (2003) Trends in blizzards at selected locations on the Canadian Prairies. *Natural Hazards*, **29**, 123–138.

Leah, T. (no date) *GIS in Avalanche Hazard Management*. Icelandic Meteorological Office. http://proceedings.esri.com/library/userconf/proc01/professional/papers/pap439/p439.htm [accessed 25 February 2010].

Lemke, P., Ren, J., Alley, R. *et al.* (2007) Observations: change in snow, ice and frozen ground, in *Climate Change 2007: The Physical Science Basis* (eds, S. Solomon, D. Qin, M. Manning *et al.*) *Contribution of Working Group I to the Fourth Assessment Report of the Intergovernmental Panel on Climate Change*, pp. 337–384. Cambridge University Press, Cambridge, UK.

Lenton, T., Held, H., Kriegler, E. *et al.* (2008) Tipping elements in the Earth's climate system, *Proceedings of the National Academy of Sciences*, **105**, 1786–1793.

Le Roy Ladurie, E. (1972) *Times of Feast, Times of Famine: a History of Climate Since the Year 1000*. George Allen and Unwin, UK.

Lewkowicz, A.G. and Harris, C. (2005) Frequency and magnitude of active-layer detachment failures in discontinuous and continuous permafrost. Northern Canada. *Permafrost and Periglacial Processes*, **16**, 115–130.

Lied, K. and Bakkehøi, S. (1980) Empirical calculations of snow avalanche runout distance based on topographic parameters. *Journal of Glaciology*, **26**, 165–178.

Lliboutry, L., Arnao, B., Morales, A. *et al.* (1977) Glaciological problems set in the control of dangerous lakes in Cordillera Blanca, Peru. Part 1: Historical failures of moraine dams, their causes and prevention. *Journal of Glaciology*, **18**, 239–254.

Lobdell, J.E. and Dekin, A.A. (1984) The frozen family from the Utqiagvik site, Barrow, Alaska – Papers from a symposium – Introduction. *Arctic Anthropology*, **21**, 1–4.

Lott, N. (1993) The Big One! A review of the March 12–14, 1993 'Storm of the Century'. *Technical report 93-01*, National Climatic Centre, NOAA. http://www1.ncdc.noaa.gov/pub/data/techrpts/tr9301/tr9301.pdf [accessed 27 February 2010].

Lott, N. and Ross, T. (1994) Weather in the Southeast: the February ice storm and the July flooding. *National Climatic Data Centre Technical Report 94–03*. http://www1.ncdc.noaa.gov/pub/data/techrpts/tr9403/tr9403.pdf [accessed 25 February 2010].

Lowe, J.J. and Walker, M.J.C. (1997) *Reconstructing Quaternary Environments*, 2nd edn. Addison Wesley Longman Ltd, Harlow, UK.

Luterbacher, U., Kuzmichenok, V., Shalpykova, G. and Wiegandt, E. (2008) Glaciers and Efficient Water Use in Central Asia, in *Darkening Peaks: Glacier Retreat*,

Science, and Society (eds B. Orlove, E. Wiegandt, and B. H. Luckman), pp. 249–257. University of California Press, Berkeley, California, USA.

Lüthje, M., Pedersen, L.T., Reeh, N. and Greuell, W. (2006) Modelling the evolution of supraglacial lakes on the West Greenland ice-sheet margin. *Journal of Glaciology*, **52**, 608–618.

Makarov, V.I., Kadkina, E.L., Pikulev, V.P. and Kolesnikova, O.V. (2000) The evolution of town Norilsk natural-technical system, in *Proceedings, International Workshop on Permafrost Engineering* (ed. K. Senneset), Longyearbyen, Svalbard, Norway, 18–21 June. Norwegian University of Science and Technology (NTNU)/ University Courese on Svalbard (UNIS), pp. 225–243.

MANICE (2005) *Manual of Standard Procedures for Observing and Reporting Ice Conditions*. http://www.ice-glaces.ec.gc.ca/App/WsvPageDsp.cfm?Lang=eng&lnid=23&ScndLvl=no&ID=172 [accessed 2 April 2010].

Marchant, H.J. (1992) Possible impacts of climate change on the organisms of the Antarctic marine ecosystem, in *Impact of Climate Change on Antarctica*. Australian Government Publishing Service, Canberra, Australia.

Margreth, S. and Funk, M. (1999) Hazard mapping for ice and combined snow/ice avalanches – two case studies from the Swiss and Italian Alps. *Cold Regions Science and Technology*, **30**, 159–173.

Marko, J.R., Birch, J.R. and Wilson, M.A. (1982) A study of long-term satellite-tracked iceberg drifts in Baffin Bay and Davis Strait. *Arctic*, **35**, 234–240.

Marko, J.R., Fissel, D.B. and Miller, J.D. (1988) Iceberg movement prediction off the Canadian east coast, in *Natural and Man-Made Hazards* (eds M.I. El-Sabh and T.S. Murty) pp. 435–462. D. Reidel, Dordrecht, The Netherlands.

Marko, J.R., Fissel, D.B., Wadhams, P. *et al.* (1994) Iceberg severity off eastern North America: its relationship to sea-ice variability and climate-change. *Journal of Climate*, **7**, 1335–1351.

Mars, J.C. and Houseknecht, D.W. (2007) Quantitative remote sensing study indicates doubling of coastal erosion in past 50 years along a segment of the Arctic coast of Alaska. *Geology*, **35**, 583–586. http://energy.usgs.gov/alaska/ak_coastalerosion.html [accessed 9 March 2010].

Marsh, P. and Hey, M. (1989) The flooding hydrology of the Mackenzie delta lakes near Inuvik, N.W.T., Canada. *Arctic*, **42**, 41–49.

McCabe, G.J., Clark, M.P. and Serreze, M.C. (2001) Trends in Northern Hemisphere surface cyclone frequency and intensity. *Journal of Climate*, **14**, 2763–2768.

McClintock, J., McKenna, R. and Woodworth-Lynas, C. (2007) *Grand Banks Iceberg Management*. PERD/CHC Report 20–84 for PERD/CHC, National Research Council Canada, Ottawa, Ontario, by AMEC Earth & Environmental, St. John's, NL, R.F. McKenna & Associates, Wakefield, QC, and PETRA International Ltd., Cupids, NL.

McClung, D.M. and Mears, A.I. (1991) Extreme value prediction of snow avalanche runout. *Cold Regions Science and Technology*, **19**, 163–175.

McClung, D. and Schaerer, P. (1993) *The Avalanche Handbook*. The Mountaineers, Seattle, USA.

McClung. D.M., Stethem, C.J., Schaerer, P.A. and Jamieson, J.B. (eds) (2002) *Guidelines for Snow Avalanche Risk Determination and Mapping in Canada*. Canadian Avalanche Association, Revelstoke, BC, Canada.

McCreedy, J. (2004) Ice storm 1998: lessons learned. *6th Canadian Urban Forest Conference*. Kelowna, B.C., Canada.

McGillivray, D.G., Agnew, T.A., McKay, G.G. *et al.* (1993) Impacts of climate change on the Beaufort sea-ice regime: Implications for the Arctic petroleum industry. *Climate Change Digest CCD 93–01*, Environment Canada, Downsview, Ontario, Canada.

Mears, A.I. (1992) *Snow-Avalanche Hazard Analysis for Land-Use Planning and Engineering*. Colorado Geological Survey Bulletin 49.

Mercer, J.H. (1978) West Antarctic Ice Sheet and CO_2 greenhouse effect – threat of disaster. *Nature*, **271**, 321–325.

Mergen, B. (1997) *Snow in America*. Smithsonian Institution Press, Washington DC, USA.

Montpelier-Vermont (no date) http://www.montpelier-vt.org/community/351/Flood-of-1992.html?id=Z6Ncw3N5 [accessed 13 July 2010].

Mool, P.K., Bajracharya, S.R. and Joshi, S.P. (2001a) *Inventory of glaciers, glacial lakes and glacial lake outburst floods: monitoring and early warning systems in the Hindu Kush-Himalayan region: Nepal*. ICIMOD, Kathmandu, Nepal.

Mool, P.K., Wangda, D. and Bajracharya, S.R. (2001b) *Inventory of glaciers, glacial lakes and glacial lake outburst floods: monitoring and early warning systems in the Hindu Kush-Himalayan region: Bhutan*. ICIMOD, Kathmandu, Nepal.

Moore, S.E. and Huntington. H.P. (2008) Arctic marine mammals and climate change: impacts and resilience. *Ecological Applications*, **18**, Supplement, S157–S165.

Morris, E.M. and Vaughan, D.G. (2003) Spatial and temporal variation of surface temperature on the Antarctic Peninsula and the limit of viability of ice shelves. *Antarctic Peninsula Climate Variability: Historical and Palaeoenvironmental Perspectives*, **79**, 61–68.

Müller, S.W. (1943) Permafrost or permanently frozen ground and related engineering problems. *Special Report, Strategic Engineering Study, Intelligence Branch, Office, Chief of Engineers, no. 62*. Second printing, 1945 (reprinted in 1947, J.W. Edwards, Ann Arbor, Michigan, USA).

NAIWMC (2010) (http://www.naiwmc.org/ [accessed 3 February 2010].

Nansen, F. (1897) *Farthest North, Volumes I and II*. Archibald Constable & Co., London, UK.

NASA (2001) http://www.nas.nasa.gov/About/Education/Ozone/radiation.html [accessed 15 April 2010].

National Park Service (2010) http://www.nps.gov/glac/parknews/park-seeks-comments-on-draft-eis-addressing-avalanche-hazard-reduction-for-burlington-northern-santa-fe-railway.htm [accessed 27 November 2010].

Nature (2008a) All eyes north. Editorial. *Nature*, **452**, 781.

Nature (2008b) Change in the weather, Editorial, 18 June 2008. Nature, **453**, 957–958. doi:10.1038/453957b http://www.nature.com/nature/journal/v453/n7198/full/453957b.html [accessed 17 April 2010].

Nelson, F.E., Anisimov, O.A. and Shiklomanov, N.I. (2002) Climate change and hazard zonation in the circum-Arctic permafrost regions. *Natural Hazards*, **26**, 203–225.

NERC-BAS (2007) http://www.antarctica.ac.uk/about_antarctica/geography/ice/streams.php [accessed 26 March 2010].

Newell, J.P. (1993) Exceptionally large icebergs and ice islands in Eastern Canadian waters: a review of sightings from 1900 to present. *Arctic*, **46**, 205–211.

NFIC (2010) http://indiancountrynews.net/index.php?option=com_content&task=view&id=26&Itemid=63 [accessed 12 March 2010].

NGDC (2008) www.ngdc.noaa.gov/ [accessed 18 December 2008].

NIC (2009) http://www.natice.noaa.gov/products/south_icebergs.html [accessed 27 November 2010].

NIWA (2009) http://www.niwa.co.nz/our-science/oceans/news/all/iceberg-spotted [accessed 1 March 2010].

NOAA (2005) http://www.wrh.noaa.gov/ggw/newsletter/winter_05/IceJams2005.pdf [accessed April 22 2010].

NOAA (2006) http://celebrating200years.noaa.gov/events/storm/welcome.html#storm [accessed 3 February 2010].

NOAA (2008) http://www.cpc.noaa.gov/products/expert_assessment/threats.shtml [accessed 4 February 2010].

NOAA (2009) http://oceanservice.noaa.gov/facts/iceberg.html [accessed 9 July 2010].

NOAA (2010) http://www.wrh.noaa.gov/tfx/hydro/IJAD/JamTypes.php?wfo=tfx [accessed 22 February 2010].

NOAA Economics (2010) (http://www.economics.noaa.gov/?file=bibliography #noaa.2000c) [accessed 2 February 2010].

NOAA-NCDC (1998) http://www.ncdc.noaa.gov/oa/climate/sd/annsum1998.pdf [accessed 17 February 2010].

NOAA-NCDC (1999) http://www.ncdc.noaa.gov/oa/climate/sd/annsum1999.pdf [accessed 17 February 2010].

NOAA-NCDC (2010) http://lwf.ncdc.noaa.gov/snow-and-ice/nesis.php#overview [accessed 1 February 2010].

NOAA NEWS (2010) http://www.noaanews.noaa.gov/stories/images/global.pdf [accessed 5 February 2010].

NOAA-NWS (2010) http://www.wrh.noaa.gov/tfx/icejam/icejam.php?state=MT [accessed 20 April 2010].

Noetzli, J., Hoelzle, M. and Haeberli, W. (2003) Mountain permafrost and recent alpine rockfall events: a GIS-based approach to determine critical factors, in *Proceedings of the Eighth International Conference on Permafrost, vol. 2* (eds M. Phillips, S.M. Springman, and L.U. Arenson). Swets and Zeitlinger, Lisse, Zurich, pp. 827–832.

NRC-CHC (2007) http://www.nrc-cnrc.gc.ca/eng/news/nrc/2007/03/07/iceberg-drifts.html [accessed 27 November 2010].

NSF (2002) *Emperor Penguin Colony Struggling with Iceberg Blockade*. http://www.nsf.gov/od/lpa/news/02/pr0291.htm [accessed 1 March 2010].

NSIDC (1995) *International Ice Patrol (IIP) iceberg sightings database*. National Snow and Ice Data Center/World Data Center for Glaciology, Boulder, Colorado USA. Digital media. http://nsidc.org/data/g00807.html [accessed 7 March 2010].

NSIDC (no date) http://nsidc.org/quickfacts/icesheets.html [accessed 26 March 2010].

NSIDC (2006) http://nsidc.org/data/docs/noaa/g02172_nic_charts_climo_grid/index.html [accessed 2 April 2010].

NSIDC (2009) http://nsidc.org/seaice/characteristics/difference.html [accessed 12 March 2010].
NSIDC (2010a) http://nsidc.org/snow/blizzard/plains.html [accessed 1 February 2010].
NSIDC (2010b) http://nsidc.org/snow/blizzard/images/fupr0112.gif [accessed 26 February 2010].
NSIDC (2010c) http://nsidc.org/snow/blizzard/images/fsfc0112.gif [accessed 26 February 2010].
NSIDC (2010d) http://nsidc.org/data/seaice_index/images/daily_images/N_daily_extent_hires.png [accessed 06 April 2010].
NSIDC (2010e) http://nsidc.org/sotc/sea_ice.html [accessed 6 April 2010].
NSIDC-IICWG (2009) http://nsidc.org/noaa/iicwg/ [accessed 27 March 2009].
Nunalogistics (2003) http://www.nunalogistics.com/projects/winter_road/ [accessed 23 February 2010].
Nunatsiaqonline (2009) www.nunatsiaqonline.ca/.../scrapping_**utilidor**_not_popular_with_resolute_residents/ [accessed 12 December 2009].
NWAC (no date) US Avalanche Danger Scale. http://www.nwac.us/media/uploads/pdfs/US%20Avalanche%20Danger%20Scale.pdf [accessed 28 February 2010].
NWS-NOAA (2000) http://www.nws.noaa.gov/om/hazstats/sum00.pdf [accessed 17 April 2010].
NWS-NOAA (2006) http://www.erh.noaa.gov/buf/storm101206.html [accessed 27 February 2010].
NWS-NOAA (2008) http://www.erh.noaa.gov/btv/events/IceStorm1998/ice98.shtml [accessed 27 February 2010].
NWS-NOAA (2010a) National Weather Service. http://www.nws.noaa.gov/glossary/index.php?letter=b [accessed 23 February 2010].
NWS-NOAA (2010b) http://www.nws.noaa.gov/om/hazstats.shtml [accessed 17 April 2010].
OcCC (2007) *Climate Change and Switzerland 2050: Expected Impacts on Environment Society and Economy.* http://www.meteoswiss.admin.ch/web/en/climate/climate_reports.html [accessed 20 April 2010].
O'Connor, J.E., Grant, G.E., and Costa, J.E. (2002) The geology and geography of floods, in *Ancient Floods, Modern Hazards: Principles and Application of Paleoflood Hydrology* (eds P.K. House, R.H. Webb, V.R. Baker, and D.R. Levish), American Geophysical Union Water Science and Application Series, No. 5, pp. 359–385.
Oklahoma County (2006) http://www.oklahomacounty.org/district1/news/snowroute.htm [accessed 4 February 2010].
Orlove, B. (2009) The past, the present, and some possible futures of adaptation, in *Adapting to Climate Change: Thresholds, Values, Governance* (eds W. N. Adger, I. Lorenzoni and K. O'Brien), pp. 131–163. Cambridge University Press, Cambridge, UK.
Orlove, B., Wiegandt, E. and Luckman B. H. (eds) (2008) *Darkening Peaks: Glacier Retreat, Science, and Society.* University of California Press, Berkeley, California, USA.
Osborn, T. (2000) *11: North Atlantic Oscillation.* Climatic Research Unit: Information sheets. http://www.cru.uea.ac.uk/cru/info/nao/ [accessed 15 April 2010].

Overland, J. Wang, M. and Walsh J. (2009) Atmosphere, in *Arctic Report Card 2009* (eds J. Richter-Menge and J.E. Overland), http://www.arctic.noaa.gov/reportcard. [accessed 10 March 2010].

Parry, M.L., Canziani, O.F., Palutikof, J.P.,*et al*. (eds) (2007) *Contribution of Working Group II to the Fourth Assessment Report of the Intergovernmental Panel on Climate Change*. Cambridge University Press, Cambridge, UK.

Patent Storm (no date) http://www.patentstorm.us/patents/4297056/description.html [accessed 8 December 2009].

Patzelt, G. (ed.) (1983) Die Berg- und Gletscherstürze von Huascaran, Cordillera Blanca, Peru. *Hochgebirgsforschung* 6.

Paul, F., Kääb, A. and Haeberli, W. (2007) Recent glacier changes in the Alps observed by satellite: consequences for future monitoring strategies. *Global and Planetary Change*, **56**, 111–122.

Paul F., Kääb A., Rott H. *et al.* (2009) GlobGlacier: A new ESA project to map the world's glaciers and ice caps from space. *EARSeL eProceedings*, **8**, 11–26.

Payne, A.J., Vieli, A., Shepherd, A. *et al.* (2004) Recent dramatic thinning of largest West Antarctic ice stream triggered by oceans. *Geophysical Research Letters*, **31**, L23401, doi:10.1029/2004GL021284.

Perla, R.I. (1980) Avalanche release, motion and impact, in *Dynamics of Snow and Ice masses* (ed. S.C. Colbeck), pp. 397–462. Academic Press, New York, USA.

Perla, R.I. and Martinelli, M., Jr. (1976, revised 1978) *Avalanche Handbook*. USDA Agricultural Handbook 489, US Government Printing Office, Washington DC, USA.

Permafrost outreach (2009) http:uafpermafrost.blogspot.com/2007_05_01_archive.html [accessed 8 December 2009].

PERMOS (2009) http://www.permos.ch/downloads/permafrostmap_bafu.pdf [accessed 21 November 2009).

Perovich, D., Kwok, R., Meier, W. *et al.* (2009) Sea Ice Cover, in *Arctic Report Card 2009* (eds J. Richter-Menge and J.E. Overland), http://www.arctic.noaa.gov/reportcard/seaice.html [accessed 10 March 2010].

Peterson, B.J., Holmes, R.M., McClelland, J.W. *et al.* (2002) Increasing river discharge to the Arctic Ocean. *Science*, **298**, 2172–2173.

Pfeffer, W.T., Harper, J.T. and O'Neel, S. (2008) Kinematic constraints on glacier contributions to 21st-century sea-level rise. *Science*, **321**, 1340–1343.

Phillips, M., Ladner, F., Muller, M. *et al.* (2007) Monitoring and reconstruction of a chairlift midway station in creeping permafrost terrain, Grachen, Swiss Alps. *Cold Regions Science and Technology*, **47**, 32–42.

Pielmeier, C. and Marshall, H.P. (2009) Rutschblock-scale snowpack stability derived from multiple quality-controlled Snow MicroPen measurements. *Cold Regions Science and Technology*, **59**, 178–184.

Post A. and Lachapelle, E.R. (revised edition) (2000) *Glacier Ice*, University of Washington Press, Seattle, USA.

Price, S.F., Payne, A.J., Catania, G.A. and Neumann, T.A. (2008) Seasonal acceleration of inland ice via longitudinal coupling to marginal ice. *Journal of Glaciology*, **54**, 213–219.

Priddle, J., Smetacek, V. and Bathmann, U. (1992) Antarctic marine primary production, biogeochemical carbon cycles and climatic-change. *Philosophical*

Transactions of the Royal Society of London, Series B-Biological Sciences, **338**, 289–297.

Pritchard, H. and Vaughan, D. G. (2007) Widespread acceleration of tidewater glaciers on the Antarctic Peninsula. *Journal of Geophysical. Research*, **112**, F03S29, doi:10.1029/2006JF000597 [accessed 17 March 2010].

Pritchard, H. D., Arthern, R. J., Vaughan, D. G. and Edwards, L. A. (2009) Extensive dynamic thinning on the margins of the Greenland and Antarctic ice sheets. *Nature*, **461**, 971–975, doi:10.1038/nature08471 [accessed 17 March 2010].

Prokop, A. (2008) Assessing the applicability of terrestrial laser scanning for spatial snow depth measurements. *Cold Regions Science and Technology*, **54**, 155–163.

Public Safety Canada (2009) http://www.publicsafety.gc.ca/prg/em/ndms/aboutsnac-eng.aspxcodes [accessed 4 February 2010].

Quincey, D.J., Lucas, R.M., Richardson, S.D. *et al.* (2005) Optical remote sensing techniques in high-mountain environments: application to glacial hazards. *Progress in Physical Geography*, **29**, 475–505.

Rahmstorf, S. (2007) A semi-empirical approach to projecting future sea-level rise. *Science*, **315**, 368–370.

Rana, B., Shrestha, A.B., Reynolds, J.M. *et al.* (2000) Hazard assessment of the Tsho Rolpa Glacier Lake and ongoing remediation measures. *Journal of the Nepal Geological Society*, **22**, 563–570.

Rapidmoc (2010) http://www.noc.soton.ac.uk/rapidmoc/ [accessed 25 March 10].

Regalado, A. (2005) The Ukukus wonder why a sacred glacier melts in Peru's Andes: it could portend world's end so mountain worshippers are stewarding the ice. *Wall Street Journal*. 17 June 2005, p. A1.

Regehr, E.V., Lunn, N.J., Amstrup, S.C. and Stirling. I. (2007) Survival and population size of polar bears in western Hudson Bay in relation to earlier sea ice breakup. *Journal of Wildlife Management*, **71**, 2673–2683.

Rennermalm, A.K., Smith, L.C., Stroeve, J.C. and Chu, V.W. (2009) Does sea ice influence Greenland ice sheet surface-melt? *Environmental Research Letters*, **4**, 024011, DOI:10.1088/1748-9326/4/2/024011 [accessed 16 April 2010].

Reuters News Agency (2007) http://www.reuters.com/article/idUSN1234898520070430 [accessed 22 April 2010].

Reynolds International (2008) http://www.reynolds-international.co.uk/mountain_hazards_group/pdf/Chapters_1_4.pdf [accessed 17 April 2010].

Reynolds, J.M. (1992) The identification and mitigation of glacier-related hazards: examples from the Cordillera Blanca, Peru, in *Geohazards* (eds G.J.H. McCall, D.J.C. Laming and S.C. Scott) pp. 143–157. Chapman and Hall, London, UK.

Reynolds, J.M. (1999) Glacial hazard assessment at Tsho Rolpa, Rolwaling, central Nepal. *Quarterly Journal of Engineering Geology*, **32**, 209–214.

Reynolds, J.M. and Pokhrel, P.A. (2001) The remediation of the Tsho Rolpa Glacier Lake, Rolwaling, Nepal: a case history. *Geophysical Research Abstracts*, **3**, 457.

Reynolds, J.M. and Taylor, P.J. (2004) Review of Mool *et al.*, 2001a and 2001b in *Mountain Research and Development*, **24**(3), August 2004.

Rice, R., Jr., Decker, R., Jensen, N. *et al.* (2002) Avalanche hazard reduction for transportation corridors using real-time detection and alarms. *Cold Regions Science and Technology*, **34**, 31–42.

Richardson, S.D. and Reynolds, J.M. (2000) An overview of glacial hazards in the Himalayas. *Quaternary International*, **65/66**, 31–47.

Richter-Menge, J. and Overland, J.E. (eds). (2009) *Arctic Report Card 2009*, http://www.arctic.noaa.gov/reportcard. [accessed 10 March 2010].

Rignot, E., Casassa, G., Gogineni, P. *et al.* (2004) Accelerated ice discharge from the Antarctic Peninsula following the collapse of Larsen B ice shelf. *Geophysical Research Letters*, **31**, L18401, doi:10.1029/2004GL020697.

Rignot, E., Bamber, J.L., Van Den Broeke, M.R. *et al.* (2008) Recent Antarctic ice mass loss from radar interferometry and regional climate modelling. *Nature Geoscience*, **1**, 106–110.

Riseborough, D., Shiklamonov, N.I., Etzelmuller, B. *et al.* (2008) Recent advances in permafrost modelling. *Permafrost and Periglacial Processes*, **19**, 137–156.

RMS (2008) *The 1998 Ice Storm: 10-year Retrospective*. Risk Management Solutions Inc. Special Report.

Robe, R.Q., Maier, D.C. and Russell, W.E. (1980) Long-term drift of icebergs in Baffin Bay and the Labrador Sea. *Cold Regions Science and Technology*, **1**, 183–193.

Rolwaling (no date) http://rolwaling.tripod.com/2k/2k-tr-fix.html [accessed 14 March 10].

Ross, J. (1835) *Narrative of a Second Voyage in Search of a Northwest Passage, and of a Residence in Arctic Regions During the Years 1892-1893*. Webster, London, UK.

Röthlisberger, H. (1981) Eislawinen und Ausbrüche von Gletscherseen. *Jahrbuch der Schweizerischen Naturforschenden Gesellschaft, wissenschaftlicher Teil* 1978, 170–212.

Rott, H., Rack, W., Skvarca, P. and de Angelis, H. (2002) Northern Larsen Ice Shelf, Antarctica further retreat after collapse. *Annals of Glaciology*, **34**, 277–282.

Rudkin, P (2005) *PERD Comprehensive Iceberg Management Database*. ftp://ftp2.chc.nrc.ca/CRTreports/PERD/Iceberg_management_05.pdf [accessed 18 April 2010].

Rudkin, P., Young, C., Barron, P., Jr. and Timco, G. (2005) Analysis and results of 30 years of iceberg management. *Proceedings 18th International Conference on Port and Ocean Engineering under Arctic Conditions*, POAC'05, Potsdam, NY, USA, **2**, 595–604.

Ryder, J.M. (1971a) The stratigraphy and morphology of paraglacial alluvial fans in south-central British Columbia. *Canadian Journal of Earth Sciences*, **8**, 279–298.

Ryder, J.M. (1971b) Some aspects of the morphometry of paraglacial alluvial fans in south-central British Columbia. *Canadian Journal of Earth Sciences*, **8**, 1252–1264.

Sale, R. and Potapov, E. (2010) *The Scramble for The Arctic*. Frances Lincoln Limited, London, UK.

Salm, B., Burkard, A. and Gubler, H.U. (1990) Berechnung von Fliesslawinen. Eine Anleitung für den Praktiker mit Beispielen. *Mitteilungen des Eidgenössisches. Institutes für Schnee- und Lawinenforschung*, Nr. 47.

Salzmann, N., Kääb, A., Huggel, C. *et al.* (2004) Assessment of the hazard potential of ice avalanches using remote sensing and GIS-modelling. *Norwegian Journal of Geography*, **58**, 74–84.

Sass, O. (2005a) Spatial patterns of rockfall intensity in the northern Alps. *Zeitschrift für Geomorphologie*, **138**, 51–65. NF., Suppl.-Bd.

Sass, O. (2005b) Rock moisture measurements: techniques, results and implications for weathering. *Earth Surface Processes and Landforms*, **30**, 359–374.

Scambos, T. A., Bohlander, J. A., Shuman, C. A. and Skvarca, P. (2004) Glacier acceleration and thinning after ice shelf collapse in the Larsen B embayment, Antarctica. *Geophysical Research Letters*, **31**, L18402, doi:10.1029/2004GL020670 [accessed 17 March 2010].

Scambos, T., Hulbe, C. and Fahnestock, M. (2003) Climate-induced ice shelf disintegration in the Antarctic Peninsula. *Antarctic Peninsula Climate Variability: Historical and Paleoenvironmental Perspectives*, **79**, 79–92.

Scambos, T., Hulbe, C., Fahnestock, M. and Bohlander, J. (2000) The link between climate warming and break-up of ice shelves in the Antarctic Peninsula. *Journal of Glaciology*, **46**, 516–530.

Schaerer, P.A. and Salway, A.A. (1980) Seismic and impact-pressure monitoring of flowing avalanches. *Journal of Glaciology*, **26**, 179–187.

Schneebeli, M. and Johnson, J.B. (1998) A constant-speed penetrometer for high resolution snow stratigraphy. *Annals of Glaciology*, **26**, 107–111.

Schoof, C. (2007) Ice sheet grounding line dynamics: steady states, stability, and hysteresis. *Journal of Geophysical Research-Earth Surface*, **112**, F03S28.

Schwartz, R.M. and Schmidlin, T.W. (2002) Climatology of Blizzards in the Conterminous United States, 1959–2000. *Journal of Climate*, **15**, 1765–1772.

Schwert, D. (2009) http://www.ndsu.edu/fargo_geology/documents/geologists_perspective_2003.pdf [accessed 13 July 2010].

Scott, E.D., Hayward, C.T., Kubichek, R.F. *et al.* (2007) Single and multiple sensor identification of avalanche-generated infrasound. *Cold Regions Science and Technology*, **47**, 159–170.

Seliverstov, Y., Glazovskaya, T., Shnyparkov, A. *et al.* (2008) Assessment and mapping of snow avalanche risk in Russia. *Annals of Glaciology*, **49**, 205–209.

Senneset, K.G. (ed.) (2000) *Proceedings of the International Workshop on Permafrost Engineering*, Svalbard, June, 2000.

Serreze, M.C., Carse, F., Barry, R.G. and Rogers, J.C. (1997) Icelandic low cyclone activity: climatological features, linkages with the NAG, and relationships with recent changes in the Northern Hemisphere circulation. *Journal of Climate*, **10**, 453–464.

Serreze, M.C. and Francis, J.A. (2006) The Arctic amplification debate. *Climatic Change*, **76**, 241–264.

Shackleton, N.J., Berger, A. and Peltier, W.A. (1990) An alternative astronomical calibration of the lower Pleistocene timescale based on ODP Site 677. *Transactions of the Royal Society of Edinburgh: Earth Sciences*, **81**, 251–261.

Shackleton, N.J. and Opdyke, N.D. (1973) Oxygen isotope and palaeomagnetic stratigraphy of equatorial Pacific core V28–239: oxygen isotope temperatures and ice volumes on a 10^5 and 10^6 year scale. *Quaternary Research*, **3**, 39–55.

Shapiro, L.H., Bates, H.F. and Harrison, W.D. (1978) Mechanics of origin of pressure ridges, shear ridges and hummock fields in landfast ice, in *Principal investigator's report for the year ending March 1978*, Environmental Research Laboratories, Boulder, Colorado, USA.

Shaw, J., Taylor, R.B., Solomon, S. *et al.* (1998) Potential impacts of global sea-level rise on Canadian coasts. *Canadian Geographer-Geographe Canadien*, **42**, 365–379.

Shepherd, A., Wingham, D., Payne, T. and Skvarca, P. (2003) Larsen ice shelf has progressively thinned. *Science*, **302**, 856–859.

Shepherd, A., Wingham, D. and Rignot, E. (2004) Warm ocean is eroding West Antarctic Ice Sheet. *Geophysical Research Letters*, **31**, L23402. DOI:10.1029/2004GL021106.

Shrestha, A.B., Wake, C.P., Mayewski, P.A. and Dibb, J.E. (1999) Maximum temperature trends in the Himalaya and its vicinity: An analysis based on temperature records from Nepal for the period 1971-94. *Journal of Climate*, **12**, 2775–2787.

Simpkins, M. (2009) Marine mammals, in *Arctic Report Card 2009* (eds J. Richter-Menge and J.E. Overland), http://www.arctic.noaa.gov/reportcard/seaice.html [accessed 10 March 2010].

Skvarca, P., Rack, W., Rott, H. and Donangelo, T.I.Y. (1998) Evidence of recent climatic warming on the eastern Antarctic Peninsula. *Annals of Glaciology*, **27**, 628–632.

Smale, D.A., Brown, K.M., Barnes, D., Keiron, K.A., Fraser, P. P. and Clarke, A. (2008) Ice Scour Disturbance in Antarctic Waters. *Science*, **321**, 371, DOI:10.1126/science.1158647 [accessed 7 March 2010].

SLF (no date) http://www.slf.ch/praevention/lawinenunfaelle/unfallstatistik/index_DE?redir=1& [accessed 16 February 2010].

SLF (2008) http://www.slf.ch/lawineninfo/zusatzinfos/interpretationshilfe/zusatzprodukte/icons/icon_karte_e.gif [accessed 28 February 2010].

SLF (2010) European Avalanche Danger Scale. http://www.slf.ch/lawineninfo/zusatzinfos/interpretationshilfe/lawinengefahrenbegriffe/europaeische_skala/index_EN [accessed 28 February 2010].

Smale, D.A., Brown, K.M., Barnes, D. *et al.* (2008) Ice Scour Disturbance in Antarctic Waters. *Science*, **321**, 371, DOI:10.1126/science.1158647 [accessed 7 March 2010].

Smiraglia, C., Diolaiuti, G., Pelfini, M. *et al.* (2008) Glacier changes and their impact on mountain tourism: Two case studies from the Italian Alps, in *Darkening Peaks* (eds B. Orlove, E. Wiegandt, and B.H. Luckman), pp. 206–215. University of California Press, Berkley, USA.

Smith, K. and Ward, R. (1998) *Floods: Physical Processes and Human Impacts*. John Wiley & Sons, Ltd, Chichester, UK.

Smith, R.C., Fraser, W.R., Stammerjohn S.E. and Vernet, M. (2003) Palmer long-term ecological research on the Antarctic marine ecosystem, in *Antarctic Peninsula Climate Variability: Historical and Paleoenvironmental Perspectives* (eds E. Domack, A. Leventer, A. Burnett *et al.*), Antarctic Research Series 79, AGU, Washington DC, USA, pp. 131–144.

Smith, L.C., Sheng, Y., MacDonald, G.M. and Hinzman, L.D. (2005) Disappearing Arctic lakes. *Science*, **308**, 1429.

Sneed, W.A. and Hamilton, G.S. (2007) Evolution of melt pond volume on the surface of the Greenland Ice Sheet. *Geophysical Research Letters*, **34**, L03501.

Solomon, S. (2002) *Tuktoyaktuk Erosion Risk assessment Report*. Geological Survey of Canada, Natural Resources, Canada.

Solomon, S., Qin, D., Manning, M. *et al.* (eds) (2007) *Climate Change 2007: The Physical Science Basis Contribution of Working Group I to the Fourth Assessment Report of the Intergovernmental Panel on Climate Change.* Cambridge University Press, Cambridge, UK.

Spencer, R.F. (1968) *The North Alaska Eskimo: A Study in Ecology and Society.* Bulletin 171, Smithsonian Institute, Bureau of American Ethnology.

Squire, V.A. and Moore, S.C. (1980) Direct measurement of the attenuation of ocean waves by pack ice. *Nature*, **283**, 365–368.

Squires, M. F. and Lawrimore, J. H. (2006) *Development of an Operational Snowfall Impact Scale.* 22nd International Conference on Interactive Information and Processing Systems (IIPS) for Meteorology, Atlanta, Georgia, USA. http://lwf.ncdc.noaa.gov/snow-and-ice/docs/squires.pdf) [accessed 17 April 2010].

SRBC (2010) http://www.susquehannafloodforecasting.org/about.html [accessed 27 November 2010].

SSF NPOLAR (2007) http://www.ssf.npolar.no/pages/news116.htm [accessed 26 January 2010].

Steffen, K., Nghiem, S.V., Huff, R. and Neumann, G. (2004) The melt anomaly of 2002 on the Greenland Ice Sheet from active and passive microwave satellite observations. *Geophysical Research Letters*, **31**, L20402, DOI:10.1029/2004GL020444.

Stern, N. (2007) *Stern Review on the Economics of Climate Change.* Cambridge University Press. Cambridge, UK.

Stirling, I., Lunn, and Iacozza. J. (1999) Long-term trends in the population ecology of polar bears in western Hudson Bay in relation to climate change. *Arctic*, **52**, 294–306.

Stoffel, M., Lièvre, I., Conus, D. *et al.* (2005) 400 years of debris-flow activity and triggering weather conditions: Ritigraben, Valais, Switzerland. *Arctic, Antarctic and Alpine Research*, **37**, 387–395.

Stroeve, J., Holland, M.M., Meier, W. *et al.* (2007) Arctic sea ice decline: Faster than forecast. *Geophysical Research Letters*, **34**, L09501.

Summitpost (2003) http://www.summitpost.org/trip-report/168940/mont-blanc-trip-report-dome-de-go-ter-route-august-2003.html [accessed 10 April 2010].

Swisseduc (2006) http://www.swisseduc.ch/glaciers/earth_icy_planet/glaciers13-en.html?id=3 [accessed 26 January 2010].

Swissinfo.ch (2009) http://www.swissinfo.ch/eng/multimedia/video/Summer_ skiing_ stops.html?cid=989530&itemId=642126 [accessed 27 August 2009].

Tangborn, W. and Post, A. (1998) Iceberg prediction model to reduce navigation hazards: Columbia Glacier, Alaska. *Ice in Surface Waters*, **1**, 231–236.

Tarasov, L. and Peltier, W.R. (2003) Greenland glacial history, borehole constraints, and Eemian extent. *Journal of Geophysical Research-Solid Earth*, **108**, Article Number: 2143, DOI:10.1029/2001JB001731.

Tarnocai, C., Kimble, J. and Broll, G. (2003) Determining carbon stocks in cryosols using the northern and mid latitudes soil database, in *Proceedings of the Eighth International Conference on Permafrost 2* (eds M. Phillips, S.M. Springman and L.U. Arenson), Balkema, Lisse, The Netherlands, pp. 1129–1134.

Tedesco, M. (2007) Snowmelt detection over the Greenland ice sheet from SSM/I brightness temperature daily variations. *Geophysical Research Letters*, **34**, L02504.

Thalparpan, P. (2000) Lawinenverbauungen im permafrost. Eidg. Institut für Schnee und Lawinenforschung, Davos.

The Avalanche Review (2003) *Journal of the American Avalanche Association*, **22**. http://www.avalanche.org/~aaap [accessed 13 February 2010].

The Diesel Gypsy (2009) http://www.thedieselgypsy.com/Ice%20Roads-3B-Denison.htm [accessed 16 July 2010].

The Guardian (2003) http://www.tuvaluislands.com/news/archived/2003/2003-07-19.htm [accessed 26 March 2010].

The Guardian (2005) http://browse.guardian.co.uk/search?search=gurschen+glacier&sitesearch-radio=guardian&go-guardian=
Search [accessed 19 April 2005).

The Guardian (2007a) http://www.guardian.co.uk/world/2007/nov/23/antarctica [accessed 1 March 2010].

The Guardian (2007b) http://browse.guardian.co.uk/search?search=M%2FS+Explorer&sitesearch-radio=News&go-guardian=Search [accessed 9 July 2010].

The Guardian (2008) http://www.guardian.co.uk/world/2008/mar/10/eu.climatechange [accessed 20 May 09].

The Guardian (2009) http://www.guardian.co.uk/world/2009/nov/15/antarctica-trapped-ship-penguin-cruise [accessed 6 April 2010].

The Guardian (2010) Climate scientists hit out at 'sloppy' melting glaciers error. http://www.guardian.co.uk/environment/2010/feb/08/climate-scientists-melting-glaciers/print. [accessed 9 February 2010].

The Iceland Reporter (incorporating News from Iceland) (1995) Number 237, November 1995, Iceland Review, Reykjavik, Iceland.

The Independent (2008) http://www.independent.co.uk/environment/climate-change/has-the-arctic-melt-passed-the-point-of-no-return-1128197.html [accessed 16 April 2010].

The Satellite Encyclopaedia (2010) http://www.tbs-satellite.com/tse/ [accessed 13 April 2010].

Thinkquest (no date) http://library.thinkquest.org/C003603/english/avalanches/casestudies.shtml [accessed 14 February 2010].

Thomas, D.N. and Dieckmann, G.S. (2003) *Sea Ice: An Introduction to its Physics, Chemistry, Biology and Geology*. Blackwell Science, Oxford, UK.

Thompson, L.G., Mosely-Thompson, E., Davis, M.E. *et al.* (2002) Kilimanjaro ice core records: Evidence of Holocene climate change in tropical Africa. *Science*, **298**, 589–593.

Topham, H.W. (1889) A visit to the glaciers of Alaska and Mount Saint Elias. *Proceedings of the Royal Geographical Society*, **11**, 424–433.

Transport Canada (2003) http://www.tc.gc.ca/eng/roadsafety/tp-tp3322-2003-menu-630.htm [accessed 22 April 2010].

Treks (1998) http://p6.hostingprod.com/@treks.org/lakeng98.htm [accessed 29 December 2009].

Trillium (1997) *Tuktoyaktuk.Erosion Control Using Monolithic Concrete Slabs*. Trillium Engineering and Hydrographs Inc.

Trivers, G. (1994) International Ice Patrol's Iceberg Season Severity, in *App. C, Report of the International Ice Patrol in the North Atlantic, Bulletin No. 80, 1994 Season*, CG-188-49. [accessed 18 April 2010].

Tufnell, L. (1984) *Glacier Hazards*. Longman, Harlow, UK.

Turner, J., Bindschadler, R., Convey, P. *et al.* (eds) (2009) *Arctic Climate Change and the Environment*. Scientific Committee on Antarctic Research (SCAR), Scott Polar Research Institute, Cambridge, UK.

UAF Thermosyphons (no date) http://www.alaska.edu/uaf/cem/me/news/thompson_drive_thermosyphons.xml [accessed 24 January 2010].

UMA (1994) *Tuktoyaktuk Shore Protection Study, Phases 2 and 3*. UMA Engineering Ltd.

UNMSM (2004) Calor intenso y largas sequías. Especials, Perú. http://www.unmsm.edu.pe/Destacados/contenido.php?mver=11.

USACE (no date) http://www.mvp-wc.usace.army.mil/ice/graphics/iceload1.png [accessed 12 01 09].

USCG-IIP (no date) http://www.uscg-iip.org/cms/[accessed 1 March 2010].

USCG-IIP (2008) http://www.uscg-iip.org/pdf/Annual_Report_2008.pdf [accessed 18 April 2010].

USCG-IIP (2009a) http://www.uscg-iip.org/pdf/AOS_2009.pdf [accessed 1 April 2009].

USCG-IIP (2009b) http://www.uscg-iip.org/General/history.shtml [accessed 27 November 2010].

USCG-IIP (2009c) http://www.uscg-iip.org/pdf/Annual_Report_2009.pdf [accessed 1 March 2010].

USCG-SOLAS (2009) http://docs.google.com/viewer?a=v&q=cache:IoXG6yHT_hUJ:www.navcen.uscg.gov/pdf/marcomms/imo/SOLAS_V.pdf+MSC/73/21/Add.2+Annex+7&hl=en&gl=uk&pid=bl&srcid=ADGEESjZ3E25kh8610UoTSAry-7hqeMSIseR7F9on63qNUXNxu1rcKJ1FGm2JUm6jLvpPSc8y4XWgwW09tU-LjYv4Or_ARXeuayHvQteJJ7ciGvh7wRVNcBSwAeE507qTQt32jLzylbp&sig=AHIEtbQQB-vCRlJ5Dib3pqc8Z6PRN6Q1JQ [accessed 27 November 2009].

Van der Veen, C.J. (2002) Calving glaciers. *Progress in Physical Geography*, **26**, 96–122.

Van Everdingen, R.O. (ed.) (1998) *Multi-Language Glossary of Permafrost and Related Ground-Ice Terms*. International Permafrost Association.

Vásquez, O.C. (2004) El Fenómeno El Niño en Perú y Bolivia: Experiencias en Participación Local. Memoria del Encuentro Binacional Experiencias de prevención de desastres y manejo de emergencias ante el Fenómeno El Niño, Chiclayo, Peru. ITDG.

Vaughan, D.G. (2008) West Antarctic Ice Sheet collapse: the fall and rise of a paradigm. *Climatic Change*, **91**, 65–79, 10.1007/s10584-008-9448-3.

Vaughan, D.G. and Doake, C.S.M. (1996) Recent atmospheric warming and retreat of ice shelves on the Antarctic Peninsula. *Nature*, **379**, 328–331.

Vaughan, D.G., Marshall, G.J., Connolley, W.M. *et al.* (2003) Recent rapid regional climate warming on the Antarctic Peninsula. *Climatic Change* **60**, 243274.

Vaughan, D.G. and Spouge, J.R. (2002) Risk estimation of collapse of the West Antarctic Ice Sheet. *Climatic Change*, **52**, 65–91.

Venkatesh, S. and El-Tahan, M. (1988) Iceberg life expectancies in the Grand Banks and Labrador Sea. *Cold Regions Science and Technology*, **15**, 1–11.

Venkatesh, S., Murphy, D.L. and Wright, G.F. (1994) On the deterioration of icebergs in the marginal ice-zone. *Atmosphere-Ocean*, **32**, 469–484.

Verbit, S., Comfort, G. and Timco, G. (2006) Development of a database for iceberg sightings off Canada's east coast. *Proceedings of the 18th International Symposium on Ice*, **2** 89–96.

Vincent, W.F. and Belzile, C. (2003) Biological UV exposure in the polar oceans: Arctic-Antarctic comparisons, in *Antarctic Biology in a Global Context* (eds A.H.L. Huiskes, W.W.C. Gieskes, J. Rozema *et al.*), *Proceedings of the 8th SCAR International Biology Symposium*, 176–181.

Vivian, R. (1966) La catastrophe du glacier d'Allalin. *Revue de Géographie Alpine*, **54**, 97–112.

Vonder Mühll, D. and Haeberli, W. (1990) Thermal characteristics of the permafrost within an active rock glacier (Murtèl/Corvatssch, Grisons, Swiss Alps). *Journal of Glaciology*, **36**, 151–158.

Vonder Mühll, D., Hauck, C., Gubler, H., McDonald, R. and Russill, N. (2001) New geophysical methods of investigating the nature and distribution of mountain permafrost with special reference to radiometry techniques. *Permafrost and Periglacial Processes*, **12**, 27–38.

Vonder Mühll, D., Nötzli, J., Makowski, K, and Delaloye, R. (2004) *Permafrost in Switzerland 2000/2001 and 2001/2002.* Glaciological report (permafrost) Vol. 2/3.

Vonder Mühll, D., Nötzli, J. and Roer, I. (2008) PERMOS – a comprehensive monitoring network of mountain permafrost in the Swiss Alps in *Proceedings of the Ninth International Conference on Permafrost Vol. 2* (eds D.L. Kane and K.M. Hinkel), Institute of Northern Engineering, University of Alaska Fairbanks, USA, pp. 1869–1874.

Vuichard, D. and Zimmermann, M. (1986) The Langmoche flash-flood, Khumbu Himal, Nepal. *Mountain Research and Development*, **6**, 90–94.

Vuichard, D. and Zimmermann, M. (1987) The 1985 catastrophic drainange of a moraine-dammed lake, Khumbu Himal, Nepal: causes and consequences. *Mountain Research and Development*, **7**, 91–110.

Watanabe, T., Ives, J.D. and Hammond, J.E. (1994) Rapid growth of a glacial lake in Khumbu-Himal, Himalaya: prospects for a catastrophic flood, *Mountain Research and Development*, **14**, 329–340.

Weinthal, E. (2006) Water conflict and cooperation in Central Asia. *Human Development Report Office Occasional Paper 32.* United Nations Development Programme, New York, USA.

Weller, G. and Lange, M. (1999) Impacts of global climate change in the Arctic regions: an initial assessment. Discussion paper, Workshop on the Impacts of Global Change, 25–26 April, 1999, Tromso, Norway, International Arctic Science Committee, Oslo, Norway.

White, K.D. (1996) A new ice jam database. *Journal of the American Water Resources Association*, **32**, 341–348.

White, K.D. and Kay, R.L. (1996) Ice jam flooding and mitigation, Lower Platte River Basin, Nebraska. *Cold Regions Research and Engineering Laboratory Special Report 96-1*, Hanover, New Hampshire, USA.

White, K.D., Tuthill, A.M. and Furman, L. (2007) Studies of ice jam flooding in the United States, in *Extreme Hydrological Events: New Concepts for Security* (eds O.F. Vasiliev, P.H.A.J.M. van gelder, E.J. Plate and M.V. Bolgov), NATO Science Series, Vol. **78**.Springer, Netherlands, pp. 255–268.

White Rose (2000) *Development Application Volume 5 Part Two (Concept Safety Analysis) Appendix A Target Levels of Safety*. http://www.huskyenergy.ca/downloads/AreasOfOperations/EastCoast/DevelopmentApplication/Vol5_AppendixA.pdf [accessed 18 April 2010].

Whittaker, L.M. and Horn, L.H. (1981) Geographical and seasonal distribution of North American cyclogenesis, 1958–1977. *Monthly Weather Review*, **109**, 2312–2322.

Wiegandt, ER. and Lugon, R. (2008) Challenges of living with glaciers in the Swiss Alps, past and present, in *Darkening Peaks* (eds B. Orlove, E. Wiegandt and B.H. Luckman), pp. 33–48. University of California Press, Berkley, USA.

Willerslev, E., Cappellini, E., Boomsma, W. *et al.* (2007) Ancient biomolecules from deep ice cores reveal a forested Southern Greenland. *Science*, **317**, 111–114.

Windupradio (1998) http://windupradio.com/hickson1.gif [accessed 26 February 2010].

Winton, M. (2006) Does the Arctic sea ice have a tipping point? *Geophysical Research Letters*, **33**, L23504, doi:10.1029/2006GL028017.

Wireless Estimator (2006) http://www.wirelessestimator.com/generaldoc.cfm?ContentID=6 [accessed 17 April 2010].

Witze, A. (2008) Losing Greenland: is the Arctic's biggest ice sheet in irreversible meltdown? *Nature*, **452**, 798–802. doi:10.1038/452798a.

Wolfe, S.A. (1998) Living with frozen ground. A field guide to permafrost in Yellowknife, Northwest Territories. *Geological Survey of Canada miscellaneous report 64*.

WWF (2008) http://wwf.panda.org/what_we_do/where_we_work/arctic/publications/?uNewsID=122260 [accessed 7 July 2010].

Yamada, T. (1998) *Glacier Lake and its Outburst Flood in the Nepal Himalaya*. Data Centre for Glacier Research, Japanese Society for Snow and Ice, Tokyo, Monograph 1.

Yamada, T. and Sharma, C.K. (1993) Glacier lakes and outburst floods in the Nepal Himalaya, in (ed. G.J. Young) *Snow and Glacier Hydrology*. Publication No. 218, International Association of Hydrological Sciences, Wallingford, UK.

Zemp. M., Haeberli, W. Hoelzle, M. and Paul, F. (2006) Alpine glaciers to disappear within decades? *Geophysical Research Letters*, **33**.L13504, doi:10.1029/2006 GL026319.

Zielinski, G.A. (2002) A classification scheme for winter storms in the eastern and central United States with an emphasis on 'Nor'easters'. *Bulletin of the American Meteorological Society*, **83**, 37–51.

Zimmermann, M. and Haeberli, W. (1992) Climatic change and debris flow activity in high mountain areas: a case study in the Swiss Alps, in *Catena Supplement 22* (eds M. Boer and E. Koster), pp. 59–72.

Zufelt, J.E. (1993) *Ice Motion Detector System.* USA Cold Regions Research and Engineering Laboratory, Ice Engineering Information Exchange Bulletin, No. 4.

Zwally, H.J., Abdalati, W., Herring, T. *et al.* (2002) Surface melt-induced acceleration of Greenland ice-sheet flow. *Science*, **297**, 218–222.

Glossary

Adiabatic lapse rate The rate at which air temperature and pressure changes with altitude. If the pressure of a mass of air is lowered it expands and does mechanical work on the surrounding air which requires energy. This is removed from the heat energy of the air mass and the temperature of the air falls (Atkinson in Goudie, 1990a).

Alas A Yakut (Siberia) term for a circular or oval, steep-sided, flat-bottomed depression developed by the progressive thawing of permafrost. It is a form of thermokarst (see below) (French, 2007).

Albedo A measure of the reflectivity of a surface and defined as the total radiation reflected by the surface divided by the total incident radiation. Numerical values range from 0 to 1. Oceans have an albedo of 0.07 while highly reflective ice varies from 0.4 to 0.6 (Henderson-Sellers in Goudie, 1990a).

Altimetry The measurement of altitude.

Antarctic bottom water (ABW) ABW is cold, dense, saline water derived from the Weddell and Ross Seas around Antarctica which sinks down the Antarctic continental margin into the ocean current system of the Atlantic Ocean to the north.

Antarctic circumpolar current (ACC) The current that flows from west to east around the Southern Ocean above the level of the ABW (see above).

Archipelago A sea containing many islands.

Atmosphere The gaseous envelope of air surrounding the Earth. Its circulation produces climate and weather.

Bathymetry The measurement of water depth.

Biosphere The zone at the interface of the Earth's crust and the atmosphere containing living matter (Simmons in Goudie, 1990a).

Cold Region Hazards and Risks, First Edition.   Colin A. Whiteman.
© 2011 John Wiley & Sons, Ltd. Published 2011 by John Wiley & Sons, Ltd.

BP An abbreviation of the term 'Before Present', used in relation to long geological time, for example 11 500 BP. The 'present' is taken as 1950 calendar years AD (Lowe and Walker, 1997).

Brittle deformation or failure Permanent deformation or failure in which material (such as ice) breaks along a fracture which might develop into a crevasse in a glacial context (Benn and Evans, 1998).

Canton A political subdivision. In this context, one of the states of Switzerland.

Cold-based (polar) glacier A type of glacier with ice that is below the pressure melting point so that it is frozen to its bed and can therefore achieve little, if any, erosion (Benn and Evans, 1998).

Congelation ice A coherent layer of frozen sea water (Hansom and Gordon, 1998).

Coriolis force (Geostrophic force) An apparent force on moving particles (ocean water or air) in a frame of reference which itself is moving. The force 'deflects' air particles to the right in the Northern Hemisphere and to the left in the Southern Hemisphere (Atkinson in Goudie, 1990a).

Creep A term used in relation to glaciers to describe deformation resulting from movement within or between individual ice crystals (Benn and Evans, 1998). Creep also describes the slow movement of soil particles downslope in a ratchet-style movement assisted, in cold regions, by the growth and decay of ice crystals.

Cryosphere The zone of the Earth's crust, ocean and atmosphere containing ice.

Cyclogenesis The formation of cylones, generally in preferred areas such as the western Atlantic and western Pacific oceans (Reynolds in Goudie, 1990a).

Cyclone (depression) A region of relatively low atmospheric pressure in which the low-level winds spiral, anticlockwise in the Northern Hemisphere and clockwise in the Southern Hemisphere. They are frequently responsible for variable quantities and types of precipitation (Reynolds in Goudie, 1990a).

Debris flow A type of mass movement of rock and sediment usually initiated on steep slopes, often waterlogged by melting snow or ice. Although initially heavily saturated with water, this often drains quickly, the sediment becomes more viscous and comes to rest as a debris fan (French, 2007).

Dendrochronology The study of tree rings for the purpose of dating objects or events.

Devensian A term used to describe the most recent cold-climate stage that ended about 11 500 BP.

Ductile deformation Permanent deformation in which material undergoes flow or creep (see above), often resulting in folding (Benn and Evans, 1998).

Elastic deformation A temporary change in the shape of a material which lasts only as long as stress is applied (Benn and Evans, 1998).

Fetch The distance from the nearest coast (or edge of the ice pack) in the direction of the wind. The length of exposed sea that wind can blow across to create waves at a coast.

Frost heave Frost heave is the displacement of bedrock, soil, foundations, buildings and road surfaces, for example due to the 9% expansion in volume when water freezes to ice. Frost heave also occurs when lenses of ice (see segregation ice below) form in the soil as a result of cryosuction (French, 2007).

Frost susceptible Soils and sediments are said to be frost susceptible when they readily allow ice lenses to form. Frost susceptible soils have a grain size distribution of 0.01 mm diameter or less (French, 2007).

Geocryogenetic A descriptive term referring to landscape features with an origin associated with the presence of ice in its various forms.

Glacial stage A period of time, usually about 100 000 years in length, during which temperatures are generally lower than during interglacials (see below) including the present time. Glacial stages are usually associated with the expansion of ice sheets, glaciers and permafrost.

Grounding line The position at which flotation occurs in a marine (tidewater) glacier (see below).

Ground-penetrating radar (GPR) GPR is a geophysical method for determining sub-surface conditions including rock layering and the presence of objects. In simple terms, the equipment emits electromagnetic pulses which travel through the ground at different rates related to the nature of the material and rebound from discontinuities where material changes. The signal is shown as a series of traces depicting these discontinuities and object boundaries.

Gulf Stream The warm surface ocean current which moves northwards along the west side of the North Atlantic Ocean. In its northern position it is sometimes referred to as the North Atlantic Drift. It is mainly responsible for keeping the north-eastern area of the Atlantic Ocean relatively ice free.

Heinrich Event Massive iceberg discharge from the Laurentide (North American), Fennoscandian and British ice sheets resulting in the deposition of carbonate-rich layers of ice-rafted debris into mid-latitude regions of the North Atlantic Ocean (Lowe and Walker, 1997).

Holocene A term meaning 'wholly recent' that is used to describe the present warm interval, since 11 500 years Before Present, in which we live (Lowe and Walker, 1997).

Hydrocarbon A naturally occurring organic compound of hydrogen and carbon only. Oil and petroleum gas are examples of hydrocarbons.

Hydrosphere The zone of the Earth formed of water.

Hypothermia A medical condition of very low, life-threatening body temperature.

Ice floe A piece of floating sea, lake or river ice which is not attached to the land. Arctic and Antarctic ice floes can vary from tens of metres to several kilometres across. They may occur as discrete floes or form a continuous mass as, for example, Arctic sea ice (Sugden in Goudie, 1990a).

Ice island A very large tabular iceberg normally derived from a disintegrating ice shelf.

Ice wedge (polygon) An ice wedge is a wedge-shaped body of ice that fills an open crack in the ground. It is caused by the cracking of the ground when permafrost becomes very cold and contracts, a process known as thermal contraction cracking. Ice wedges occur in a polygonal network. The expression of this vertical network at the ground surface produces ice-wedge (or tundra) polygons, one of the most characteristic features of permafrost terrain (French, 2007).

Interglacial A warm episode between glacials (see above) during which temperatures in mid- and high-latitude regions were occasionally higher than those of the present day (Lowe and Walker, 1997).

Levee An embankment formed at the edge of a river or debris flow.

Lithosphere The part of the Earth system composed of rock.

Little Ice Age The widespread period of climatic cooling during the Holocene which lasted approximately from the fourteenth to the nineteenth centuries AD.

Atlantic meridional overturning circulation (MOC) The Atlantic MOC carries warm upper waters, such as the Gulf Stream, into far-northern latitudes where cooling and sinking takes place, and returns the cold deep waters southward across the Equator. The heat transport of the Atlantic MOC makes a substantial contribution to the moderate climate of maritime and continental Europe and any slowdown in the overturning circulation would have profound implications for climate change (Bryden *et al.*, 2005).

Marine (tidewater) glacier That part of a glacier that is forced to float by water that is deep enough to make the glacier buoyant. Such a glacier may extend to become an ice shelf.

Moraine A generally ridge-shaped accumulation of sediment deposited around the margin of a glacier or ice sheet. (Note, in some texts the term 'moraine' is used to refer to glacial sediments in general, in addition to landforms).

Moulin A vertical or steeply dipping hole in glacier ice, usually formed where a crevasse opens across the line of a surface meltwater stream, and which carries meltwater into and possibly through a glacier (moulin (French) = 'mill') (Benn and Evans, 1998).

Nilas The term which describes a thin elastic crust of ice up to 10 cm in thickness, easily bending on waves and swells and under pressure growing in a pattern of interlocking 'fingers' (finger rafting). Nilas is a stage in the development of sea ice (Environment Canada, 2003b).

North Atlantic Oscillation (NAO) The NAO is a north–south shift (or vice versa) in the track of storms and depressions across the North Atlantic Ocean, related to fluctuations in pressure strength over Iceland and the Azores. The change in the mean atmospheric circulation drives patterns of warming and cooling over much of the Northern Hemisphere and can significantly affect temperature and pressure over the Arctic Ocean (Osborn, 2000).

Ozone A form of oxygen containing three atoms in the molecule (O_3). The ozone layer contains high concentrations of ozone which absorbs harmful solar ultraviolet radiation. In recent years the ozone layer over Antarctica and the Arctic has lost significant amounts of ozone and these areas are referred to as 'ozone holes'.

Pleistocene A term meaning 'most recent' which refers to a period of time which ended around 11 500 years ago, at the end of the last glaciation. It is frequently used interchangeably with the term 'Quaternary' (see below) (Lowe and Walker, 1997).

Polynya An area of open water within sea ice.

Salinity A term meaning 'saltiness', often used with reference to sea water which has a mean salinity of around 35 parts per 1000. Salinity affects the density of water and its freezing properties. Fresh water is less dense than sea water and generally freezes more quickly.

Quaternary The most recent major subdivision of the geological record extending up to the present day (Lowe and Walker, 1997).

Riffle A shallow section of river bed composed of gravel and cobbles which alternate with deeper sections referred to as pools (Richards in Goudie, 1990a).

Run-out zone The area on a slope, usually near or at the bottom of the slope, where an avalanche decelerates and most of the snow and other debris is deposited.

Salinization The process of significantly increasing the saltiness of soils and sediments.

Saltation The bouncing or hopping motion of objects (grains, pebbles, boulders) transported by a fluid (air or water) after being ejected from the ground by lift forces. The impact of an object returning to the ground initiates further particle movement and a chain reaction of accelerated transport typical of an avalanche (Richards in Goudie, 1990a).

Segregation ice This is ice which forms as a lens when water is drawn up (cryosuction) between grains of soil or sediment towards the level at which freezing is taking place. As long as there is a supply of water the lens of segregation ice will continue to grow. Cryosuction depends on surface tension and soil particle volume (French, 2007).

Serac A block or pinnacle of ice, isolated between open crevasses and usually associated with ice falls where a glacier is accelerating down a steep slope.

Southern Hemisphere annular mode (SAM) A term which refers to the state of atmospheric circulation in the Southern Hemisphere. It involves north–south switches in the position of main wind systems and changes in pressure around the south-polar region.

Start zone The area on a slope where an avalanche is initiated.

Storm surge High tide and wave activity, often well in excess of normal high-tide level, induced by storm-force winds.

Supercooled water Water that is cooled below its normal freezing point without changing to ice. This may occur under certain pressure conditions or if the water is contaminated by salts (van Everdingen, 1998).

Talik A Russian term used to denote areas of ground in (closed) or around (open) permafrost that remain unfrozen, although within a permafrost region (van Everdingen, 1998).

Thermohaline circulation (THC) A term which refers to the contribution made to the circulation of ocean-water masses by the internal factors of salinity (density) and water-temperature gradients. Other drivers of the **ocean conveyor**, the global system of ocean currents, are meltwater and freshwater influxes (Lowe and Walker, 1997).

Thermokarst A term used to describe the uneven landscape resulting from processes associated with the thaw of especially ice-rich permafrost that lead to collapse, subsidence, erosion and instability of the ground surface. Thermokarst is likely to increase under the present warming climatic regime (French, 2007).

Thermosyphon A device designed to transfer heat from the ground to the air to avoid the melting of permafrost and the consequent subsidence of the ground and the disturbance of surfaces and buildings.

Utilidor A system of continuously-insulated elongated aluminium boxes, carrying water and waste pipes between buildings and a central location, above ground with permafrost to avoid the permafrost being melted and the ground and pipework disturbed (French, 2007).

Ultraviolet (UV) radiation High-frequency, short-wave solar radiation that is largely absorbed by ozone at the top of the atmosphere. UV-B light can be harmful to skin and marine organisms (NASA, 2001).

Warm-based (temperate) glacier A type of glacier with ice that is above the pressure melting point (except for a thin surface layer subject to seasonal temperature cycles) so that water is available to facilitate sliding and erosion at its bed (Benn and Evans, 1998).

Acronyms

AARI	Arctic and Antarctic Research Institute
ACC	Antarctic circumpolar current
ACE	Air convection embankment
ACECRC	Antarctic Climate and Ecosystems Cooperative Research Centre (Australia)
ACIA	Arctic Climate Impact Assessment
AFI	Air freezing index
AMAP	Arctic Monitoring and Assessment Programme
AP	Antarctic Peninsula
ASMA	Antarctic Specially Managed Area
ASPA	Antarctic Specially Protected Area
ATI	Air thawing index
BAPS	iceBerg Analysis and Prediction System
BAS	British Antarctic Survey
BBC	British Broadcasting Corporation (UK)
CAIC	Colorado Avalanche Information Centre (US)
CAFF	Conservation of Arctic Flora and Fauna
CBC	Canadian Broadcasting Corporation
CC	Cryospheric Commission
CIS	Canadian Ice Service
CRREL	Cold Regions Research and Engineering Laboratory (US)
DEM	Digital elevation model
DTM	Digital terrain model
EAIS	East Antarctica Ice Sheet

Cold Region Hazards and Risks, First Edition.   Colin A. Whiteman.
© 2011 John Wiley & Sons, Ltd. Published 2011 by John Wiley & Sons, Ltd.

ENRA	Estonian National Road Administration
ERT	Electrical resistivity tomography
EU	European Union
FAR	Fourth Assessment Report (of IPCC)
FLAR	Forward-looking airborne radar
FPSO	Floating production and storage offloading
GBS	Gravity based structure
GCM	General circulation model
GHG	Greenhouse gas
GIS	Greenland Ice Sheet, geographical information system
GLOF	Glacial lake outburst flood
GPR	Ground-penetrating radar
GPS	Global positioning system
GST	Ground surface temperature
GTN-P	Global Terrestrial Network for Permafrost
HEP	Hydroelectric power
HVRI	Hazards and Vulnerability Research Institute (US)
IAATO	International Association of Antarctic Tour Operators
IASC	International Arctic Science Committee
ICESat	Ice, cloud and land elevation satellite
ICIMOD	International Centre for Integrated Mountain Development
IICWG	International Ice Charting Working Group
IIP	International Ice Patrol
IPCC	Intergovernmental Panel on Climate Change
IISD	International Institute for Sustainable Development
IPA	International Permafrost Association
LAKI	Limit of all known ice
LIA	Little Ice Age
MAAT	Mean annual air temperature
MANICE	Manual of Standard Procedures for Observing and Reporting Ice Conditions (Canada)
MOC	Meridional Overturning Circulation
NAIWMC	North American Interstate Weather Modification Council
NAO	North Atlantic Oscillation
NAPA	National Adaptation Programme of Action
NERC	Natural Environment Research Council
NESIS	Northeast snowfall impact scale
NIC	National Ice Centre

NIWA	National Institute of Water and Atmospheric Research (New Zealand)
NOAA	National Oceanic and Atmospheric Administration
NRCan	National Resources Canada
NRC-CHC	National Research Council–Canadian Hydraulics Centre
NSIDC	National Snow and Ice Data Centre
NWS	National Weather Service
NWT	Northwest Territories
PACE	Permafrost and Climate in Europe
PERD	Program of Energy Research and Development
PERMOS	Swiss Permafrost Monitoring Network
RADARSAT	Radar satellite
SAM	Southern Hemisphere Annular Mode
SAR	Synthetic aperture radar
SCAR	Scientific Committee on Antarctic Research
SCNAT	Swiss Academy of Sciences
SLAR	Side-looking airborne radar
SLF	Swiss Federal Institute for Snow and Avalanche Research
SOLAS	Safety of life at sea
TAR	Third Assessment Report (of IPCC)
THC	Thermohaline circulation
UNEP	United Nations Environment Programme
USCG	United States Coastguard
USGS	United States Geological Survey
WAIS	West Antarctic Ice Sheet
WGI	Working Group 1 (IPCC, 2007)
WGII	Working Group 2 (IPCC, 2007)
WGIII	Working Group 3 (IPCC, 2007)
WWF	World Wildlife Fund

Index

AARI (Arctic and Antarctic Research Institute, 41
ACIA, 6,
  Scientific report, 42
acceleration of terrestrial ice, 8
active layer, see permafrost
adaptive capacity, 10, 69
Adelie Island, Antarctica, 88
Adelie penguin, 66
adfreeze bonding, 164
air convection embankment ('ventilated shoulder'), 182, 184
aircraft visual reconnaissance, 93
air freezing index (AFI), 164
air-lifting mechanisms, 208
air thawing index (ATI), 164
alas, 168
Alaska, 15, 16, 27, 28, 29, 30, 31, 36, 41, 42, 76, 112, 119, 120, 157, 158, 159, 164, 168, 174, 175, 177, 179, 180, 181, 182, 183, 184, 185, 186, 188, 265
  Division of Homeland Security and Emergency Management, 28
  Department of Natural Resources, 188
  Highway, 121, 158
albedo, 14, 58, 128, 242, 246
  positive feedback, 22, 38
  reduction, 14

Alberta, 279, 298
  'Alberta Clipper' storms, 279
alien species impact, 37, 66–67
Allalin Glacier, Switzerland, 105, 113, 118
Allemansrätt (Swedish – 'right to roam'), 104
alpine regions, 5
alpine tourism, 10
Alps, The, 165, 209, 213, 215, 228
Alsek River, Yukon, Canada, 121
Altels Glacier, Switzerland, 105
altimetry, 19, 56, 61
AMAP, 42
Amery Ice Shelf, 94
Amundsen Sea, 58, 60, 61, 94
Andermatt, Switzerland, 127
Andes, 1, 114, 119, 122, 125, 127
animal migration routes, disruption, 37
Archimedes principle, 50
Antarctica, see ice sheets
Antarctic Bottom Water, 66, 90
Antarctic Circumpolar Current, 58
Antarctic Peninsula, see ice sheets,
Antarctic Specially Managed Area (ASMA), 67
Antarctic Specially Protected Area (ASPA), 67
Antarctic Treaty System, 67

Cold Region Hazards and Risks, First Edition. Colin A. Whiteman.
© 2011 John Wiley & Sons, Ltd. Published 2011 by John Wiley & Sons, Ltd.

AP, see ice sheets
  tourist landings, 62, 69
  response to climate change, 62
Arctic Amplification, ice albedo
    feedback, 38
Arctic and Antarctic Research Institute,
    see AARI
Arctic Climate Impact Assessment, see
    ACIA,
Arctic Council, 6, 42
Arctic ecosystems, 35–37
Arctic inhabitants, communities, 6, 23
Arctic Monitoring and Assessment
    Programme (see AMAP)
Arctic Ocean, 5, 8, 13, 21, 34, 37, 42, 64, 75
  area of ice on, 13
  basin, relief map, 16
  circulation map, 15
  exchange of water, 16
  ice-free, -less, 13, 15, 300
  predictions of ice-free state, 13
  water, 13
Arctic sea ice, 6, 13–43, 300
  absence, 15
  access, 15
  age distribution in February, 20
  albedo of associated surfaces, 38
  area, 21
  break-up, 26
  calving, break-off of shorefast ice, 25
  causes of movement, 21
  changes, 19, 21–23
  definition, 17
  distribution, 16, 19, 21–23
  drift, 24
  effects of diminishing extent, 25, 29–38
  entrapment, 23
  export, 19, 38
  extent, 14, 16, 22
  factors affecting, 21
  freeze-up, 26
  hazards, 23
  loss, 14, 41
  hazards, 13, 23
  impacts, direct and indirect, 23–38
    entrapment, 24
    climate change, 37–38
  coastal erosion, 29–34
  ecological, 35–37
  exploration and exploitation of
    minerals, 34–35
  tourist shipping, 34–35
  iceberg severity, 35
  icesheet stability, 35
  ivu (override), see ivu
  reduced access, 24–27
  influence of weather, 19
  Inuit elders' comments, 25–26
  marine mammals, 36
  melt season, 23
  mitigation of impacts, 38–42
  ice-breaking, 39
  information services, 39–41
  movement, 21
  newspaper headlines, 13
  processes of formation, 17
  reflector of solar radiation, 6
  seasonal changes, 21–22
  summer (September) ice
    minimum, 13, 16, 22
  summer minimum trend, 23
  thermodynamic equilibrium, 18
  thickness, 6, 17–21
  tipping point, 22, 39
  trend, 22
  unpredictability, 25–26
  volume, 21, 22
  winter (March) ice maximum, 16, 22
Arctic, The, 5, 35, 37
  a focus of interest, 37
Argentière Glacier, France, 118
Arkansas, USA, 282
artificial snow, 127
Asia, 13, 117
Association for Hydraulic Research, 248
Atlantic, Ocean, 1, 22
Atlantic Oceanographic and
    Meteorological
    Laboratory, 67
atmospheric manipulation, 290
Austria, 1, 133, 208, 213, 214, 215, 225, 236
avalanche, see snow avalanche
    and ice avalanche
    avalanche-prone area, 10

forecasting, 10
modification, 10
types (see snow avalanche and ice avalanche)
Axel Heiberg Island, Canada, 119
Ayles Ice Shelf, 87
Azores, The, 79

Baffin Bay, 19, 21, 37, 51, 75, 77, 78, 81, 92
Baffin Current, 78
Baffin Island, 16, 35, 51, 76
Bangladesh, 7, 63
Banks Island, Canada, 26
Barents Sea, 21, 77, 88, 96
Barrow (Alaska, USA), 27, 28, 29, 31, 41, 175
BAS, 49, 62
Bathurst Inlet, Canada, 268
beach nourishment, 33
Beaufort Island, Antarctica, 65
Beaufort Sea, 33, 87, 88, 246
    Coastlands, 29, 30, 34
    Hydrocarbon exploration, 32
    Hydrocarbon field, 27
bedrock ramp, 109
Bellingshausen Sea, Antarctica, 60, 94
benthic resources, 37
Bering Strait, 19, 21
    Development Council, 32
Bermuda, 80
Bernese Alps, Switzerland, 105, 107
Bhutan, 151, 153, 154
Bilibino nuclear power station, 175, 176
biodiversity, 130
    loss, 37
biosphere, 5
Björnöya Island, Norway, 21
Black Rapids ('Galloping') Glacier, Alaska, 120
blizzards, 2, 4, 273–298
    'Alberta Clipper', 279
    Buffalo, New York State, USA, 282, 293, *Buffalo News*, 293
    definition, 274–275
    distribution, 279
    frequency, 277, 278
    impact of climate warming, 296–298

    impacts, 281–291
        damage to property, 284, 286
        economic losses, 285, 287
        examples from N. America, 281, 282
        fatalities, 281, 282, 284, 285–286, 287, 288, 291
        indirect effects, 285–286
        power loss (outage), 285, 289
        transport and traffic disruption, 285, 287
    mitigation measures, 289–296
        effective forecasting, 290–291
        effective procedures, 290, 293–294
        forward planning, 290, 291–293
        public response, 290, 294–295
    Mongolia, 282
    Northeast Snowfall Impact Scale (NESIS), 294, 295–296, categories, 295–296, 297
    probability in USA, 278
    'snow bursts', 281
    'Storm of the Century' (USA, 1993), 281, 282, 286, 290,
    summary of impacts, 284
    weather systems and processes, 273, 276–279
breakup, see river ice
brine, 17
British Antarctic Survey, see BAS
British Columbia, Canada, 119, 123
buoys, remotely-sensed, 24
buried glacier ice, 168
Bylot Island, Canada, 76

CAFF, 42
calving glaciers, 49, 55, 76, 77, 86
Canada, 1, 15, 34, 35, 42, 157, 158, 162, 164, 168, 173, 175, 178, 179, 180, 181, 182, 184, 186, 187, 188, 265
Canadian Archipelago, 15, 19, 21, 34, 36, 75, 76
Canadian Broadcasting Corporation (CBC), 268, 269–272

Canadian Coast Guard, 92
Canadian Committee on River Ice
    Processes and the
    Environment, 240
Canadian Cordillera, 123
Canadian Hydraulics Centre, 95
Canadian Ice Service, see CIS
Canadian Panel on Energy Research and
    Development, 200
Cape Race, Newfoundland, 99
Cape Crozier, Antarctica, 6, 89
Cape York, Greenland, 75
carbon dioxide, 172, 173, 181,
    export, 173, 174
    budget, 173
Caucasus Mountains, Russia, 113, 119
Central Asia, 127
Chamonix, France, 118
Chesapeake Bay, USA, 80
Chile, 76, 77
China, 125, 157, 174, 237
Chukchi Sea, 88
Circum-Arctic Map of Permafrost and
    Ground Ice, 161, 175, 198
circumpolar deep water, 58
CIS, 39, 40, 68, 71, 73, 92, 95, 98
climate
    change, 5, 23
    global, 15
    pattern, 15
    rapid change, 2
    regional, 14
coastal erosion, 29–34, 41
    coastal ground-ice content, 29
    impact of fetch, 29
    sea level rise, 29
    vulnerability of Beaufort Sea
        coastlands, 29
coastal retreat, 31
    rate, 30, 32
coasts, 10
coefficient of soil salinity, 175
cold-based glacier, see glacier
cold regions engineering and
    technology, 173
Cold Regions Research and Engineering
    Laboratory, see CRREL

Colombia, 126
Columbia Glacier, Alaska, 76,
complex avalanche, 113–114,
    process combinations, 113
Conservation of Arctic Flora and Fauna
    Committee (see CAFF)
continuous permafrost, see permafrost
Cordillera Blanca, 114
coriolis force, 21, 77
cornice, 210
Cotacachi Glacier, Ecuador, 125
CRREL, 158, 174, 240, 241, 253, 254, 261,
    263
crevasse, 8
Cryosat-2 (see satellites)
cryosphere, 5, 101
    cryospheric research, 7
    cryospheric system monitoring, 7
cryotechnological solutions to permafrost
    problems, 174
cultural importance of glaciers, 127
cyclogenesis, 298

Dalton Highway, Alaska, USA, 175
Danube River, 239
Daugaard-Jensen Gletscher, 75
Davis Strait, 19, 21, 35, 51, 75, 77, 99
Dawson City, Yukon, Canada, 158, 179
debris flow, 7, 143
deformable wet sediment slurries, 49, 54
deformation of basal till, 52
Dempster Highway, Yukon,
    Canada, 175
dendrochronology (Tree-ring
    dating), 219
Des Bossons Glacier, France, 118
design event, 9
Devensian Glaciation, 122
Devon Island, Canada, 76
Diavik Diamond Mine, NWT,
    Canada, 265
digital terrain model (DTM), 228
Dig Tsho GLOF, see GLOF
discontinuous permafrost, see permafrost
Disko Bay, Greenland, 75
Drangajökull, Iceland, 118,
    destruction of farms, 118

driving stress, 52
dry adiabatic lapse rate, 189
Dutch Delta Commission, 63

EAIS (East Antarctic Ice Sheet), 35, 301
early warning systems, 147, 150, 151, 153
East Antarctic Ice Sheet, see EAIS
Eastern Ross Sea, Antarctica, 94
Eastern Weddell Sea, Antarctica, 94
ecosystem, 2, 35–37, 62, 66–67, 69
ecotourism, 15
Eemian (Ipswichian) Interglacial, 52
'Egg Code', 39, 41
   map, 40
Ekati diamond mine, NWT, Canada, 265
electrical resistivity tomography
   (ERT), 195, 196
Ellesmere Island, Canada, 75, 76, 86, 87,
   88, 90, 119
emperor penguin, 65, 66, 89
endemic species, loss, 66
equilibrium line (of glaciers and ice
   sheets), 55
Estonia, 267
Estonian ice roads, 267
Estonian National Road
   Administration, 267
EU, 34
   Guidance on the Arctic, 34,
   Fifth Framework Programme, 102
Eurasia, 15, 157, 161, 166, 175, 200, 237
Europe, 13, 214, 215, 221, 231, 233
European Alps, 1, 6, 101, 102, 111, 122,
   125, 189, 192, 200
European Avalanche Danger
   Scale, 234–235
European Permafrost Conference, 6
European Union (EU), see EU
evacuation, 10
exploratory drilling, 27
exploratory wells, 31

factor of safety, 165, 194, 207
Fastnet, Ireland, 80 fetch (of waves), 23,
   29, 32, 33
Filchner-Ronne ice shelf, 50, 59
Finland, 237, 267
firn, 111, 127

First Nations, 269, Pehdzeh Ki, 270, 271
fishing, 25, 37
Fitchburg, Massachusetts, USA, *'Sentinel
   and Enterprise'*
   newspaper, 293
fixed installations, 35, 71, 81, 86, 95, 96, 98
fjord bathymetry, 86
FLAR (Forward Looking Airborne
   Radar), 85, 92
Flemish Cap, 77
floating glacier tongues, 50
Floating Production Storage Off-loading
   Platform (FPSO), 97
Fond-du-Lac (settlement),
   Saskatchewan, Canada, 269
Former Soviet Union, see FSU
forty eighth (N) parallel (48°N), line of
   latitude, 73, 82, 85, 99
Forward-looking airborne radar , see
   FLAR
Fram (Norwegian polar exploration
   ship), 24
Fram Channel, 21
Fram Strait, 16
France, 6, 213, 215, 224
Franz Joseph Glacier, 124
Franz Joseph Land, 76
Freezing-degree-days (FDD), 18
freezing point, 17
freeze-thaw, 170,
   cycle, 189
freeze-up, see river ice
fresh-water ice, 14,
   density, low, 67
frost heave ('jacking'), 162, 163, 164,
   165, 166, 170, 178, 184, 185
frost-susceptible sediments, 168
'frozen family', 28
FSU, 188, 157
Fujita Scale (tornadoes), 295

Galtür, Austria, 1, 203, 204, 208,
   223, 225
General Circulation Model (GCM),
   22, 175
general public, 2
geocryogenic hazards, 2, 10, 11,
   300, 301

geocryogenic systems, 301,
  rates of change, 7
Geographical Information Systems (GIS), 295
Geoinformatics, 129
geomorphological mapping, 114
geophysical methods, 162, 195, 196, 197, 200, 262
geotextiles, 33, fleece, 128
geothermal,
  heat, 48,
  gradient, 159,
  modelling, 199
Giéto Glacier disaster, 129
Gietroz Glacier, Switzerland, 118
glacial stage, 5, 300
Glacier(s), 4,
  ablation, 8
  accumulation, 8
  alpine glaciers, 5
  backstress, 60
  basal meltwater, 49
  basal shear, 8,
  basal sliding, 8, 49
  buttressing effect, 135
  calving, 7
  cold-based, 111
  covering, 127
  creep, 8
  crevasses, 7, 101, 102–104, 105, 108, 111, 113, 130,
    mitigation, 103–104
  cultural significance, 127
  dam, 8
  deformable wet sediment slurries, 49
  deforming subglacial sediments, 8
  firn, 111
  'hanging', 105, 124, 139, 140
  hazards, 101–131
    inherent, 102–116, see also crevasses,
      seracs, ice avalanches
    advancing, 8, 117
      impacts, 118
    retreating, 121–128
      impacts,
        slope instability, 122–124,
        lakes, 124
        ice avalanches, 124
        loss of ice, 125–127
        others, 128
      monitoring, 124
    surging 101, 118–121,
      cyclicity, 118, 120,
      predictability, 120,
      secondary effects, 120–121
    hazard mitigation, 108–109, 129–130
    hazard monitoring, 102, 129
    hazard zoning, 130
    impacts as 'divine retribution' ('act of God'), 129
    insulating membrane, 6
    inventory of glacier-related hazardous events, 129–130
    mass balance, 8, 86, 101, 102, 116–117, 121, 130
    meltwater, 128,
      storage, 134–135,
      system, 134
    Norwegian, 103, 118,
      crevasse fatalities, 103
    rate of flow, 8
    retreating margin, 8
    risk assessment, 102,
    risk perception, 129, 130
    serac, 8, 101, 104, 105, 130,
      fatalities, 104,
      mitigation, 104
    surface runoff, 8
    warm-based, 111
glacier lake outburst flood, see GLOF
Glaciorisk, 102, 108, 129
global coastal communities, 52
ground penetrating radar, see GPR
global positioning systems, see GPS
Global Terrestrial Network for Permafrost (GTN-P), 199
global warming, 45, 60, 62, 64, 85, 166
GLOF, 5, 124, 133–155, 301
  adaptation strategies, 153
  Dig Tsho GLOF, 136, 139–142, 147, 152, impacts 141–142
  displacement (impulse) wave, 135, 136, 137, 142, 149

failure threshold, 143,
frequency, 143
historical records, 143,
impacts, 141–142
inventories of potential GLOF
    sites, 152–153, remote
    sensing, 153
mapping, 152–153
mean recurrence interval, 143
mitigation, 150–154
   hard (engineering)
       measures, 150–151, 152
   soft measures, 151, 152, 153, 154
monitoring, 133, 144, 148, 153
pressure threshold, 134, 136, 143
remedial engineering works, 142, and
    see Tsho Rolpa
risk assessment, 142–144, 149
risk management, 150
supraglacial morainic debris, 143
terrain model, 135
trigger mechanisms, 136–141,
GPR (ground penetrating radar), 195, 262
GPS (global positioning system), 27, 54, 262
Grächen, Switzerland, 192
Grand Banks, 35, 73, 75, 77, 78, 79, 80, 81, 84, 85, 86, 87, 92, 95, 96, 97, 98, 99
graupel, 210
grave goods, 31
gravel pads, 33, 178, 179
gravity base structure (GBS), 86, 96, 97
Great Lakes, North America, 279, 281
Greenland, 15, 16, 21, 35, 45, 46, 47, 48, 49, 51, 56, 61, 71, 72, 75, 76, 77, 78, 79, 89, 90, 94, 98, 99, 157
  glaciers, 55
  iceberg source area, 65, 68, 75, 76
  record snow melt, 54
Greenland Ice Sheet, see ice sheets
Greenland Sea, 21
'*Gridabase*', 102, 103, 108, 110, 129
Grimsvötn, Iceland, 134
Groins, 33
ground-ice content, 41, 163, 189
grounding line, 49, 55, 60, 61, 64
ground settlement index, 175

ground subsidence, 168, 170, 175
growlers, 81, 99
Gulf of Mexico, 282
Gulf Stream (North Atlantic Drift, Current), 13, 21, 77
Gurschen Glacier, Switzerland, 127
Gutz Glacier (Gutzgletscher), Switzerland, 105, 107–109, 111

hail, 210, 274
hanging glacier, see glacier
harvesting platform, 25
Haynes Junction, Yukon, Canada, 121
hazard(s), 8–10,
  awareness, 5, 299,
  characteristics, 3,
  classification, 2,
  global context, 4
'heat island' effect, 170
heat pump, 180, 182
heat transfer, 195
Heinrich events, 72
HEP (hydroelectric power), 1
  projects, 152,
  station, 154
Hibernia,
  drilling platform, 86, 97, 98
  oil field 86
Himalaya, 1, 104, 122, 124, 125, 133, 136, 141, 142, 143, 145, 149, 151, 152, 154, 155, 189, 300
historical records, 10
Holocene,
  climate change, 117,
  Period, 122
Huascarán, Peru, 1
Huascarán avalanche, 112, 113, 114–115, 130
Hudson Bay, Canada, 89
  railway, 175
human societies, 2
hunting season, 14
hydrocarbon, exploration,
    exploitation, 30, 32, 37,
    resources, 34
hydrosphere, 5

IASC, 42
ice
  ablation, 49, 55, 104, 105, 107, 121, 127, 134
  advance, 29, 47, 58, 59, 101, 102, 105, 116, 117, 118, 120, 121, 129, 130
  avalanche, 1, 2, 5, 104–113, 124, 143,
    typology, 105–111,
    initiation processes, 111,
    starting volume, 111–112,
    probability, 108, 112–113,
    run-out, 108, 112
  black ice, 261
  bonding, 7, 165
  brittle behaviour, 7
  cold ice, 7
  crystals, 7
  deformation, 7
  dome, 46, 47, 48
  ductile behaviour, 7
  elastic behaviour, 7
  falls, 104
  floating, 2, 4
  floe, 18, 21, 25, 26, 243, 244, 248, 249, 250, 258
  formation processes, 241–244,
    hydro-meteorological conditions, 243
  Froude number, 242
  islands (large icebergs), 87
  load-bearing capacity, 25, 27
  observing year, 82
  override (see ivu)
  physical properties of, 7
  rafting, 17
  resilience, 18
  retreat, 5
  shelf, 46, 49–50
  store of water, 10
  stream, 4, 7, 46, 47, 49,
  white ice, 261
  iceberg, 2, 4, 8, 65, 68, 71–99
    Analysis and Prediction System (BAPS), 98
    armada, 72, 89
    avoidance, 97–98
    bergy bits, 96
    categories of impact and risk, 81
      ship-iceberg collisions, 82–86, 87
      iceberg impact on fixed/moored installations, 8, 86–88
      seabed scouring of submarine installations, 88
      impact on ecosystems, 88–89
      impact on global systems, 88–90
    calving rate, 77
    characteristics, 72–74
      composition, 72–73
      size, classifications, 73
      shape, classification, 74
    controls of iceberg movement, 75
    databases, 82, 90, 91, 94–95
      Comprehensive Iceberg Management Database, 95, 96
      Iceberg Shape Database, 95
      Iceberg Sighting Database, 94
      Ship-Iceberg collision Database, 95
    decay processes, 79
    density, 72, 77
    deterioration processes, 79
      rates, 79, 80
    distribution, 76–78
    drift rates, 77
    factors influencing occurrence, 77
    factors determining survival, 79
    fatalities due to icebergs, 82
    flux (frequency of occurrence), 47, 56, 62, 65, 68, 77
    hazard to fixed installations, 8
    hazard to shipping, 8
    hydrocarbon exploration and exploitation, 86
    'Iceberg Alley', 77, 95
    impact and risk, 35, 72, 81–90
    inter-annual variability, 86
    international cooperation, 91
    large tabular, 6, 8, 72, 89
      archiving, 94
      B15A, 65, 88
      C16, 65,
      C19, 88
    life expectancy (deterioration), 77, 79–81
    management, 82, 86, 97

monitoring, 65, 71
mitigation strategies, 90–98
   detection, monitoring, databases and
      research, 91–95
   threat evaluation and
      prediction, 95–96
   ice management, 96–97
   avoidance, 97–98
number crossing 48°N, 84, 85
   reasons for increase, 84–85
prediction models, 95–96
reconnaissance, 85, 93
   methods, 92
risk assessment, 65
Russian contexts, 87–88
scour troughs, 72
season, 86
season 2008 on the Grand
      Banks, 98–99
ships' archives, 80
size, see characteristics
severity, 85
shape, see characteristics
sources, 72, 73, 75–76
towing, 96–97
trajectory (route), 77, map, 78
ice-bonding, 165
ice-breaker, 6
ice-capable vessel, 37
ice-dammed lakes, 143
ice-free arctic, 15, 21, 22, 24, 25, 33, 37
ice-cored moraine, see moraine
ice-covered detachment surface, 167
ice jam, 2, 8, 238–259
   affected hydrograph, 248
   breaking front, 249
   breakup, (Spring), see river ice
   contributory factors, 240, 249
   database (CRREL), 240
   distribution, in USA, 238
      in 2000 and 2001, 247
   effect of climate change, 258–259
   freeze-up, (Autumn, Fall),
      see river ice
   grounded jam, 249
   hanging dam, 244, 249, 250
   impacts, 240, 250–253
      classification, 251

'daylighting', 251, 253
   economic (indirect), 251
   fatalities, 237, 240, 251, 252, 257, 258
   flooding, 239, 240, 248, 250, 251, 252,
      253, 255, 256, 258
   geomorphological change,
      erosion, 251, 253
   habitat change, 251, 253
   ice blocks, 8, 253
   structural damage, 251
mega-floods, 239
mitigation, 253–259
   methods, 253–257,
   classification (structural,
      non-structural), 254
   strategies (anticipation,
      reaction), 257–258
processes, 249–250
recurring, 252
severity, 250
sites, 249
   prime sites, 251
strength, 250
Iceland, 118, 119, 189, 195
ice-observing year, 82
ice properties, 7
   elastic behaviour, 7
   ductile behaviour, 7
   brittle behaviour, 7
ice-resistant drilling platforms, 88
ice-rich permafrost (see permafrost)
ice (winter) road, 2, 259–272
   construction methods, 261, 262–263
   hazards, 267–272
      accidents 267–268
      fatalities, 267
      season length 268–272
   ice forming processes, 261
   ice types and strength, 260–261
   'Ice Road Truckers', 2, 237,
      truckers safety, 262
   Kitikmeot Inuit Association,
      Canada, 268
   loading criteria, 263–264
   location of networks, 265
   movement of vehicles and people, 8
   N'Dulee Ice Road, Canada, 269, 271
   Northwest European, 267

ice (winter) road (*Continued*)
  quality control, 265
  regulations (Canadian), 268
  season, 237, 262, 265, 267, 268, 272
  Tibbitt to Contwoyto Winter Road, NWT, Canada, 265, 266, 267, 268
  trucking vehicles, 260
  Tulita to Deline Ice Road, NWT, Canada, 268
  Wrigley to Tulita Ice Road, NWT, Canada, 268
ICESat (see satellites)
ice sheet(s), 4, 45–70, 101
  Antarctica, 6, 8, 35, 37, 45, 46, 49, 50, 51, 56–62, 63, 66, 67, 68, 69, 71, 76
    alien species, 67, 69
    climate change effects, 62
    contribution to sea level rise, 49, 63
    ecosystems, 62
    iceberg monitoring, 68
    impacts, 47
    isolation, 46
    loss of ice, 56, 57
    mass balance, 63
    mass gain, 64
    net accumulation zone, 48
    ozone hole, 62
    permafrost, 157
    plateau, 48
    sea ice, 51, 66
    size, 56,
    stability, 63
    structure and flow, 48
    warming trend, 56
  Antarctic Ice Sheet, 45, 46, 61, 68
    rate of change, 61
  Antarctic Peninsula (AP), 6, 46, 50, 56, 58, 76
    climate change effects, 62
    change in marine glaciers, 59
    ice shelves, 58
    loss of ice from, 56, 57
    rate of warming, 56
    source of icebergs, 76
  AP, see Antarctic Peninsula
  backstress, 60
  contrast between Antarctica and Greenland, 46, 48
  deglaciation, 63, 64
  EIAS, East Antarctic Ice sheet, 45, 46, 50, 56
  Greenland Ice Sheet, 5, 35, 45, 46, 52–55,
    acceleration of marginal glaciers, 54
    contribution to sea level rise, 52
    deglaciation
      total, 64
      partial, 65
    driving stress, 52
    glacier flow models, 54–55
    mass balance, 54
    mean rate of surface change, 53
    melting, 52–54
      index, 53
    meltwater penetration, 54
    rate of surface change, 53
    size, 52
    stability, 63
  impacts of ice sheet loss, 62–68
    sea-level rise, 62–65
    iceberg flux, 65
    ecosystem change, 66–67
    climate and ocean circulation change, 67–68
  location, 46
  mitigation of impacts, 68–69
  model inadequacy, 55, 64, 65
  rapid decay and iceberg flux, 72
  rate of flow, 8
  stability, 35
    and sea ice, 35, 51
  structure and flow pattern, 48
  systems, 47–51
  terrestrial, 4
  timescale for change, 47, 56
  WAIS, West Antarctic Ice Sheet, 6, 46, 47, 49, 50, 56, 59, 60, 61, 63, 64, 65, 68, 69
    vulnerability to rapid change, 46
Ice shelf, 2, 4, 6, 8, 46, 49–50, 88
  abrupt collapse, 50, 56
  basal crevasses, 58

basal melting, 58
buttress effect for glaciers, 60, 64
characteristics, 49–50
disintegration, fragmentation, 6
dynamics, 58
Filchner-Ronne, 50, 59
grounding line, point, 50
influence of sea ice, 51
Larsen A, 50, 56, 57, 76, 88
Larsen B, 50, 57, 58, 76, 88
loss, 56, 57, causes, 58
modelling, 58
ponding and surface melting, 58
progressive retreat, 56
Ross, 50, 59, 65, 88
thermal limit of viability, 50, 56
thinning, 50, 58
Wilkie, 50, 76
Wordie, 50, 88

Ice storms, 2, 4, 273–298
definition, 274–275
distribution in USA, 279
freezing drizzle, 276, 283
freezing rain, 275, 276, 279, 280, 282, 283, 287, 288
in Montreal, 292
freezing spray in Alaskan Peninsula, 280
frequency, 297
ice loading, 273, 289, 293
ice pellets, 276, 282, 283
ice storm '98 (1998), 273, 274, 275, 286, 292, 294,
summary of impacts, 287,
distribution and timing of precipitation, 283
impacts,
damage, 281
economic losses, 285, 286–287
fatalities, 274, 284, 285–286, 287, 288, 289
electricity transmission lines, 273, 292, 295,
pylons, 284, 287, 288, 292
power loss (outage), 285, 289
transport and traffic disruption, 285, 287

mitigation, 290
probability, 291
return period, 291, 292
weather systems (cyclonic) and processes, 276–282,
air temperature 280–282
pressure patterns, 276–279
ice types
anchor, 244, 249, 250
border, 243, 244, 250
brash, 248
congelation, 17
frazil, 17, 243, 244, 246, 248
grease, 17
landfast (shorefast), 17
model, 18
nilas, 17
pack, 17, 32, 33
pancake, 17, 243
skim, 243
ice wedge, 163
polygon, 162, 163
Icy Bay, Alaska, USA, 120
IIP, 68, 71, 72, 73, 74, 79, 80, 82, 84, 92, 93, 94, 95, 98, 99
collaborating countries, 92
annual report, 92
IISD, 26
Ikaahuk (Sachs Harbour), Canada, 26
Imja Glacier, Nepal, 144,
Lake, 144,
lake expansion, 144
Indian Ocean, 69, 76
indigenous communities, 13, 24, 25
indigenous population, 6
Institute for Ocean Technology, Canada, 82
insolation angle, 189
interferometric synthetic-aperture radar, 60
interglacial stage, 5, 300
Intergovernmental Panel on Climate Change (IPCC), 5
International Arctic Science Committee, see IASC
International Friendship Association, 147, 148

International Ice Charting Working
    Group (IICWG), 94
International Institute for Sustainable
    Development, see IISD
International Permafrost Association,
    see IPA,
Intergovernmental Panel on Climate
    Change, see IPCC
International Ice Patrol, see IIP
International Centre for Integrated
    Mountain Development
    (ICIMOD), 152, 153
Interstate Weather Modification Council,
    USA, 289
Inuit, 13, 14, 26
  Circumpolar Conference, 41
  comments on sea ice, 25
  culture, 25, 41
  economy, 25
  hunting and fishing, 25, 41
  hunting season, 14, 23
  loss of cultural and historical sites, 31
  loss of culture and identity, 41
  Observations of Climate Change
      Project, 26
  traditional knowledge, 6, 23
Inupiaq (Inuit) 27, 31
  cemetery, 31
  village, 31
Inuvialuit, 26
Inuvik, NWT, Canada, 175, 178, 186, 187,
    188
IPA, 161, 175, 198
IPCC, 47
  confidence rating, 47
  Fourth Assessment Report (FAR), 63,
      300
  probability of predictions, 7
  Third Assessment Report (TAR), 63
Iqaluit, Nunavut, Canada, 281
isostatic adjustment, 46
Italy, 209, 213, 215
ivu (ice override), 13, 27–29
  driving forces, 27, 28
  examples of ivu impacts, 28
  fatalities, 28
  'frozen family', 28

penetration distance, 28
processes, shove, pile-up, ride-up, 28
timing, 27

Jakobshavn Isbrae, 55
Japan, 133, 227
Japan-Nepal collaboration, 133
jökulhlaup, 134
Jostedalsbreen, Norway, 118
Journal of Quaternary Science, 300
'Journey to the Edge of the World', 34

Kapitan Khlebnikov (tourist vessel), 24
Kara Sea, 77, 96
Karakoram Mountains, Asia, 117, 119
Kazakhstan, 127
kettle hole, 135
Khimti hydroelectric power
    complex, 146, 147
Khumbu Himal, 138, 139, 143, 153
Kolka Glacier, Russia, 114
Kolka-Karmadon avalanche,
    Russia, 113–114
Kotzebue, Alaska, 28
krill, 66
Kubaka Gold Mine, Russia, 184, 185
Kyrgyzstan, 127

Labrador, 81, 99
Labrador Current, 77, 79, 246
Labrador Sea, 73, 77, 79, 80, 81, 92
lacustrine surface meltwater, 55
Lake Erie, New York State, USA, 282,
    285
LAKI, 93, 98
Landfast ice, see ice types
Landslide, landslip, 7, 135, 143
Land-use
  planning, 10
  regulations, 10
  zoning, 10
Langmoche Glacier, Nepal, 140, 141,
    Lake, 151,
    disaster, 152
Larsen A Ice Shelf, see ice shelf
Larsen B Ice Shelf, see ice shelf
Laurentide Ice Sheet, 89
Lena River, Siberia, 239, 240, 250, 252

Lensk, Russia, 239
less developed region, 1
light transmission, 14
'Limit of All Known Ice', see LAKI,
lithosphere, 5
Little Ice Age (LIA), 5, 101, 117, 118, 121, 122, 136, 239
 moraines, 143
 trimlines, 123
Lomonosov Ridge, Arctic Ocean, 15
Longard tubes (geotextiles), 33
longshore protection, 33
Lowell Glacier (Nàlùdi), Yukon, 121
Lower Platte River Basin, Nebraska, USA, recurring ice jams, 252
Lupin mine, NWT, Canada, 265

MAAT, 111, 194, 195
Mackenzie River, Canada, 32, 33, 246, 269, 270, 271
Mackenzie Delta, 186, 251
Macquarie Island, New Zealand, 77
MANICE, 39
Manitoba, Canada, 298
marine-based subsistence culture, 13
marine ecosystem, 14
marine (tidewater) glacier(s), 7, 58, 59, 65, 76
 retreat and its causes, 58, 59–61,
 processes, 60
 advance, 58
 as source of icebergs, 72
marine safety, 71
mass balance, 56
material strength, 8
Matterhorn, Switzerland, 6, 192,
 Lion Ridge, 191, 192
Mattmarksee, Switzerland, 105, 106, 113, 118
Max Plank Institute of Meteorology, 67
mean annual air temperature, see MAAT
media, 2, 5, 6
Melbern Glacier, BC, Canada, 123
melting index, 53
meltwater, see glacier meltwater
Melville Bay, Greenland, 75

Mer de Glace, France, 118
meridional overturning circulation (see MOC)
Meteorological Service of Canada, 92
methane, 172, 173, 174
Miami University, USA, 67
migration
 of residents, 9, 47, 69
 of tourists, 9
mineral exploration, 15
Missouri River, 239
MOC (Meridional Overturning Circulation), 24, 37, 38, 47, 67, 68, 69
monolithic slabs, 33
monsoon, 134
Montana, USA, 252, 256, 258,
 'Ice Jam Awareness Day', 258
Mont Blanc, 6, 192
Montpelier, Vermont, USA, 251
Montreal, Canada, 251
Montroc, France, 231
moraine, 1,
 ice-cored, 124, 136, 142, 143
 lateral, 119, 122, 123
 looped medial, 119,
 supraglacial, 143
 terminal, 143
moraine-dammed lakes, 136, 139, 143, 144
moulin, 54, 134
Mount Cook, New Zealand,
 National Park, 123
 Region, 124
Mount Everest (Chomolungma, Devadhunka, Chingopamari), 104, 136
Mount Fletcher, New Zealand, rock avalanche, 123
Mount Kilimanjaro, Tanzania, 125, 126
Mount Pinatubo, Philippines, 52
M/S Explorer, sinking, 71, 81, 86
multispectral ASTER satellite imagery, 54

Namche Bazaar hydroelectricity plant, Nepal, 136, 141
NAO, 81

narwhal, 21, 37
National Climatic Data Centre
    (USA), 294, 295
National Electricity Safety Code
    (USA), 293
National Guard Armoury, USA, 294
National Ice Centre, USA, see NIC
National Oceanographic and
    Atmospheric
    Administration, see NOAA
National Petroleum Reserve, Alaska,
    USA, 30
National Research Council, Canada, 82
National Resources Canada, 94
National Snow and Ice Data Centre,
    USA, see NSIDC,
National Weather Service, USA,
    see NWS
Natural Environment Research Council,
    UK, see NERC,
neglect of cold region hazards, 1
Nepal, 138, 139, 143, 144, 145, 146, 147,
    148, 151, 152, 153, 154
Nepalese Department of Hydrology and
    Meteorology, (DHM), 147
NERC, 49, 67
NESIS, see blizzards
Net accumulation zone, 48
Netherlands, The, 133, 148,
Netherlands-Nepal Friendship
    Association, 147, 148
New Brunswick, Canada, 284, 288
Newfoundland, Canada, 75, 81
New York, USA, 282
New Zealand, 76, 77, 122, 123, 124
NIC, 76, 88, 92, 94
Nigardsbreen, Norway, tax
    reductions, 118
NOAA, 19, 62
Noril'sk, Russia, 157, 170, 171
North America, 13, 71, 157, 158, 161,
    175, 200, 214, 215, 229, 230
North American Interstate Weather
    Modification Council
    (USA), 289
North Atlantic Drift, Current,
    see Gulf Stream

North Atlantic Ocean, 16, 21, 67,
    71, 75, 92,
North Atlantic Oscillation, see NAO
Northeast shipping route, 34
Northeast Snowfall Impact Scale
    (NESIS), USA,
    see blizzards
Northern circumpolar Arctic
    tundra, 172
northern latitudes, regions, 2
northern resource development
    support vessels, 37
north magnetic pole, 24
Northwest Passage, 13, 22, 23,
    24, 268
northwest shipping route, 34
North Pole, 21, 25, 34
North Slope Borough, Alaska, 28
Northwest Territories, see NWT
Norway, 174, 189, 195, 197, 215, 267
Norway-Russia Arctic Off-shore
    Workshop, 96
Novaya Zemlya, Russia, 76
NSIDC, 41
Nunavut (Canada), 15, 188
Nuwuk, Alaska, 28
NWS, 255, 256, 257, 291, 295
NWT, 15, 267, 268, 269, 270, 271, 291,
    Department of Transport,
    NWT, Canada, 269, 271

ocean water
    solar radiation-absorbing, 14
offshore breakwater, 33
oil spillage threat, 37
Oklahoma City and County, USA, 294
Ontario, Canada, 283, 284, 287, 288
open water, 18, 21, 26, 34, 37
    seasons, 33
optical attenuation of water, 54
Ottawa River, Canada, 255
Mongolia, 282
outlet glacier, 47, 49
over-fishing in the Arctic, 37
Overseas Ohio, 86
ozone hole, 37
    depletion, 37, 68

PACE (Permafrost and Climate in
    Europe Project), 192, 194,
    196, 197,
  boreholes, 198
Pacific Ocean, 16, 21, 22, 69, 76
paraglaciation, 122
Parks Canada, 179
passive microwave satellite platforms, 52
pattern of ice drift, 24
Peace River, Alberta, Canada, ice jam v
    summer flood
    discharges, 240
penguin, 35, 65, 66
PERD, 94, 95, 96
Permafrost, 2, 5
  active layer, 162, 163, 164, 165, 168,
    170, 171, 173, 178, 189, 195
    thickening, 162, 170, 171, 174, 175,
    184, 185, 188, 191
  awareness, 165, 192
  bearing capacity, 166, 170
  boundary movement, 175
  continuous, 161, 162, 169, 175
  creep, 164, 165, 189, 194, 196,
    rates, 185, 190, 192, 194
  definition, 159
  degradation, factors, 168
    susceptibility to, 175
    potential hazards, 190
  discontinuous, 161, 162, 163, 165, 168,
    169, 175, 178, 180, 188
  distortion of buildings and
    infrastructure, 158
  distribution and
    characteristics, 159–162, 163,
    165, 166 , 189, 194, 195, 197,
    198, 200, controls on
    distribution, 160–161
  engineering design, 158, 164
  engineering solutions, 178, 179, 191
  extent, 158
  'glue', 194
  ground ice as a major cause of
    permafrost hazards, 165
  ground-thermal regime, 160
  hazard management, 165
  hazardousness, 163–166

  enhancing factors, 163–164
  hazard potential maps, 175, 177, 178
  hazard zonation maps, 170, 175,
    176, 191
  heat balance, 159
  heave (frost, 'jacking'), 8, 162, 184
  human infrastructure, 175, 191
  ice-bonding, 165
  ice-content, 162, 163, 175, 189, 195
  ice-rich, 7, 162, 168, 171, 173, 178, 179,
    181, 197
  impact of climate change, 165
  infrastructure, 158, 163, 164, 166, 177,
    188, 194, 199
  isolated, 161, 162
  load-bearing capacity, 184
  lowland-permafrost, 165–166, 168
    climate change, 174
    geotechnical engineering, 178–188
    hazard zonation maps, 170, 175, 176,
      191
    impacts, 166–173
    mapping, 174–178
    mitigation, 173–188
    reasons for hazardousness, 168
    regulation, 188
  mountain-permafrost, 6, 162, 163,
    165–166, 188–200
    bedrock destabilisation, factors, 191
    factors affecting distribution, 189
    geotechnical engineering
      solutions, 178
    hazards, impacts, 188–194
      relation to slope angle, 189–191
    lower limit,
      Norway, 189
      rise, 192
      Switzerland, 192
    mitigation strategies
      engineering, 199
      Canadian guidelines, 200
      design life of structure, 199
      modelling and mapping, 194–195
      monitoring, 195–199
      regulation, 200
    rock failure, falls, 189–191
    slope stability, 191, 192

Permafrost (*Continued*)
    topography, importance of, 189
    tourist facilities (chair lifts, huts etc.), 194
  passive cooling systems, 192
  piling, piles,
    methods, 184–185
    creep 185
  pipe insulation, 185–188
  railways on, 175
  refrigeration, 179
  remote sensing, 162
  saline, 185
  segregation ice, 162, 163, 165
  sensitivity, 164, 165, 166
  sporadic, 161, 162, 165
  subdivisions, 161
  supercooling, 181
  susceptibility to permafrost degradation, 175
  table, 162, 174, 182
  talik, 159, 169, 170
  temperature, 163, 164
    changes, 170, 193
    monitoring, 180
  thaw sensitivity maps, 175
  thaw settlement, subsidence, 165, 171
  thaw-stable foundation pad, 178
    gravels, 179
  thermal, conductivity, 171, 189
    equilibrium, 168, 178, 179
    regime, 171
    stability, 179, 191, 199
  thermosyphons (-probes, -piles), 179, 180, 182, design, 180
    hairpin, 181, 182
    loop system, 180
  thickness, controlling factors, 162
  transport networks, 174, 175, 191
  warm permafrost, 165, 180, 185, 191
  volumetric expansion, 165
  utilidor, 186–188
Permafrost and Climate in Europe (PACE) Project, see PACE
PERMOS, 195, 196, 199, 200
Peru, 1, 114, 126, 142
Philippines, 52

phytoplanktonic bloom, 66
Pine Island Glacier, Antarctica, 61
Pleistocene glaciers and ice sheets, 117
Point Barrow, Alaska, 28, 31
polar bear, 14, 25, 26, 35, 36
politicians, 2
polynya, 21, 65, 66
  biological significance, 21, 65, 66
  causes, 21
ponded lakes and their seasonality 116
population density, 1, 9
pore water pressure, 165
positive feedback, 14, 38
Prairie Ecotone, western, 298
precautionary principle, 47
pressure gradient, 210
primary production, 66
Prince Edward Island, Canada, 284, 288
Prince William Sound, Alaska, 86
proglacial lake, 124, 136
Programme of Energy Research and Development, (Canada), see PERD
property-casualty insurance industry, 277
Prudhoe Bay, Alaska, 29
Prudhoe oil field, 28

Qalluvik, Alaska, 31
Quaternary Period, 143
Quebec, Canada, 274, 283, 284, 287, 288,
  Quebec Province, 292

RADARSAT, see satellites
radiative melting, 54
recurrence interval, 9
Red River, N. Dakota, USA, 250
reflective ice, 14
refugia, 36
remote sensing, 7, 9, 45, 52, 111, 113, 129, 130, 162
remotely sensed buoys, 24
rescue services (from sea ice), 25, 27
resettlement, 34
resilience, 10
Resolute, Nunavut, Canada, utilidor replacement, 188

responsible authorities, 10
retrogressive thaw slumps, 142
Rhine River, Germany, 239
Rhone River, France, 118
Richardson Highway, Alaska, 120
Rideau River, Ottawa, Canada, annual ice management, 255
ringed seal, 25, 36
risk, definition, 9
river ice, (see also ice jams and ice roads), 4, 237–272
  black ice, 244, 246, 250, 261
  breakup, 241, 249, 250, 255, 256, 259,
    processes, 246–248
    timing in Canada, 245
  controlling factors, 260–261
  distribution, 237, 238
  freeze-up, 241, 243, 244, 246, 248, 249, 250, 257, 258, 259
    timing in Canada, 245
  hydrothermal melting, 246
  hydro-meteorological conditions, 243, 244
  mean January temperature (-5°C), 237
    USA, 238
    Eurasia, 239
  river flow regime, 241, 244
  stages of development, 241–244
  white ice, 246, 261
rockface stability, 7
rockfall, 6, 7
rock glacier, 165, 166, 190, 192
  Muragl, 166
Murtèl-Corvatch, Switzerland, 196
Rocky Mountains ('Rockies'), 122, 125, 215, 246
Rogers Pass, British Columbia, Canada, 225, 227, 232
  train disaster, 213, 217
Ross Ice Shelf, see ice shelf
Ross Island, Antarctica, 65
Ross River community school, Yukon, Canada, 180, 182
Rotlisberger channels, 134
Rowaling Valley, 146, 147, 153, 154
Ruitor Glacier, 118

Russia, 15, 41, 75, 76, 87–88, 94, 96, 158, 159, 162, 164, 168, 169, 170, 171, 172, 173, 174, 175, 184, 185, 188, 200, 237, 265
  Arctic cities, 170, 172

Sabai Tsho Glacier, Nepal, 136, 138
Sabai Tsho Glof, 136, 138, 139
Sachs Harbour (Ikaahuk), Canada, 26
Safety of Life at Sea, see SOLAS
Saffir-Simpson Scale (hurricanes), 295
Salang Pass, Afghanistan, 213, 214–215
salinity, 14
saline ocean water, 17, 67
sandbagging, 33
Saas-Visp River, Switzerland, 105, 118
St. Elias Mountains, Yukon, Canada, 121
St. Lawrence River, Canada, 251
SAR, 85
Saskatchewan, Canada, 298
satellites
  ASTER, 54
  Cryosat-2, 6, 7
  ERS, 7
  ICESat, 7, 20, 21, 61
  Landsat, 7, 30
  RADARSAT, 7, 85, 92
  SAR, 92,
  remote sensing capabilities, 7
Scandinavia, 13, 15, 157, 189, 215, 265
SCAR, Reports, 63, 64
scientific journals, 7
scour troughs, see ice berg
sea ice, 2, 4, 51, 99
  access, 6
  age
    new ice, 18, 21
    young ice, 18
    first-year ice, 18, 25, 26
    multi-year ice, 18, 19, 23, 25, 26
    old ice, 17
  asset for travel, 13
  buffer, 51
  changing hazards, 14
  concentration, 59
  density, 35
  decreasing trend, 22, 23, 37

sea ice (*Continued*)
  depletion, 35
  differences between the Arctic and Antarctica, 51
  excess of, 14
  extent, 10
  failure, 23
  fatalities, 13
  formation processes, 17
  girdle in Antarctica
  hazards, 18, 22, 23
  impacts, 23–38
  influence of winds, 8
  influence of sea currents, 8
  keel, 21
  lead, 18, 25
  light transmission, 14
  movement, 21
  over-ride (see ivu)
  pattern of change, 21, 22
  paucity of, 14
  pressure ridge (stamukhi), 17, 21, 23
  protective role, ice sheets, icebergs, 35, 51
  rate of thickening, 17
  regional variations, 37
  satellite monitoring, 22
  season, seasonal, 4, 19
  stability, 25
  strength, 25, 27
  surface relief, 13
  stage of development, age, 18
  thickness, 6, 17–18
  thickness model, 19
  thinning of ice cover, 14
  unpredictability, 25
  volume, 19, 22
Sea level, 2, 45
  global, 7
  rise, 7, 128
segregation ice, see permafrost
seismic line scar, 159
serac, see glacier
Severnaya Zemlya, Russia, 76
Shakwak Valley, Yukon, Canada, 121
shallow 'skin' flows, 165
Shepard Glacier, Montana, USA, 125

ship-iceberg collisions, 71, 72, 81, 82–86, 90
  database, 82, 83, 95
  reasons for decrease, 82
shipping
  access, 15
  lanes, 71
  routes, 34
Shishmaref, Alaska, 31, 42
shorefast ice, see ice types
Shtokman gas-condensate field, Russia, 88
Siberia, 15, 16, 19, 158, 160, 164, 166, 168, 169, 170, 171, 175, 200,
  floodplains and rivers, 166
side-looking airborne radar, see SLAR
skidoo, 14, 25
skiing industry, 6, 128
ski tourism, 127
SLAR, Side-Looking Airborne Radar, 85, 92
slope stability, 7
  at Tasman Glacier, NZ, 123–124,
Smith Sound, Greenland, 75
snow,
  accumulation, 64,
  properties, 210, 222, 224
snow avalanche, 2, 4, 203–236
  air blast, 207, 214
  air-lifting mechanisms, 208
  catastrophies, 212–213
  causal factors, 207–212
    aspect, 211, 234
    critical slope angle, 211
    meteorology, 207–211
    terrain, 204, 207, 209, 211–212, 218, 219, 220, 222, 225, 228, 229, 235, 236
    weather patterns ('grosswetterlagen'), 208
  classification criteria, 204–205
  corridor avalanche management system, 232
  danger forecast, Switzerland, 236
  danger scales, 233,
    Europe, 233–235,
    USA, 233

defences, 206
definition, 204
flow dynamics, saltation, rolling, sliding, 204, 222,
  density, 214
hard engineering, 204
hazard maps, 229
hazard ratings, 233
hazard zoning, see mitigation
impact pressure, 213, 214, 218, 227, 228, 229, 230
impacts, 212–218,
  costs, 214
  damage, 214
  fatalities, 213, 214–218,
    annual, global *recorded* death toll, 204, 215
    change over time, 215–218
    distribution, 214–215
management, 231, 232, 233
metamorphism of snowpacks, 210–211
  constructive metamorphism, 210
  depth hoar (facetted snow), 210, 211
  destructive metamorphism, 211
  recrystalisation, 210
  surface hoar, 209, 210, 211, 212
mitigation methods, 217, 218–233,
  artillery, 227, 231
  automatic detection,
  avoidance, 227–233
  classification, 219
  design, 227
  detection and alarm, 217, 231–233
  education, 217,
  evacuation, 219, 228, 231
  forecasting, 217, 224
  forest control, 226–227
  information, 219–224,
  mapping, 219, 220, 228, 230
  modification and control, 225–227
  terrain engineering, 225, 226
  warning, 217, 232–233
  zoning, 217, 228–231
    Swiss scheme, 229
    American scheme, 230

movement mechanisms, 205
off-piste, 204, 216
piste, 204,
radiation recrystalisation, 210
return period, 204, 223
runout distance, 204, 205 218, 220, 221–223, 225, 228
  modelling, 219, 220, 222
  zone, 220, 225, 228, 231, 235
snowpack, structure, stability, 5, 203, 204, 205, 207, 210, 211, 212, 219, 223–224, 225, 226, 233, 234
statistics,
  Canada, 217
    by activity in Canada, 218
  Switzerland, 215, 216
  USA, 216
    by age in USA, 217
structure, 205–206
start zone, 205, 211, 223, 225, 232
three-day (H_{72}) precipitation and avalanche probability, 209
track, 204, 207 219, 220, 221, 222, 223, 225
types, loose snow, slab, 205
unprecedented snow avalanche events, 203
vegetation as frequency indicator, 221
velocity estimates, 207
'Year of Terror' (1951), 228, 236
snowmobile, 25, 27
Snowshoe, West Virginia, USA, 282
solar radiation, 189
SOLAS, 91, 93
sonar, 19
Southampton University, UK, 67
Southern Alps NZ, 122
Southern Hemisphere Annular Mode, 58
Southern Ocean, 58, 64, 66, 68, 76, 88
South Pole, 46
Spanish Institute of Oceanography, 67
sporadic permafrost, see permafrost
stamukhi, 17, 21, 23

stress release, 122
stress threshold, 136
subglacial topography, 49, 60
submarine, 19
submarine installations, rigs, pipelines and cables, 72
sub-tropics, 2
summer ice minimum. see sea ice
summer of 2003, 165, 192
supercooled water, 242, 244, 280
supraglacial lakes, 142
surface melt-water, 55
surface reflectivity, see albedo
surge cycle, see glacier
surging glacier, see glacier
Svalbard, 16, 21, 76, 119
Sweden, 237, 267
Swiss Alps, 105, 125, 166, 192
   seasonal floods, 116,
   lower level of permafrost, climate modelling, 192
Swiss Camp (Greenland), 54
Swiss cantons, 129
Swiss Federal Government, 192
Swiss Federal Institute for Snow and Avalanche Research (SLF), 108
Swiss Federal Office for the Environment, 195
Switzerland, 2, 6, 105, 106, 107, 110, 112, 118, 127, 129, 208, 209, 212, 213, 215, 217, 221, 223, 224, 228, 229, 233, 300
Synthetic Aperture Radar, see SAR

Tajikistan, 127
Talik, see permafrost
Tasman Glacier, NZ, 123, 124
temperate maritime environment, 2
temperature gradient, 210
temperature of the earth, 8
temperature threshold, 8
Terra Nova oil field and drilling platform, 86
thaw settlement, subsidence, 158, 165, 170, 190

thaw-stable sediments, 178, 179
thaw unconformity, 170
THC, 67, 90
thermistor string, 197
thermohaline circulation, see THC
thermokarst, 168, 170, 171, 172, 175
   depressions, 169, lakes, 169
The Scramble for the Arctic, 35
Thompson Drive, Fairbanks, Alaska, 181, 182, 183, 184
threat of oil spillage, 37
Thwaites Glacier, Antarctica, 61
tidewater glacier, see marine glacier
Tien Shan Mountains, Asia, 119
Tim Williams Glacier, BC., Canada, 123
tipping point, 56, 64, 69, 300
Titanic, 1, 71, 82, 90, 91, 92, 95
tourism, 130
tourist cruise ships, impact, 37
toxic drilling wastes, 31
traditional knowledge (see Inuit)
Trakarding Glacier, Nepal, 147
Trans-Alaskan Pipeline, 120, 176
Transantarctic Mountains, 56
Trans-Siberia Railway, 175
travelling platform, 14
trimline, 120, 123
tropics, 2
Tsho Rolpa, Nepal, 144–149,
   expansion, 145
   mitigation measures,
     hard, 146
     soft, 147
   threat to Khimti HEP complex, 146
two-way radio, 27
Tuktoyaktuk NWT, Canada, 29, 32, 33, 34, 42
   coastal protection methods, 33–34, 42
   erosion problems, 32–34
   Lowlands, 162, 168
   natural harbour, 32, 33
   vulnerability, 32
tundra, 172
Tyumen, Russia, 170

UK, 133,
  Geology UK, 149
ultraviolet radiation, see UV radiation
Unalakleet, Alaska, Development
        Plan, 32
UNEP, 153
United Kingdom, see UK
United Nations, 133
United States Coast Guard, (formerly US
        Revenue) see USCG
United States Geological Survey,
        see USGS
Upernavik, Greenland, 75
USA, 2, 34, 41, 42, 157, 158, 173, 174, 184,
        212, 214, 215, 216, 222, 224,
        227, 231, 233, 238, 239, 244,
        246, 247, 250, 251, 252, 255,
        256, 257, 258, 265, 274, 275,
        276, 278, 279, 281, 282, 284,
        285, 286, 287, 288, 291, 292,
        294, 295, 297
US Army Corps of Engineers
        (USACE), 158
US Army Engineering and Research
        Development Centre
        (ERDC), 255
USCG, 92
USGS, 30, 34, 68, 95, 255, 256
  water stage gauges, 256, 257
US National Weather Service
        see NWS
utilidor, see permafrost
Utkiave, Alaska, 28
Utqiagvik, Alaska, 28
UV (ultraviolet) radiation, 37, 66
Uzbekistan, 127

Valais Alps, Switzerland, 105
Val d'Isère, Switzerland, 213
Vatnajökull, Iceland, 118, 134
  destruction of farms, 118
Vedretta Piana Glacier, Italy, 127
Vestfjord Gletscher, Greenland, 75
Victoria Island, Canada, 26
volcanic activity, 135
volumetric change, 8
Vortuka, Russia, 170

vulnerable communities, 10
vulnerable economies, 7
WAIS (West Antarctic Ice Sheet),
        see ice sheets
walrus,
  redistribution and trampling
        deaths, 36
  depletion of near-shore benthic
        resources, 36
warm-based glacier, see glacier
warning procedures, 10
water properties, 7
'water towers' (mountain glaciers),
        127, 300
wave attack, 35
wave fetch, 23, 32
wave height (amplitude), 33
water year, 246
Weather Mitigation Research
        Programme, USA, 289
weather pattern, 15, 208
Weddell seal, 66
Weddell Sea, Antarctica, 24, 94
West Antarctic Ice Sheet,
        see WAIS
Western Europe, climate
        change, 89
Western Ross Sea, 94
West Siberian Plain, 176
West Spitzbergen, 76
whaling, 24
  camps, 25
White Rose,
  drilling platform, 86, 97, 98
  oil field, 86, 99
Whymper Glacier, Switzerland, 105
Wilkes Land, Antarctica, 94
Wilkie Ice Shelf, see ice shelf
winter hunting seasons, 6
'winter of terror', see snow avalanche
winter storms, 273–298, see also blizzards
        and ice storms,
Wollaston Lake (settlement),
        Saskatchewan,
        Canada, 269
Wordie Ice Shelf, see ice shelves
Wrigley, NWT, 269, 270, 271

WWF (World Wildlife Fund), 37, 42
Yakutsk, Russia, 170, 171, 239, 240
Yakutia Lowlands, Russia, 170
Yamal Peninsula, Russia, 162, 168
Yellowknife, NWT, Canada, 178, 265
 thermosyphon, 180

Yellowstone River, Montana, USA, ice jam, 252
Yukon Territory, Canada, 15, 119, 121
Yungay, Peru, 130

zero annual amplitude, 162
zero degree Celsius threshold, 201

WITHDRAWN

129.95   7/19/12.

LONGWOOD PUBLIC LIBRARY
800 Middle Country Road
Middle Island, NY 11953
(631) 924-6400
mylpl.net

LIBRARY HOURS

Monday-Friday	9:30 a.m. - 9:00 p.m.
Saturday	9:30 a.m. - 5:00 p.m.
Sunday (Sept-June)	1:00 p.m. - 5:00 p.m.